Beam and
Fiber Optics

Beam and Fiber Optics

J. A. Arnaud

Crawford Hill Laboratory
Bell Laboratories
Holmdel, New Jersey

ACADEMIC PRESS New York San Francisco London 1976

A Subsidiary of Harcourt Brace Jovanovich, Publishers

ACADEMIC PRESS, INC.
111 Fifth Avenue, New York, New York 10003

United Kingdom Edition published by
ACADEMIC PRESS, INC. (LONDON) LTD.
24/28 Oval Road, London NW1

Library of Congress Cataloging in Publication Data

Arnaud, Jacques A   (date)
    Beam and fiber optics.

    (Quantum electronics series)
    Includes bibliographical references and index.
    1.   Beam optics.   2.   Fiber optics.   3.   Optical
communication.   I.   Title.
QC389.A76   535'.89   75-13108
ISBN 0-12-063250-0

To Marianne and Marie-Louise

The most striking feature of waves is, without doubt, their capability of carrying energy over long distances. Besides the energy, they carry also, more or less imperfectly, information.

Lighthill, *J. Inst. Math. Appl.* **1**, 28 (1965).

Contents

3 WAVE EQUATIONS

5 PIECEWISE HOMOGENEOUS MEDIA

Preface

Beam optics is a specialized field of optics giving consideration to waves that have small angular divergence. In contrast to conventional radio waves, which spread out almost uniformly from the radiating antenna, optical beams are confined to the neighborhood of some axis with the help of discrete focusing elements such as lenses or continuous changes in the refractive index of the medium. Many theoretical aspects of beam propagation have been investigated during the first half of this century in connection with radio waves in atmospheric ducts. Important new results were obtained in the 1960s, stimulated by the invention of coherent sources of electromagnetic radiation of short wavelength (lasers, backward-wave oscillators, impatt oscillators) and the growing need for communication systems of high capacity. Further impetus followed the discovery that glass fibers could be fabricated with losses as small as 1 dB/km. Dielectric waveguides, previously restricted to special applications at microwave frequencies, now appear to be one of the most promising approaches to terrestrial communication. The feasibility of intercontinental communication by artificial satellites, on the other hand, stimulated studies of quasi-optical devices that can perform operations of beam guiding and filtering at the ground stations. These are examples of the applications of beam or fiber optics techniques to communication that are discussed in this book.

Intuitive arguments are given for explaining the physical phenomena involved in beam propagation. For instance, beam guidance can be interpreted as a balance between the force of diffraction, which tends to make optical beams diverge, and the focusing action of lenses, which tends to make them converge. Our main concern is to clarify the concepts in wave and geometrical optics that are most relevant to a deeper understanding of beam optics. The book also provides the necessary algebraic details in simple form. In particular, the laws of beam propagation through unaberrated optical systems are shown to be formally the same as the laws of gaussian ray optics. We did not attempt to discuss every aspect of beam optics on the same footing. Some topics are discussed in depth; others are discussed superficially. In the latter case, the reader is directed to relevant textbooks or recent papers. Duplication of existing texts has been mostly avoided, but sufficient background material is given to make the book useful to unspecialized readers.

The book is divided into five chapters. The first chapter gives a broad view of the subject matter almost free of mathematical derivations. In the last sections of Chapter 1, a comparison is made between the laws of mechanics and the laws of optics. Chapter 2 is essentially self-contained. Its purpose is to present the laws of propagation of gaussian beams through freespace, unaberrated lenses, or lenslike media and resonators. The simplest configurations (two-dimensional with isotropic media) are first considered, but a few advanced problems are also treated. We discuss the use of gaussian beams at millimeter wavelengths. In Chapter 3, various wave equations relevant to beam optics are given, and their relationship is discussed. The importance of the Lorentz reciprocity theorem for problems of coupling between beams or fibers is emphasized. The geometrical optics limit of these equations is the subject of Chapter 4. The propagation of optical pulses in dispersive inhomogeneous (graded-index) fibers is considered from the point of view of Hamiltonian optics. The final chapter is devoted to piecewise homogeneous dielectric waveguides, such as the dielectric slab and the dielectric rod. A new method is given to evaluate the bending loss of open waveguides. This method requires that only the field of the straight waveguide be known.

This book should be useful to students, professors, and research engineers in the field of electromagnetic communication. Chapter 2 is essentially a self-contained course on beam optics. Research engineers studying communication by glass fibers should find Chapters 4 and 5 useful. The most important characteristics of glass fibers for communication are summarized in the last section of the book.

Acknowledgments

The subject of this book was taught for two years by the author as an "in-hour" course of the Bell Laboratories, in 1972 and 1973. I wish to express my thanks to the Bell Laboratories for authorizing me to write this book. Discussions with members of the Bell Laboratories have been most stimulating and are gratefully acknowledged. Many thanks are due to Mrs. B. Griffin and Miss J. Fernandes for their dedication and skill in typing the manuscript.

Notations and Definitions

The system of units used is the international system of units (meter, kilogram-mass, second, ampere). The field of traveling waves is denoted $\exp[i(kz - \omega t)]$. As the wave propagates in the $+z$-direction, its phase thus advances. $k \equiv 2\pi/\lambda$ denotes the wavenumber and λ the wavelength in the medium. The free space wavelength is denoted λ_0. When we discuss transformations through optical systems, primed quantities are used in the object space, and unprimed quantities in the image space, where most of the transformations are carried out. The radius of a gaussian beam is defined at the $1/e$ point of the beam irradiance rather than at the $1/e$ point of the field modulus. If the former radius is denoted ξ and the latter w, we have $w = \sqrt{2}\,\xi$. The ξ-notation is selected because it provides more symmetrical expressions between far and near fields than the w-notation. Furthermore, ξ, rather than w, corresponds to the classical turning point. The ξ-notation is always used in quantum mechanics. The field of scalar modes in circularly symmetric fibers is denoted $\psi_{\mu\alpha}$, where $\mu = 0, \pm1, \pm2, \ldots$ is the azimuthal mode number and $\alpha = 0, 1, 2, \ldots$ the radial mode number. For step-index fibers, the notation $LP_{|\mu|, \alpha + 1}$ (where LP stands for "linearly polarized") has been used in place of $\psi_{\mu\alpha}$ in some recent works. This alternative notation, however, is not consistent with that commonly used for graded-index fibers. The normalized frequency of a fiber, which we denote F, is often denoted V. Note also that the numerical aperture (NA) of a fiber is defined from the radiation pattern *in air*. Vectors and matrices are distinguished by boldface, usually with lowercase and capital letters, respectively. Scalar products are denoted $\mathbf{a} \cdot \mathbf{b}$, $\tilde{\mathbf{a}}\mathbf{b}$, or simply, when no confusion with a tensor product is possible, $\mathbf{ab}$.

Latin Letters

a: radius of the core of a fiber; radius of an aperture; normal-mode amplitude; asymmetry parameter

a: element of a ray matrix

a_m: coefficient of spectral expansion

a: potential vector

A: element of a ray matrix; area

A: 4-potential vector = $(\mathbf{a}, iV)$

$Ai(x)$: Airy function

b: normalized phase velocity

b: element of a ray matrix

B: element of a ray matrix; radiance

B: magnetic field; binormal vector

c: velocity of light in free space; transverse coupling

c: element of a ray matrix

c_{ab}: coupling

C: element of a ray matrix; capacitance per unit length; fiber axis curvature, = $1/$radius of curvature $\equiv 1/\rho$

$\mathcal{C}$: capacitance

d: distance between two mirrors; half slab thickness

d: element of a ray matrix

D: element of a ray matrix; diffraction walk-off parameter; spacing

D: electric induction; spectral matrix of an operator

ds: elementary ray length

dC: vector perpendicular to a planar contour, with magnitude equal to the elementary arc length

dS: vector perpendicular to a surface, with magnitude equal to the elementary area

$D = (\omega/k)(dk/d\omega)$: dispersion parameter

$\mathbf{D}_1, \mathbf{D}_2$: differentiation matrices

D_κ, D_n: inhomogeneous material dispersion parameters

e: = 2.718 . . . ; electron charge

$\bar{e}$: electron charge divided by $\hbar$

E: magnitude of the electric field; pulse energy

E: electric field

f: focal length; frequency; force; ray density in phase space

F: Fresnel number; finesse; normalized frequency = $(k^2 - k_s^2)^{1/2} \times (a$ or $d)$ (denoted V in other works)

F: electromagnetic field 6-vector; hamiltonian parameter

g: resonator parameter, = $1 - d/R$

g: hamiltonian parameter

$G(\mathbf{x}; \mathbf{x}')$: Green's function

G: geometrical walk-off parameter

G: Hamiltonian parameter

h: Planck's constant = $2\pi \times \hbar$

$\hbar$: = 1.054×10^{-34} joule $\times$ second

$h(z)$: curvature of the surface of wave vectors

$H(\mathbf{k}, \mathbf{x}) = 0$: hamiltonian function

H: magnetic induction

$H_m(x)$: Hermite polynomial of order m

$He_{m_1 m_2}(x_1, x_2)$:   modified Hermite polynomial in two variables.

i: $(-1)^{1/2}$; time dependence denoted: $\exp(-i\omega t)$; in subscript: imaginary part

i', i: angles of incidence and of refraction defined with respect to the normal to the surface

I: electric current; Lagrange ray invariant

$\mathrm{Im}(\ )$:   imaginary part

$I(x, t)$:   intermediate frequency current density

J: adiabatic invariant

$J_\mu(x)$;   Bessel function of the first kind of order μ

$\mathbf{J}_M$:   Maxwell current

$\mathbf{J}_F$:   Fock current with components $\mathbf{j}$ (transverse) and ρ (axial)

$k(\mathbf{r})$:   free wavenumber at some point $\mathbf{r}$

k: Fiber core free wavenumber, $= 2\pi/\lambda$

k_s: cladding or substrate free-wavenumber

k_0:   wavenumber on-axis or in the upper medium

$\mathbf{k}$:   wave vector, with components k_x, k_y, k_z or p_1, p_2, k_z

$\mathbf{K}$:   4-wave vector, with components $k_x, k_y, k_z, i\omega/c$; wave function operator; third material matrix (6×6)

K: Boltzmann constant, $= 1.38 \times 10^{-23}$ joule/°C; normalized coupling

$K_\mu(x)$:   modified Bessel function of the second kind of order μ

l: axial mode number; relative loss

L: loss in dB or loss per unit length in dB/km; inductance per unit length

$\bar{L}$:   lagrangian of the beam axis

$\mathcal{L}$:   lagrangian density; inductance

$\bar{\mathcal{L}}$:   average lagrangian density

$\mathbf{L}$:   second material matrix with submatrices $\boldsymbol{\epsilon}, \boldsymbol{\xi}, \boldsymbol{\zeta}, \boldsymbol{\mu}$

$\mathcal{L}(s_x, s_z) = 0$:   surface of ray vectors

$\mathbf{L}$:   Lagrange function

m: electron mass; transverse mode number

$\bar{m}$: $= m/\hbar$

m^*:   effective mass

m_0:   mass per unit length

m_1, m_2:   transverse mode numbers

$\mathbf{M}$:   ray matrix; material matrix (6×6) with submatrices $\mathbf{M}_{11}, \mathbf{M}_{12}, \mathbf{M}_{21}, \mathbf{M}_{22}$

n: refractive index, $= \lambda_0/\lambda = k/k_0$; mode number

N: power coupling in two dimensions, mode number density; number of transverse dimensions

N^2:   power coupling in three dimensions

$\mathbf{n}$:   hamiltonian parameter

$\mathbf{N}$:   hamiltonian parameter

NA: numerical aperture, $=$ sine of half-radiation angle in air

$\mathbf{p}$:   transverse ray momentum with components p_1, p_2 or k_x, k_y

P: power; power in the core

P_s: power in the cladding

P_t: total power

$\mathbf{P}, \mathbf{P}^\dagger$:   complex matrical ray momenta (2×2)

$q(z)$:   complex ray

q^+, q^-:   positive and negative frequency components of a complex ray

Q: mismatch parameter; complex ray

$Q^\dagger$: complex ray

$\mathbf{Q}, \mathbf{Q}^\dagger$: complex matrical rays (2×2)

r: radius in space or in a transverse plane; field reflectivity; as a subscript, denotes real part

R: wavefront mismatch parameter; mirror radius of curvature; power reflectivity, $= rr^* = 1 - T$. Also, $R \equiv r^2$, where r denotes radius; field strength parameter

Re( ): real part

R_M: maximum ray radius squared

s: normalized susceptance with dimension $1/\mathrm{length}$; integer

$\mathbf{s}$: ray vector; 4-vector in spatial phase space

$S(\mathbf{x})$: eikonal (phase)

$S(\mathbf{x}; \mathbf{x}')$: point-eikonal, = phase shift along a ray

$\mathbf{S}$: energy flux

t: time

T: power transmission, $= 1 - R$

$\mathbf{T}$: rotation matrix; Lorentz transformation matrix (6×6)

u: magnitude of group velocity; ray parameter; circular rod parameter

u_0: average axial group velocity

u_z: axial group velocity

$\mathbf{u}$: point-eikonal parameter; group velocity

U: electric potential; point-eikonal parameter

U_g: generating point-eikonal parameter (two dimensions)

$\mathbf{U}_\mathrm{g}$: generating point-eikonal parameter (three dimensions)

v: phase velocity of free waves in a medium; ray parameter; circuit rod parameter

v_z: axial phase velocity, $= \omega/k_z$

v_g: axial group velocity

$\mathbf{v}$: point-eikonal parameter

V: scalar (electric) potential; point-eikonal parameter

V_g: generating point-eikonal parameter (two dimensions)

$\mathbf{V}_\mathrm{g}$: generating point-eikonal parameter (three dimensions)

w: $\sqrt{2} \times$ gaussian beam half-width ξ

W: point-eikonal parameter (two dimensions); energy density

$\mathbf{W}$: point-eikonal parameter (three dimensions); modal matrix of operators; dispersion curvature matrix (3×3)

W_g: generating point-eikonal parameter (two dimensions)

$\mathbf{W}_\mathrm{g}$: generating point-eikonal parameter (three dimensions)

x: transverse coordinate

$\mathbf{x}$: defines a point in space with components x, y, z or x_1, x_2, x_3

$\mathbf{x}_1, \mathbf{x}_2, \mathbf{x}_3$: period vectors

X: $\equiv x^2$

$\mathbf{X}$: defines a point in space–time ($\mathbf{x}, ict$)

y: transverse coordinate

Y: admittance; also $Y \equiv y^2$

z: axial coordinate

$\bar{z}$: normalized axial coordinate

Z: Fabry–Perot response parameter; ray period

Greek Letters

α: angle of a ray to the axis; loss coefficient; integer for summation; radial wavenumber

$\boldsymbol{\alpha}$:   generating vector with components α_1, α_2

β: axial propagation constant (alternative notation, k_z); integer for summation

γ: integer for summation

$\delta(x)$:   Dirac symbolic function

δ: prism angle; also $= (1 - k_s^2/k^2)^{1/2}$

δ_{ij}:   1 if $i = j$; 0 otherwise

Δ: small variation, e.g., Δk_x, Δn; characteristic parameter of resonators; aberration; Laplacian operator; $= \delta^2/2 \approx \Delta n/n$

ϵ: small quantity; medium permittivity

ϵ_0:   free-space permittivity $= (4\pi \times 9 \times 10^9)^{-1}$ MKSA

ϵ/ϵ_0: $= n^2$, n refractive index

$\boldsymbol{\epsilon}$:   permittivity tensor (3×3)

ζ: modified axial coordinate; displacement of a surface

$\boldsymbol{\zeta}$:   material submatrix (3×3); 5-vector in the phase space (k_x, k_y, ω, x, y)

η: waveguide efficiency

θ: beam axial phase shift; normalized axial ray angular momentum

$\bar{\theta}$:   normalized phase shift

θ_0:   phase shift of axial rays

$_1$, θ_2 or θ_x, θ_y: beam phase shifts in three dimensions

κ: power-law parameter; also stands for ϵ or μ

λ: wavelength; eigenvalue

$\bar{\lambda}$:   eigenvalue

λ_0:   wavelength in free-space

Λ: wavelength in oblique directions

μ: medium permeability; azimuthal scalar wave number; complex wavefront curvature

μ_0:   free-space permeability, $= 4\pi \times 10^{-7}$ MKSA

$\boldsymbol{\mu}$:   permeability tensor (3×3)

ν: azimuthal e.m. wavenumber; characteristic angle of helical fibers; rotation angle

ξ: gaussian beam half-width, $= 1/e$ point of the beam irradiance

$\bar{\xi}$:   normalized gaussian beam half-width

ξ_0:   gaussian beam waist half-width

$\bar{\xi}_0$:   half-width of matched gaussian beams

$\boldsymbol{\xi}$:   material submatrix (3×3); 6-vector in space–time phase space

π: $= 3.14159\ldots$

ρ: wavefront curvature radius; helix curvature radius

$\bar{\rho}$:   normalized wavefront curvature radius

σ: ray parameter; rms pulse width

$\sum$:   summation sign

τ: proper time; relative time of flight of an optical pulse; spatial rate of rotation

$\bar{\tau}$:   normalized relative time of flight

φ: azimuthal coordinate

ϕ: phase shift (e.g., under total reflection)

ϕ_1, ϕ_2, ϕ_3:   phase shifts per period
ϕ: electromagnetic field 6-vector
χ: normalized transverse vector with components χ_1, χ_2
ψ: wavefunction $\approx (k_0)^{1/2} \times$ electric field transverse component
$\psi_m; \psi_{m_1, m_2}$:   wavefunctions of modes of order m and m_1, m_2, respectively
Ψ: wavefunction for off-set beams
ω: angular frequency, $= 2\pi \times$ frequency
ω_c: cutoff angular frequency
Ω: focusing strength; solid angle; path twist angle
Ω_0:   spatial period of perturbation
$\boldsymbol{\Omega}$:   focusing strength matrix
∂:   partial differentiation
$\mathbf{V}$:   gradient operator with rectangular components: $\partial/\partial x, \partial/\partial y, \partial/\partial z$
$\mathbf{V} \cdot$:   divergence operator
$\mathbf{V} \times$:   rotational operator

Other Symbols

$*$:   complex conjugation
$\tilde{}$:   transposition (omitted on first vectors of products)
$\dagger$:   adjointness
$\cdot$:   one upper dot; first derivative with respect to argument
$\cdot\cdot$:   two upper dots; second derivative with respect to argument
$!$:   factorial, $a! = 1 \times 2 \times \cdots \times a$
$\cdot$:   scalar product, $\mathbf{a} \cdot \mathbf{b} \equiv \tilde{\mathbf{a}}\mathbf{b} \equiv \mathbf{ab}$
$\times$:   vector product
$\binom{a}{b}$: $= a!/[(a - b)! \, b!]$
$\approx$:   approximately equal
$\sim$:   order of magnitude; asymptotic expansion
$\propto$:   proportional

Definitions

Angular frequency: $2\pi \times$ frequency
Bianisotropy: dependence of the electric and magnetic inductions ($\mathbf{D}, \mathbf{H}$) on both the electric and magnetic fields ($\mathbf{E}, \mathbf{B}$)
Canonical momentum: product of the wavevector ($\mathbf{k}$) and an adiabatic invariant (J)
Dispersion hypersurface (or hypersurface of 4-wave vectors): hypersurface in the k_x, k_y, k_z, ω space. Restriction to the k_x, k_y, k_z space is called the dispersion surface. Restriction to the ω, k_z space is called the dispersion curve
Distribution (f): density of rays in phase space ($\mathbf{k}, \mathbf{x}$)
Inertial coordinate system: frame of reference in which a particle not submitted to a recognized force (e.g., electrical force) has constant velocity
Inhomogeneous dispersion: spatial variation of $D \equiv v/u \equiv (\omega/k)(dk/d\omega)$
Irradiance: power radiated by an extended source per unit area, in watts/meter^2
Material dispersion: $M \equiv (\omega^2/k)(d^2k/d\omega^2)$
Radiance: power radiated by an extended source per unit projected area and unit solid angle in some given direction, in watt/(steradian $\times$ meter^2). For a lambertian source, the radiance is independent of direction, denoted B

Description of Optical Beams

The laws of diffraction in homogeneous media, the properties of typical light-guiding systems, and the essential features of open resonators are discussed in this introductory chapter in an essentially qualitative manner. A few other topics are selected for their illustrative value.

1.1 Diffraction by Apertures

The concept of light ray presumably originates from the observation of the sunlight passing through small apertures. An aperture defines a narrow pencil that can be used to perform simple experiments in geometrical optics.[1] The light, however, eventually spreads out because of the finite angular size of the source, as illustrated in Fig. 1-1. Except for very large and very small aperture sizes, this angle is independent of the diameter of the aperture. It depends only on the apparent size of the sun.

If a coherent source of light, such as that originating from a collimated laser beam, illuminates a small aperture, beam divergence is also observed. However, the divergence angle is now inversely proportional to the aperture diameter and is a consequence of the wave nature of light. An expression for the angular divergence of the beam transmitted through the aperture is easily obtained for that case by observing that the field has a significant intensity at the observation point O only if the radiation from one edge A of the aperture is approximately in phase with the radiation from the aperture center O' (see Fig. 1-2). The difference in length between

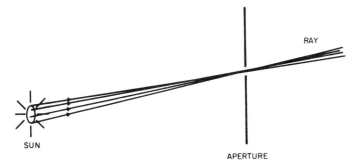

Fig. 1-1 The sunlight passing through a small aperture defines approximately a ray. The light eventually spreads out because of the finite angular size of the sun (about 0.5°). Diffraction effects are negligible in that experiment.

$O'O$ and AO for far-off observation points that are at an angle α to the axis is approximately equal to $a\alpha$, if a denotes the radius of the aperture. Thus the angular divergence of the beam is approximately

$$\alpha \sim \frac{\lambda}{a} \tag{1.1}$$

This expression is applicable only if α is a small angle, that is, if a is large compared with the wavelength.[2,3]

Figure 1-3a shows that three regions can be distinguished: the near-field (or shadow) region, the intermediate (or Fresnel) region, and the far-field (or Fraunhofer) region. The beam width departs significantly from the aperture width at a distance d of the order of $a/\alpha = a^2/\lambda$. For example, if $a = 1$ mm and $\lambda = 1$ μm, the distance d is equal to 1 m. In the millimeter wave range, for $\lambda = 1$ mm and $a = 0.3$ m, the distance d is equal to 100 m. Far from the aperture ($d \gg a^2/\lambda$), the beam wavefront is almost spherical and centered approximately at the center O' of the aperture. The beam then closely resembles a ray pencil.

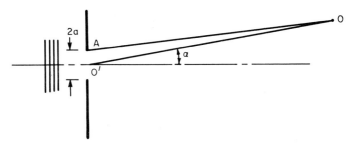

Fig. 1-2 When a coherent source of light illuminates a small aperture, the light beam spreads out because of diffraction. The divergence angle is of the order of λ/a, where λ is the wavelength and a the aperture radius.

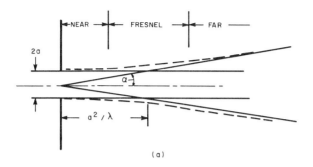

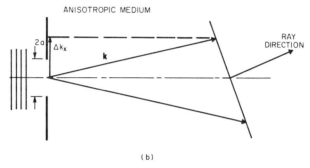

Fig. 1-3 (a) The near-field (or shadow) region, intermediate (or Fresnel) region, and far-field (or Fraunhofer) region are defined here. The intermediate distance d is equal to a^2/λ (that is, $a^2/\lambda d \approx 1$). (b) If the medium is anisotropic, we need to consider the difference in phase shift across the aperture of the various plane waves that constitute the transmitted beam. This difference is of the order of unity at the beam angular half-width. The condition $\Delta k_x a \approx 1$ is similar to the Heisenberg uncertainty relation in quantum mechanics: $\Delta pa \approx$ Planck constant.

In the expression a^2/λ given above, λ denotes the wavelength in the medium rather than in free space. When the wavelength λ decreases, either because the frequency of the source increases or because the refractive index of the medium increases, the rate of expansion of the beam is reduced.

In anisotropic media, the phase shift per unit length along a ray depends on the direction of that ray. In evaluating the phase shift for the two paths $O'O$ and AO in Fig. 1-2, one must be cautioned not to assume that the phase shifts per unit length are the same for both paths on the ground that these two paths are almost parallel when O is sufficiently far away. This would be incorrect because the difference in phase shift per unit length is multiplied by the distance $O'O$, which tends to infinity as O recedes to infinity. The correct expression for beam divergence in anisotropic media is obtained by specifying that the phase shift across the aperture is of the order of unity. Using the notation in Fig. 1-3b, we have $\Delta k_x a \sim 1$, where Δk_x denotes the change in the component of the wave vector **k** in the aperture plane. The wave vector **k** is defined as perpendicular to the wavefront. Its magnitude k, called the wavenumber, is equal to $2\pi/\lambda$. Note that the rays are not necessarily perpendicular to the aperture plane, in contrast to the case of isotropic media. In most of this chapter, only isotropic media are taken into consideration.

Consider now a point source O' and a detector O, as shown in Fig. 1-4a, and an aperture located between O' and O. If this aperture, originally widely opened, is closed down, the amplitude at O does not decrease steadily, as one might have expected, but instead oscillates. For some

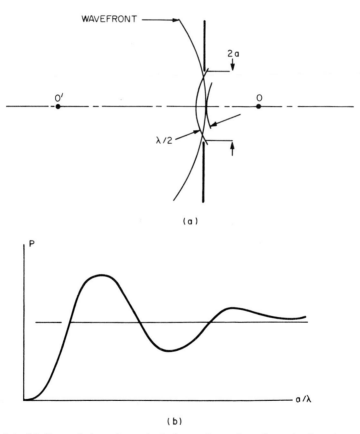

(a)

(b)

Fig. 1-4 (a) Transmission of a spherical wavefront through a circular aperture. By selecting the points of the wavefront that are approximately in phase at the observation point O, an aperture can increase the field intensity at O. (b) Variation of the power density at O as a function of the normalized aperture radius.

aperture radii, the irradiance at O is much larger than when the aperture is opened. This startling result can be understood on the basis of the Huygens principle, which states that each portion of the spherical wavefront at the aperture plane can be considered a source of secondary waves. If these wavelets are in phase at O, they reinforce each other. Otherwise, they tend to cancel out. The aperture, when given the proper radius, helps select the central portion of the wavefront (called the first Fresnel zone), which contributes to the field at O, and absorbs the outer part of the wavefront, that tends to cancel out the contribution of the first zone. The variation of the power density at O is shown in Fig. 1-4b as a function of the aperture radius a. A periodic sequence of apertures can, in fact, guide optical beams by repeatedly selecting the first Fresnel zone. The loss suffered, however, is higher than when guidance is provided by focusing lenses, an arrangement discussed in the next section.

The above explanation based on the Huygens principle suggests that, in order to increase the transmitted power, we can do better than merely absorb the outer part of the wavefront. If the phase of the field is reversed on successive rings, the wavelets add up approximately in phase at O. These successive phase reversals are accomplished by the Fresnel-zone plate. The Fresnel-zone plate is a thin sheet of dielectric material whose thickness varies by steps. At a fixed frequency, a Fresnel-zone plate behaves approximately as a focusing lens. Compared to a lens, it presents the advantage, important at microwave frequencies, of requiring a smaller amount of material. On the other hand, Fresnel-zone plates can be used only in narrow bands of frequency.

1.2 Beam Guidance by Lenses

A prism has the property of deflecting rays, as illustrated in Fig. 1-5a. The larger the prism angle, the larger is the ray deflection. If the prism angle δ is much less than 1 rad, the ray deflection α is equal to $(n - 1)\delta$, where $n \equiv \lambda_0/\lambda$ denotes the refractive index of the prism material and λ_0 is the wavelength in free space. This result can be obtained by applying the Descartes–Snell law of refraction to the two air–medium interfaces. It should be noted at that point that the law of refraction is truly a law of refraction for wavefronts rather than for rays. This distinction will become clearer when we consider in Chapter 4 refraction between anisotropic media. What really matters is the direction of the normal to the incident wavefront at the point where this wavefront encounters the interface. The prism deflection angle α given above is, in fact, more easily obtained by considering the tilt of the wavefront, which results from the change in phase shift introduced by the prism at various heights, than by a direct application of the law of refraction.

A lens can be viewed as a prism whose apex angle increases linearly as a function of the distance from axis (see Fig. 1-5b). A collimated (or parallel) manifold of rays converges approximately to a common point located at a distance f from the lens vertex, called the lens focal length. A lens, however, cannot concentrate incident beams into perfectly sharp points. The minimum radius a of the spot is given approximately (using the same

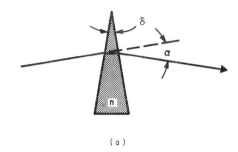

(a)

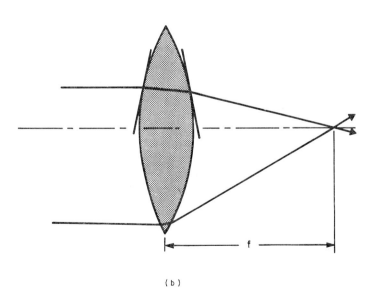

(b)

Fig. 1-5 (a) Refraction by a prism. The deflection angle for small prism angles δ is $\alpha \approx (n - 1)\delta$. (b) Refraction by a lens. Parallel rays converge at the lens focal point, a distance f from the lens vertex.

argument as before) by $a/\lambda = \alpha$, where α denotes the angle of convergence (see Fig. 1-6a). If the lens suffers from aberrations, the spot size may be considerably larger.

The previous discussion suggests that optical beams can be periodically refocused by sequences of lenses and kept confined as illustrated in Fig. 1-6b. Periodic systems of lenses were studied at the beginning of the century for the design of submarine periscopes. For that application, wave optics effects can be ignored. These effects, on the contrary, are essential in optical communication. If the lens radius is a_0, and $f \approx d/4$, where d denotes the lens spacing, the minimum spot size is, as we have seen before, of the order of $\lambda d/a_0 \ll a_0$. In most practical systems, one tries to

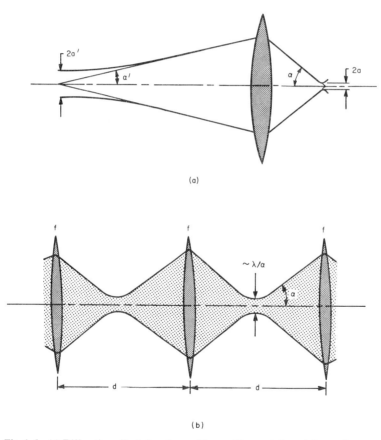

(a)

(b)

Fig. 1-6   (a) Diffraction effects in a focused beam. The spot size a is inversely proportional to the convergence angle α (or to $\sin \alpha$ if α is not small compared with unity). (b) The beam behavior for a periodic sequence of focusing lenses with small negative values of $d - 4f$.

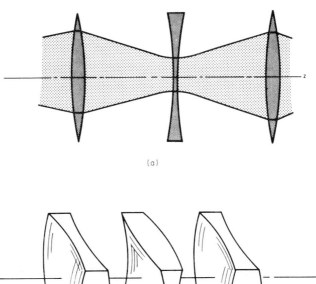

(a)

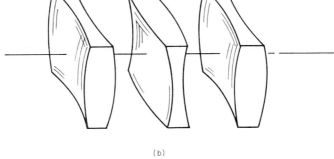

(b)

Fig. 1-7 (a) A "strong-focusing" arrangement with lenses alternately focusing and defocusing. There is a net focusing action because the beam is broader at the focusing lenses. (b) Saddle-shaped lenses can be used to accomplish the strong-focusing action in three dimensions.

maximize the lens spacing. The confocal spacing $d \sim 2f$ is usually selected. The minimum beam radius then becomes of the order of the lens radius a_0. To understand the behavior of optical beams when the minimum beam radius is comparable to the lens radius, a more precise theory of diffraction, which is given in Chapter 2, is required.[4]

Optical beams can be kept confined by sequences of alternately converging and diverging lenses, as illustrated in Fig. 1-7a.[5] To understand how this is possible, let us first observe that the ray deflection provided by a lens increases linearly with the distance from axis. This can be understood by considering the prism tangent to the lens surface at the point of incidence as we discussed earlier. On that basis, it can be seen that the force of diffraction[§] can be balanced by the net focusing power of the lens system shown in Fig. 1-7a, if the beam undulates in size, with the large

---

[§] This expression is used here only for its illustrative value.

diameters located at the focusing lenses (where the wavefronts are deflected toward the axis by relatively large amounts) and the small diameters located at the defocusing lenses (where the wavefronts are deflected away from axis by relatively small amounts). This principle was first applied to particle accelerators and called "strong focusing." This is somewhat of a misnomer because the strong-focusing mechanism is less effective than the conventional converging lens arrangement described before. Strong focusing is of interest when there is a physical impossibility in having all the lenses convergent. A three-dimensional "strong-focusing" arrangement with saddle-shaped lenses is shown in Fig. 1-7b. A related guiding system, the helical fiber, will be treated in Chapter 2.

1.3 Continuously Guiding Media

A series of lenses, in the limit of small spacings and large focal lengths, is equivalent to a continuously guiding medium having a variation of refractive index in the transverse direction. Beams can be "trapped" if the wavenumber decreases away from axis, as shown in Fig. 1-8a, b. In Fig. 1-8b we have represented a typical distribution $k^2(r) \propto n^2(r)$, the k^2-axis being oriented downward to clarify the following analogy: We can think of

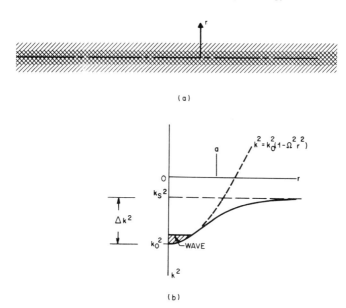

Fig. 1-8 (a) A graded-index fiber represented schematically. Beams can be trapped if the refractive index is maximum on axis. (b) The refractive-index profile shown here can sometimes be approximated by a parabolic variation.

the wave guided by the medium as a liquid (shaded area) filling up the well, the volume of the liquid being of the order of λ^2. The trapping of the wave is more efficient if the well is wide (large diameter $2a$) and deep (large change of refractive index Δn). At optical wavelengths, materials whose refractive indices vary from about 1 to 2 are available. For low-loss solids and liquids, the range is narrower, perhaps 1.4 to 1.6. Small changes of refractive index of the order of 1% are usually sufficient to guide optical waves if the duct is many wavelengths wide. This statement should be qualified with the remark that, for a given product $\Delta n \, a^2$, the sensitivity of the system to misalignments is lessened if Δn is large and a^2 small rather than the opposite. For that reason, glass fibers, which have relatively large $\Delta n \, (\sim 0.01)$, provide a more efficient guidance of optical beams than gas lenses ($\Delta n \sim 0.0001$), even if the product $\Delta n \, a^2$ is the same in both cases. The details of the refractive index profile $\Delta n(r)$, however, are rather unimportant, as far as the confinement of the beam through random bends is concerned. Omitting numerical factors, the bending loss in decibels per wavelength for a radius of curvature ρ has the form (see Chapter 5) $(\Delta n \, a/\lambda)^2 \exp[-(\Delta n \, a/\lambda)^3(\rho/\lambda)]$. Thus the bending loss depends on the product $\Delta n \, a$, rather than $\Delta n \, a^2$. Put another way, for a given $\Delta n \, a^2$, the critical bending radius that makes the argument of the exponential term in the expression of the bending loss of the order of unity is proportional to the cube of the fiber radius a.

Great simplification in the analysis occurs if it can be assumed that the law $k^2(r)$ or $n^2(r)$ is parabolic. This parabolic law, shown as a dashed line in Fig. 1-8b, constitutes an acceptable approximation to the actual law if the ray trajectories remain below the point where the curve $k^2(r)$ departs from parabolicity. Within this parabolic approximation, which will be discussed at great length in this book, the field of the fundamental mode of propagation is a gaussian $\psi \sim \exp(-r^2)$. The theory of optical propagation in square law fibers, in fact, coincides with the quantum theory of harmonic oscillators.[6] However, the usual formulation of quantum mechanics, based on the expansion of incident wave functions in stationary states, is not the most convenient in optics. A more adequate formulation rests on the observation that the irradiance of a gaussian beam of any radius launched into a square-law fiber remains gaussian. The beam radius, in general, pulsates at twice the ray oscillation frequency as shown in Fig. 1-9a. Such beams have been called "beam modes." The radius of the beam remains a constant only if the initial radius has a well-defined value that depends on the wavelength and the focusing strength of the medium. For that case, fundamental beam modes coincide with the ground states of harmonic oscillators. The difference between the optical and the mechanical cases is that in quantum mechanics the phase of the wavefunction ψ is often ignored on the ground that only the probability density $\psi^*\psi$

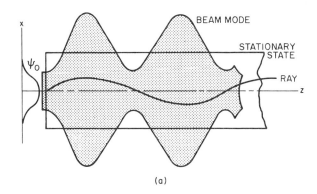

(a)

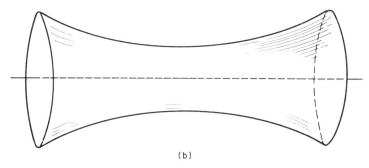

(b)

Fig. 1-9 (a) The difference between an ordinary mode (or stationary state) shown with a straight profile and a "beam mode" with undulating profile, the profile being defined as the loci of $1/e$ points from axis in beam irradiance. The period of beam oscillation is half the period of ray oscillation. (b) Beam mode in free space. The profile is an hyperboloid of revolution.

is of physical interest. In optics, we more often need to keep track of the phase along different paths. Thus the phase of ψ is important, particularly in the analysis of optical resonators. In free space, beam modes assume the form shown in Fig. 1-9b. The beam reaches its minimum radius, the so-called beam waist, only once. The study of gaussian beams provides a more precise model of optical beams than the one discussed before involving a near zone, a far zone, and a somewhat fuzzy region (Fresnel zone) between them (compare Fig. 1-9b and Figs. 1-3a and 1-6). When a gaussian beam is launched in a square-law fiber off axis, not only does its radius pulsate, but also its center oscillates about the fiber axis. The center, in fact, follows a classical ray trajectory. This is illustrated in Fig. 1-10. It can be shown that the beam profile is generated by an offset ellipse, rotating at a continuous rate Ω in the so-called phase space k_x, x, as z varies.

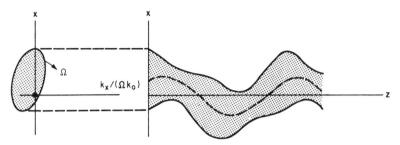

Fig. 1-10 The propagation of gaussian beams in square-law media, launched at some angle to the medium axis. The profile can be generated by an ellipse rotating at a uniform rate Ω in the "phase space" $k_x/\Omega k_0$, x.

Another simple refractive index law is when k^2 is a constant for $r < a$ (core) and, again a constant but with a lower value, for $r > a$ (cladding). An approximate (WKB) representation of the wave propagation can be given in terms of waves totally reflected at the interface. This picture is similar to the picture often given for conventional metallic waveguides. In dielectric waveguides, however, in contrast to metallic waveguides, there is a minimum angle below which the waves are totally reflected. In particular, if the discontinuity in refractive index is small, only rays at grazing incidence are totally reflected. Near the critical angle, the field rate of decay outside the core is slow.[7] The field pattern of the fundamental mode (denoted HE_{11}) is shown in Fig. 1-11 for various depths of the refractive-index well. When the well is shallow (Fig. 1-11a), most of the power flows in the cladding ($r > a$). For very deep wells, on the contrary, the field is concentrated in the core (Fig. 1-11c). Furthermore, the field tends to vanish near the core–cladding interface. This is unlike the case of round metallic waveguides, where the field has a comparable magnitude at the boundary and near axis.

A typical dispersion curve ω (angular frequency) versus β (wavenumber) is shown in Fig. 1-12 for a dielectric slab in free space and for a dielectric slab with wavenumber k supported by a substrate with wavenumber $k_s < k$. In the former case (symmetrical slab), there is no low-frequency cutoff, that is, waves can propagate at arbitrarily low frequencies. At very low frequencies, however, most of the power flows outside the slab, as is the case for the dielectric rod in Fig. 1-11a, and the propagation is very sensitive to bends. In the latter case (asymmetrical slab), a trapped mode can be sustained only above an angular frequency ω_c. Below that frequency, waves excited in the film radiate into the substrate.

The wavenumber k introduced in the previous discussions may be an "effective" rather than a true wavenumber. A thin dielectric film, as we have seen above, can sustain waves with wavenumber β in any direction in

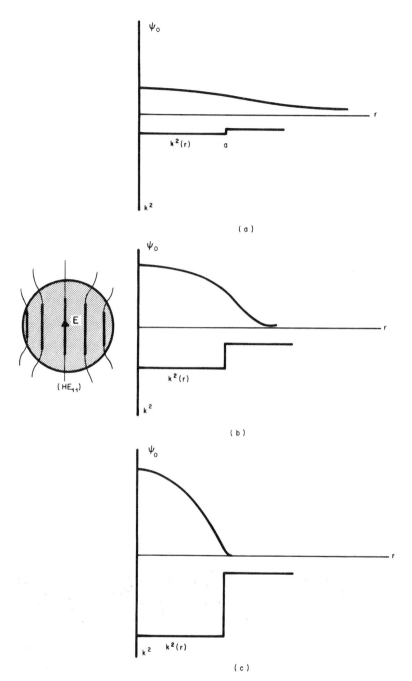

Fig. 1-11 Clad fibers. (a) Shallow well: Most of the power flows in the cladding. (b) Moderate well depth: On the left the electric field lines of the fundamental HE_{11}-mode are shown. (c) Deep well: The field tends to vanish at the boundary.

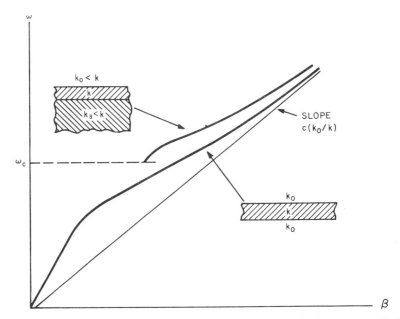

Fig. 1-12 Typical dispersion curve $\omega(\beta)$ of a dielectric slab in free space and of a dielectric slab supported by a substrate.

the plane of the film. β differs somewhat from the bulk wavenumber k of the film as Fig. 1-12 shows and is a function of the film thickness $2d$. Let β be called an "effective" wavenumber. If the film thickness varies as a function of x, β also varies, the bulk wavenumber k remaining a constant. The previous discussion is applicable to the effective wavenumber $\beta(2d)$. This is illustrated in Fig. 1-13 for prismlike and graded-index-like thin films.[8] The film thickness determines the local value of β, and thereby the wave and ray optics properties of waves guided in the plane of the film. Focusing can also be achieved with films of uniform thickness by curving the substrate as illustrated in Fig. 1-14. The substrate in Fig. 1-14a is a sphere, and the propagation is confined to the surface of that sphere.[9] It is clear that rays from the point source at O' converge to the opposite point O. This behavior is often observed with radio waves guided between the ionosphere and the surface of the earth. The sphere is only one example of a class of two-dimensional lenses called "configuration" lenses. Another example is the Rinehart–Luneburg lens shown in Fig. 1-14b. Collimated rays are focused to a point. Mathematically, the behavior of these lenses can be understood as resulting from a deformation of the metric compared to that of flat space.

Let us now discuss another class of guiding structures where the refractive-index law illustrated in Fig. 1-11 is reversed. The cladding has a higher

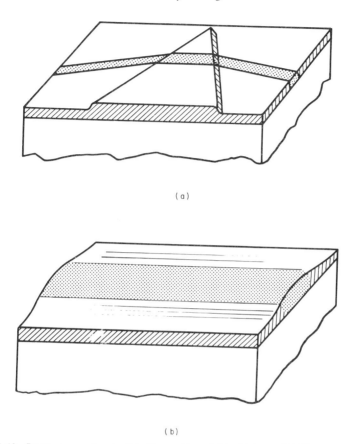

(a)

(b)

Fig. 1-13   Beam propagation in thin films with variable thickness $2d$. The propagation is characterized locally by an effective wavenumber $\beta(2d)$. (a) Prismlike thin film. (b) Continuously focusing thin film.

refractive index than the core, as shown in Fig. 1-15. Waves propagating in the core are never totally reflected. However, for waves that have an almost grazing incidence, most of the power is reflected, and only a small fraction of the incident power is transmitted to the cladding, and from there, to free space. This type of fiber, although essentially lossy, may have small radiation losses.[10] The larger the core diameter, the smaller are the losses because the rays associated with the fundamental guided wave make small angles to the axis. In the guiding systems described earlier, the exact shape of the refractive-index profile was relatively unimportant. In the present case, it is essential that the refractive index be discontinuous. Although the loss is most easily evaluated from a ray picture, it can also be obtained from a modal analysis. The modes are leaky waves, whose field intensity increases as we move away from axis as shown in Fig. 1-15b for

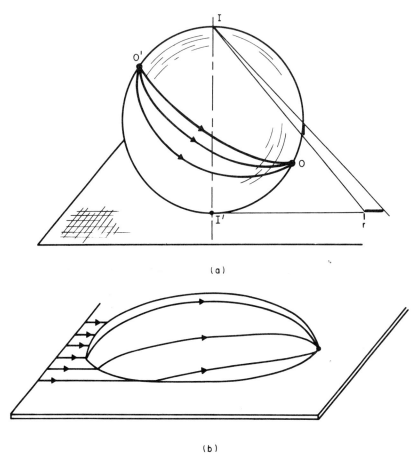

(a)

(b)

Fig. 1-14   (a) Spherical configuration lens. Rays emitted at O' are bound to follow the spherical surface. They converge at the point O opposite O'. A stereographic projection from I to the plane tangent to the sphere at I', with conservation of the optical length element, gives the Maxwell fish-eye distribution of refractive index $n = n_0(1 + r^2)^{-1}$. (b) The Rinehart–Luneburg configuration lens, which gives perfect focusing of collimated manifolds of rays.

$r > a$. This type of waveguide, with inverted refractive indices, has applications in gas lasers, the active medium (excited atoms or molecules) filling up the capillary.

Another interesting example of optical wave guidance is provided by the so-called whispering-gallery modes, which have a tendency to cling to the concave sides of curved boundaries, as shown in Fig. 1-16a. This mode of propagation was first investigated by Rayleigh in acoustics. Because there is an approximate equivalence between curvature and linear variation of

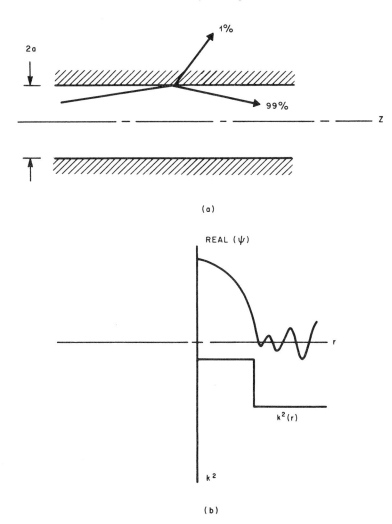

Fig. 1-15 (a) Optical waveguide with inverted refractive-index law. Although essentially lossy, this type of waveguide can have low losses (~ 2 dB/km if $a/\lambda \sim 1000$). (b) Real part of the field of a (leaky) mode in this configuration as a function of r.

refractive index,[11] a whispering gallery is equivalent to a medium having a constant transverse gradient of the wavenumber, bounded by a plane boundary. The modes are described approximately by Airy functions (Bessel functions of order $\frac{1}{3}$). Whispering-gallery modes are also important for the analysis of the bending loss of dielectric waveguides.

The equivalence between curvature and linear gradient of the wavenumber is illustrated further in Fig. 1-17 for a square-law fiber. The

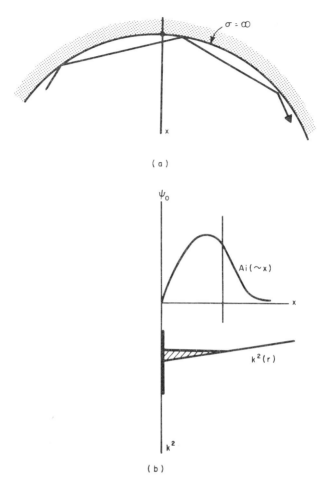

Fig. 1-16 (a) Whispering galleries are curved concave boundaries. Ray manifolds that remain confined to the close neighborhood of the boundary can be defined. (b) The field pattern is described approximately by Airy functions.

effective wavenumber law shown in Fig. 1-17b is obtained by "tipping" the well representing the straight fiber. Radiation losses thus can be understood as resulting from a phenomenon of "tunneling" from regions with low effective wavenumber to regions with large effective wavenumber. Radiation losses are always suffered in curved open waveguides because the effective medium wavenumber is, at some radius, larger than at the bottom of the well. Tunneling losses exist because the wave field does not stop abruptly as previous figures may have suggested, but decays continuously beyond the classical (or ray) limit. For the curved fiber shown in

Fig. 1-17a, strictly speaking, no trapped wave can propagate, but only leaky waves. An alternative way of explaining why radiation takes place is to point out that the phase velocity of the wave exceeds the free-space velocity at some radius, because the phase velocity increases in proportion to the radius from the curvature center. The radiation loss depends on the strength of the field at that particular radius.

Pulses of electromagnetic radiation propagating in single-mode waveguides tend to broaden if the group velocity depends on frequency because some spectral components of the pulse arrive ahead of others. In a single-mode waveguide, the group velocity varies rapidly with frequency particularly near cutoff. In the microwave band, this pulse broadening reduces the bandwidth capacity of the system. For optical waveguides, the effect is relatively unimportant if the source has a small relative bandwidth, as is the case for most laser sources. It contributes significantly to pulse broadening only when the source (e.g., a light-emitting diode) has a broad linewidth.

The most important source of pulse broadening in optics, encountered in multimode waveguides, is ray dispersion. The ray picture used before for describing the propagation in step-index fibers clearly suggests that the average length of a ray reflected repeatedly at the core–cladding interface depends on the angle that this ray makes with the fiber axis. Since many different angles are permissible when the fiber radius is large, step-index fibers suffer from large ray dispersion. Ray dispersion can be precisely defined by considering the group velocities of the various modes that can propagate in the fiber.[12] Usually, low-order modes transmit signals faster than high-order modes. Even if the source is coherent and a single mode is launched into the fiber, the power is likely to be found distributed into many modes after a certain distance because of small defects in the fiber or because of bends. Square-law fibers, or optical waveguides incorporating hard lenses, suffer from almost negligible ray dispersion (<1 nsec/km) even if they are highly multimoded. This can be understood on a simple geometrical optics basis. Rays in square-law media are sinusoids that all have the same optical length, at least as long as the ray amplitude is not too large.[13] If the material dispersion can be neglected, they also have the same group length.

In conclusion, when a large transmission capacity is desired, one may use either single-mode fibers or square-law fibers. The choice between these two alternatives depends on the source available. The most attractive source for optical communication is the injection laser. The lifetime of this type of laser now exceeds 10,000 hr at room temperature, but mode control is still far from perfect. For moderate distances and transmission capacities, square-law fibers and light-emitting diodes are most attractive.

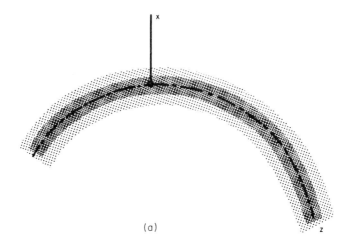

(a)

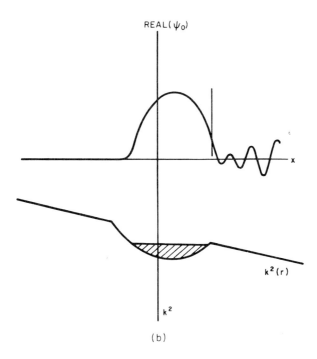

(b)

Fig. 1-17   (a) Curved graded-index fiber. (b) Because of the equivalence between curvature and constant gradient of refractive index, curving a fiber amounts to "tipping" the wavenumber law $k^2(x)$.

1.4 Fabrication of Optical Waveguides

Although it is not the main purpose of this book to discuss the technology of optical waveguides, a few indications may be useful to motivate the subsequent theoretical discussions. A brief review is given here.

The fabrication of ordinary "hard" lenses goes back to the sixteenth century. A major challenge over the past centuries in the design of hard lenses has been the reduction of geometrical optics and chromatic aberrations. Mathematically, geometrical optics aberrations result from the fact that the phaseshift between two points is not exactly a quadratic function of the point coordinates. This quadratic condition is approached, but can never be reached exactly, by suitable combinations of lenses with spherical surfaces, or by using lenses with aspherical surfaces. Chromatic aberration is due to the dependence of the refractive index n on the wavelength (or color). Because only small relative bandwidths are required in optical communication systems, chromatic aberration is usually not a problem for guiding systems using hard lenses. Because these light-guiding systems are operating close to the diffraction limit, geometrical optics aberrations can usually be neglected too. For hard lenses, the bulk loss in glass does not contribute significantly to the total loss, but it is important to prevent reflections from taking place at the air–glass interfaces by depositing antireflective coatings or orienting the lenses at the Brewster angle (for linearly polarized sources). Usually the loss per surface can be reduced with the help of antireflective coatings from 4 to 0.1%, in narrow bands. A pair of mirrors is sometimes to be preferred to a lens because mirrors do not suffer from Fresnel reflection. Scattering losses, which are due to minute irregularities of the lens or mirror surfaces, are important, particularly if many optical beams are transmitted through the same optical guide for increased transmission capacity. Scattering introduces cross talk between the different channels. The main technical difficulty that has been encountered with optical guides using discrete focusers is associated with accidental misalignments of the system. These are bound to happen in any realistic environment. Although various servoloop arrangements have been successfully implemented, the difficulty essentially remains.

Various methods have been devised to fabricate continuously guiding media with suitable refractive-index profiles. One method consists in sending a flow of cool gas in a hot tube as shown in Fig. 1-18a. The hot walls of the tube heat the outer layers of the gas, thereby reducing their density and refractive index.[13] The absolute change of refractive index is minute, yet sufficient to guide optical beams. This system, called a gas lens, suffers from geometrical optics aberrations mainly due to gravity. There is a tendency of the cooler parts of the gas to move to the lower parts of the waveguide, and this motion distorts the refractive-index profile. Further difficulties are associated with the rather large power required to heat the gas and the fact that the total change in refractive index being rather small ($\Delta n \sim 0.0001$), close lens spacing is required in bends.

It is possible to induce continuous changes in refractive index in solids such as glass by ion exchange, ion implantation, and other techniques. The most popular guiding system is perhaps the step-index fiber. Step-index fibers are usually made by pulling together two coaxial cylinders made of two glasses having different refractive indices, as shown in Fig. 1-18b.

The main sources of loss in glass fibers are the impurities and the scattering at the core–cladding interface and in the bulk. The development of techniques for making pure glasses has been a major factor in reducing the loss from some 1000 dB/km down to 1 dB/km.[14,15] An absorption loss of 0.8 dB/km has been measured for fused silica (suprasil). One part per billion of transition elements, such as iron or copper, or one part per million of the radical OH in the glass may result in absorption losses of the order of 1 dB/km. Because some of these impurities come from the crucible used to melt the glass, the best glasses are now prepared directly

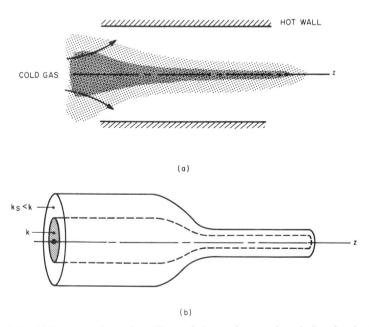

(a)

(b)

Fig. 1-18 (a) Structure of a gas lens. The gas being cooler near the axis, its refractive index is larger. (b) Clad fibers can be fabricated by pulling together two glass cylinders with different refractive indices, or by vapor phase deposition.

in tubing form by oxidizing vapors of silicon compounds (vapor phase deposition). The second source of loss is Rayleigh scattering, the scattering of light by inhomogeneities whose dimensions are small, but not negligible, compared to the wavelength. The wavelength dependence of this loss is λ_0^{-4}. Rayleigh scattering turns out to be negligible for glass for λ_0 above 1.2 μm. The inhomogeneities mainly result from the thermal motion of the molecules. In solids, the inhomogeneities are essentially frozen in at the temperature of solidification and are less pronounced if the solidification temperature is low. On that ground alone, glasses should be preferred to fused silica. In nonpolar liquids, such as tetrachloroethylene, Rayleigh scattering is the dominant source of loss (~ 40 dB/km at $\lambda_0 = 0.6328$ μm and 5 dB/km at $\lambda_0 = 1$ μm).[16,17]

The irregularities of the core–cladding interface turn out to be very small in step-index fibers, because of the strong surface tensions that smooth out the initial defects. The residual distortion of the interface scatters predominantly the higher-order modes of propagation because these modes correspond to larger ray angles with the surface. Graded-index fibers, in principle, do not suffer from this source of loss. The technology of glass fibers is rapidly evolving at the time of this writing.[18]

1.5 Transverse and Axial Coupling

When two guiding structures are put side by side, there is a certain amount of power transferred from one to the other. This arrangement, illustrated in Fig. 1-19, is essentially a directional coupler. It has application in integrated optics. If the two structures are identical, the coupling mechanism is most easily understood as resulting from the beat between symmetric and antisymmetric modes, which have slightly different propagation constants. If only one waveguide is excited at the input plane, the symmetrical and antisymmetrical modes are excited with equal amplitudes. After a certain distance, when the phases of these two modes differ by π, the power is transferred to the other waveguide. If the coupling is small, the beat length is large compared to the optical wavelength. Yet the physical size of such devices may not exceed a few tenths of a millimeter ($\sim 100\lambda$). If the substrate has electrooptics properties, the change of coupling can be effected by applying an electric field to the device.[19]

Analysis of the slab-coupled-rod waveguide in Fig. 1-20c involves the coupling between an open waveguide (an oversized dielectric rod) and a substrate (a dielectric slab) supporting radiation modes. This coupling helps reduce the number of modes that the dielectric rod would otherwise support.[20] The waveguide in Fig. 1-20a is similar to that in Fig. 1-13b. The

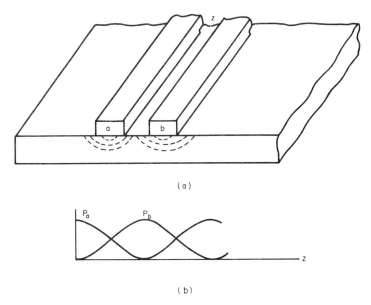

(a)

(b)

Fig. 1-19 (a) Coupled optical waveguides. (b) The beat wavelength increases as the spacing between the two waveguides increases. If the system is symmetrical and lossless, the power fed in guide *a* is eventually transferred completely to guide *b* and, from there, back to *a*.

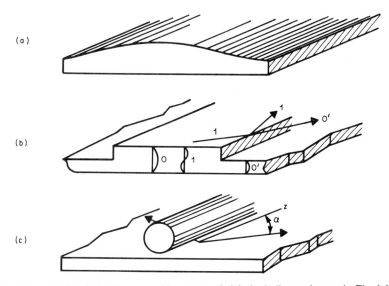

Fig. 1-20 (a) This dielectric waveguide, a tapered slab, is similar to the one in Fig. 1-13b. (b) A useful mode-selection mechanism is provided by the step shown here, which couples the various modes denoted 0, 1, ... (c) The central part is loosely coupled to a uniform slab. The high-order modes of the rod are attenuated through coupling to the faster slab modes.

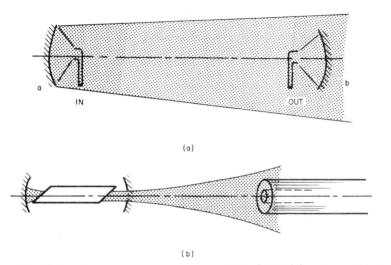

(a)

(b)

Fig. 1-21 (a) Coupling between two microwave antennas. Some of the power radiated by antenna *a* is collected by antenna *b*. (b) Coupling between a laser beam and a clad fiber.

waveguide in Fig. 1-20b establishes a transition between the waveguide in Fig. 1-20a, where the various slab modes are essentially uncoupled, and the slab-coupled rod in Fig. 1-20c, where this coupling plays an essential role.

The concept of axial coupling is involved in the evaluation of the power transferred from one microwave antenna to another (Fig. 1-21a), or in the coupling between a laser beam and an optical fiber illustrated in Fig. 1-21b. Evaluation of the response of optical resonators to given sources also involves expressions for the axial coupling between various beam modes.[21]

1.6 Optical Resonators

The simplest type of optical resonator is the plane-parallel Fabry–Perot resonator, two plane mirrors facing each other with a separation *d*, represented in Fig. 1-22a. In first approximation, the resonance condition is, as for a violin string, that an integral number *l* of half-wavelengths fits between the two mirrors (see Fig. 1-22b). The resonant wavelengths λ_l are consequently given approximately by

$$\left(\frac{2\pi}{\lambda_l} \right)d = l\pi, \qquad l = 1, 2, 3, \ldots \qquad (1.2)$$

l is called the axial mode-number.

The loss per transit is easily evaluated under the assumption that the two mirrors are in the far field of each other, as shown schematically in Fig.

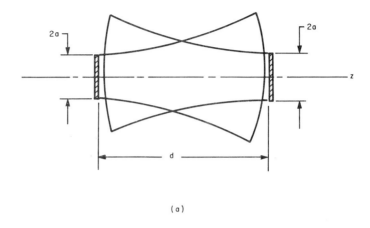

(a)

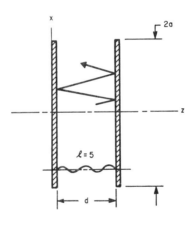

(b)

Fig. 1-22 (a) Optical resonator incorporating two plane circular mirrors facing each other. The loss per transit is easily calculated if the two mirrors are in the far field of each other ($a^2/\lambda d \ll 1$). (b) In two dimensions xz, the mode of resonance can be viewed as a high-order mode propagating in the $+x$- and $-x$-directions in a near-cutoff metallic plane waveguide.

1-22a. The angular divergence of the beam reflected by the mirror on the left side is, as we have shown before, $\alpha \sim \lambda/a$. The fractional power intercepted by the opposite mirror is therefore of the order of

$$\frac{\pi a^2}{\pi(\alpha d)^2} = \left(\frac{a^2}{\lambda d}\right)^2 \tag{1.3}$$

Thus, the loss per transit in decibels is approximately

$$\mathcal{L}_{dB} \sim 20 \log \frac{\lambda d}{a^2} \tag{1.4}$$

a very high loss in the range of application of the present theory ($a^2/\lambda d \ll 1$).

High-loss resonators are of some practical interest, owing to the very high gain of lasers such as the helium–xenon gas lasers or the dye lasers; in most cases, however, the loss per transit should not exceed a few percent if an oscillation is to be sustained. The reduction in loss is first accomplished by increasing the radii of the mirrors.

The operation of plane-parallel Fabry–Perot resonators can be understood by considering the two plane mirrors as a conventional metallic waveguide supporting high-order modes, the propagation taking place in the x-direction (see Fig. 1-22b). These waveguide modes, as is well known, can be represented by plane waves that zigzag between the two mirrors. They are in part reflected at the two open ends, thereby forming a standing-wave pattern in the x-direction. A fraction of the power is lost at both ends. Usually the mode number, that is, the number of half-wavelengths in the z-direction, is very large, perhaps 10^6. The waveguide operates very close to its cutoff for that mode, and the guided wavelength (in the x-direction) is much larger than λ; it is of the order of a, the mirror radius.[22] The loss of plane-parallel Fabry–Perot resonators is essentially the same as the loss of a periodic sequence of apertures discussed in Section 1.1. The critical parameter, here again, is the dimensionless factor $a^2/\lambda d$. If $a^2/\lambda d$ is much larger than unity, the loss per transit can be small.[23]

Still lower losses are observed when the mirrors are slightly curved. Because of this curvature, the separation between the two mirrors becomes smaller and smaller as the wave moves away from axis. At a certain distance from axis, the waveguide is below cutoff for the mode considered. Consequently the wave is reflected without even approaching the mirror edge. The rays representing the wave are bound in that case by a caustic curve. Thus the losses due to edge diffraction are small. Some loss nevertheless occurs because the fields extend slightly beyond the caustic. This type of resonator with curved mirrors is relatively insensitive to misalignment. Indeed, the tilt of one mirror merely amounts to a small displacement of the system axis, defined as the line joining the centers of curvature of the two mirrors. On the basis of the previous discussion, we are tempted to conclude that the losses become smaller and smaller as the curvature of the mirrors increases. This is not the case, however; it turns out that if the mirror radius of curvature is less than half the mirror separation, the rays, instead of being reflected back, "walk away." The resonator is then called "unstable"; it may exhibit very large losses.[24]

The following discussion is applicable only to stable resonators, where the rays, bouncing back and forth between the two mirrors, possess an

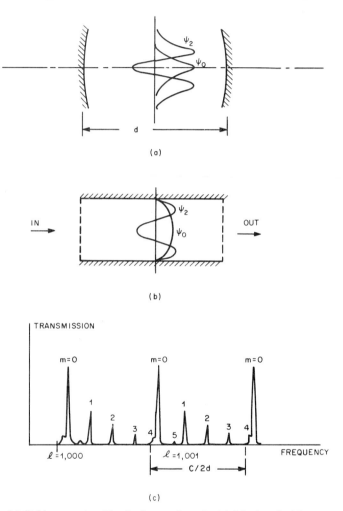

Fig. 1-23 (a) Stable resonator. The fundamental mode $\psi_0(x)$ is described by a gaussian function $\exp(-x^2)$. The fields of higher-order modes are obtained by multiplying $\exp(-x^2)$ by a Hermite polynomial. (b) For comparison, the modes of a metallic waveguide cavity are shown here. (c) A typical response of stable resonator is shown. l denotes the axial-mode number and m the transverse-mode number. Note that there is overlap between successive l series.

envelope (caustic). Assuming that the mirror edges are far from the caustic, it can be shown that the dependence on x of the fundamental mode of resonance $\psi_0(x)$ is gaussian, of the form $\exp(-x^2)$.[25] The higher-order modes $\psi_m(x)$ are described by the product of a function of Gauss and a Hermite polynomial of order m (see Fig. 1-23a). These modes essentially coincide with the modes of propagation in square-law media that we have discussed in previous sections. They differ from the modes of resonance of sections of metallic waveguides, short-circuited at both ends, shown in Fig. 1-23b, in that the irradiance patterns are not independent of z, the beam reaching its minimum size halfway between the two circular mirrors. The transverse field distribution is described by Hermite–Gauss functions instead of sine or cosine functions. In systems with rotational symmetry, the mode fields of optical resonators and metallic waveguides are described, respectively, by Laguerre–Gauss functions and Bessel functions. Thus, in open waveguides, the field extends to infinity, while in metallic waveguides, it has a finite support. However, in both cases the set of modes is orthogonal, and the number of nodes of the field coincides with the mode number m. Another important point of similarity between optical and metallic waveguides, which is useful to keep in mind, is that the phase velocity v_z always exceeds the free-space velocity (c, or c/n if the bulk refractive index is n). v_z also increases with the transverse mode number m. This is a consequence of the transverse dimension of the beam being limited. Because the wave vector has a transverse component, its axial component $k_z \equiv \omega/v_z$ is reduced. The resonance frequency of any transverse mode, $m = 0, 1, 2, \ldots$ is therefore higher than one would expect on the basis of the simple violin string model discussed earlier, and it further increases as m increases. A typical response of optical resonator is shown in Fig. 1-23c.

One of the major problems in the design of optical resonators is the elimination of unwanted modes. The frequency spacing between adjacent axial modes (e.g., $l = 1000$ and $l = 1001$) being relatively large, axial-mode selection may rest on the finite linewidth of the active medium incorporated in the cavity. In some cases, for instance for dye lasers that have a very broad linewidth, we need to introduce dispersive elements in the resonator, such as prisms or diffraction gratings. Combinations of beam splitters and auxiliary mirrors have also been proposed and successfully implemented to eliminate unwanted axial modes.[26]

The selection of a single transverse mode, usually the fundamental mode, is achieved with the help of an aperture. The round-trip loss introduced by an aperture increases with the mode number m. For the case of an aperture with gaussian transmittivity, for instance, the loss is proportional to $m + \frac{1}{2}$. Thus if the loss of the fundamental mode ($m = 0$) is 1%, the loss of the next higher-order mode ($m = 1$) is 3%. If the laser gain is equal to 2%, the oscillation is sustained only for the fundamental mode, which is the desired result.[27] A near confocal arrangement (radius of curvature $\approx$ spacing) with circular apertures appears to be near optimum as far as transverse mode selection is concerned.[28]

An important class of resonator is the class of "degenerate" resonators,[29,30] an example of which is shown in Fig. 1-24 incorporating two identical lenses and two plane mirrors located at the lens focal planes. The condition for a resonator to be degenerate is that arbitrary rays recycle, that is, retrace their path after a round trip.[31] It is not difficult to see that this is the case for the resonator shown in Fig. 1-24. It is sufficient to verify the condition for two independent rays, defined as rays that do not intersect the axis at the same point. When this geometrical optics condition is satisfied, any field configuration reproduces itself after a round trip in the resonator, except for an amplitude and phase factor. Thus, any field distribution can be considered a "mode" of the resonator, as long as aberrations remain negligible. An example of degenerate optical cavity free of aberration is the Maxwell fish-eye medium with continuous refractive-index law $n(r) = n_0(1 + r^2)^{-1}$, shown before in Fig. 1-14a (in two-dimensional form) as the "projection" of a spherical configuration lens. In such a medium, all the rays are circles. Degenerate resonators provide frequency filtering of optical waves without introducing at the same time spatial filtering, that is, without distorting the field pattern.

Let us briefly explain the mechanism of operation of laser gyroscopes. For simplicity, we consider a circular perfectly conductive boundary, as

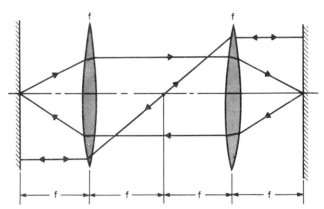

Fig. 1-24 Degenerate optical cavity. A cavity is degenerate to first order if all the rays recycle. This is verified in the figure for two independent rays. Degenerate optical cavities provide frequency filtering of optical beams without introducing spatial filtering.

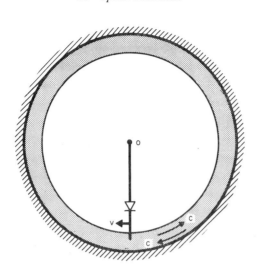

Fig. 1-25 This figure illustrates the principle of the laser gyroscope. Because of the Doppler effect, the beat frequency is $\Delta f/f = 2v/c$, if v denotes the linear velocity of the detector. The rotation of the metallic boundary is immaterial.

shown in Fig. 1-25. We have seen that waves can cling to the concave sides of curved boundaries (whispering-gallery modes). These waves propagate in the clockwise or counterclockwise direction with a velocity close to c. (The velocity is exactly equal to c only along the caustic line, but this distinction is not important here.) Let us now suppose that a detector is rotating with a tangential velocity v as shown in the figure. Because of the Doppler effect, the relative frequency difference between clockwise and counterclockwise waves that is detected is

$$\frac{\Delta f}{f} = \frac{2v}{c} \tag{1.5}$$

It is immaterial whether the perfectly conductive boundary is fixed or rotates together with the detector. Thus the above expression is applicable to laser gyroscopes, where the coupling device rotates together with the resonator itself. Rotation rates as low as 10^{-3} rad/hr have been measured.[32]

In previous sections we have discussed a few applications of beam optics at visible or infrared wavelengths. Another important field of application of beam optics is in the microwave range. Millimeter waves are strongly attenuated by rain.[33] The attenuation at 100 GHz exceeds 40 dB/mile in New Jersey 50 min/yr on the average. Yet beams of millimeter waves can be used in cities. Quasi-optical devices such as focusers and diplexers also prove useful in feeding the main reflector of an antenna transmitting information to a synchronous satellite or collecting information from it.

The subsequent sections of this chapter deal with a comparison between mechanics and optics. They can be omitted on first reading.

1.7 The Mechanical Theory of Light

There is a close analogy between the time-dependent Schrödinger equation, applicable to nonrelativistic particles, and the scalar parabolic equation applicable to radio waves propagating in atmospheric ducts or optical waves in weakly guiding optical fibers, which was first discussed by Fock.[34] The optical-wave equation follows from the substitutions $t \rightarrow z$, $m/\hbar \rightarrow k(0)$, $-eU(x)/\hbar \rightarrow k(x)$ in the Schrödinger equation, using conventional notations.[§] This analogy is for a given color of the ray or a given energy of the particle. Dynamical effects such as radiation pressure are not essential in principle to understand problems of linear propagation. They are often deemed too small to be of much interest in optics. Recent experiments, however, have shown that radiation forces are very significant for small objects, and applications have been suggested.[35] The major motivation for considering forces in this chapter is that dynamical arguments often provide new insights and simple derivations concerning propagation problems.[36-38] After a brief historical review, we shall discuss the dynamical aspects of light waves and compare them to those of matter waves.

A mechanical model for the refraction of light rays was proposed in 1637 by Descartes (see, for instance, Sabra[1]), who noted that optical-ray trajectories (defined as curves in ordinary space) may coincide with the trajectories of particles that have a nonzero rest mass. The law of refraction, discovered at about the same time by Snell, was derived, perhaps independently, by Descartes from an analysis of the trajectory of massive particles traversing a sheet that reduces their velocity by a constant factor, that is, by a factor independent of the angle of incidence. To dispel a frequent misunderstanding, let us point out that the analogy does not suggest that the velocity of the light pulses is proportional to the velocity of the equivalent massive particles. Only trajectories in space are compared in that analogy. The Descartes theory is in agreement with modern concepts, provided the quantity that he calls the "determination" of the ray is

---

[§] e, m denote the particle charge and mass, respectively, $\hbar$ the Planck constant divided by 2π, t time, and $U(x)$ the electric potential. The quantity $k(x)$ denotes the wavenumber $= 2\pi/$wavelength in the medium.

understood as the ray "canonical momentum," or wave vector. However, Descartes himself was not aware of the wave nature of light, nor of that of massive particles. The dynamical theory of light proposed later by Newton is not general because the dispersive properties of refractive media are usually different from those of massive particles in free space. Newtonian dynamics is applicable to cold plasmas and to thick dielectric slabs because these media happen to have the same dispersion as massive particles in free space. Just two centuries after Descartes' proposal, Hamilton proved that the equations for rays and the equations for massive-particle trajectories could be cast in the same general form. The Hamilton equations form an essential part of modern physics because they are more clearly suggestive of the wave nature of massive particles than the Galileo principle of inertia and because they can be generalized to physical continua. Even after Hamilton's contribution, a number of difficulties remained concerning the dynamics of optical-wave packets that have been resolved only during the last ten years. These difficulties will be discussed in some detail. Readers interested only in concrete experiments concerning the dynamics of waves may go directly to Section 1.8 and omit the rest of this section.

The similarity between the motion of optical pulses and that of massive particles rests on their respective dispersion equations. A dispersion equation is denoted, in concise form, $H(\mathbf{K}, \mathbf{X}) = 0$, where $\mathbf{X} \equiv \{\mathbf{x}, it\}$ denotes a point in space–time and $\mathbf{K} \equiv \{\mathbf{k}, i\omega\}$ denotes the 4-wave vector. (c is set equal to unity, and near linearity of the wave equation is assumed.) The function $H(\mathbf{K})$, at some $\mathbf{X}$, can be obtained for any particular system with the help of interferometric measurements. For charged particles in a potential vector $\mathbf{A}(\mathbf{X}) \equiv \{\mathbf{a}, iV\}$, where $\mathbf{a}$ denotes the usual 3-vector potential and V the scalar potential, for example, the following dispersion equation holds:

$$|\mathbf{K} - \bar{e}\mathbf{A}(\mathbf{X})|^2 + \bar{m}^2 = 0 \qquad (1.6)$$

where $\bar{e}$ and $\bar{m}$ denote constants that we shall define later. The relativistic equation of motion of a charged particle in arbitrary electric and magnetic fields follows from (1.6) with the help of the Hamilton equations (see Chapter 4), provided the radiation reaction can be neglected.

The canonical momentum of a wave packet can be shown to have the form $J\mathbf{K}$, where J denotes an adiabatic invariant, that is, a quantity that does not change when the wave packet moves through a slowly varying lossless medium. This conclusion is fundamental. It ties together the different analogies that have been proposed from special points of view. At small intensities, it is observed that the adiabatic invariant J is an integral multiple of a universal constant, denoted $\hbar$ ($=$ Planck constant divided by 2π). This fact does not follow from the classical wave theory discussed so far, but requires the second quantization. According to Dirac, the wave functions themselves should be considered noncommuting operators. Because we do not wish to go beyond the classical wave theory, we shall use reduced masses and charges, $\bar{m} = m/\hbar$ and $\bar{e} = e/\hbar$, instead of the actual particle mass m and charge e. Thus, formally, the Planck constant does not enter in the discussion.

The above general statements need to be qualified. First of all, only scalar wave equations have been considered, spin and polarization effects being omitted. A scalar approximation of the optical field is reasonable when the transverse changes of dielectric constant are small. (If the changes are piecewise continuous, the wave function and its first derivatives remain continuous.) Fortunately, this is the case for most glass fibers of current interest and for gas lenses. Polarization effects play an essential role only in special coupling problems. As far as massive particles are concerned, scalar wave equations are known to be applicable to the classical field of the π meson, an unstable massive particle responsible for the strong interaction. The electron being a more familiar particle than the π meson, it will be considered for the sake of illustration. The electron spin, as well as the photon spin, can be neglected in many experiments.

It should be noted that different functions $H(\mathbf{K}) = 0$ may represent the same dispersion (hyper) surface, the latter being the physically significant object. The dispersion surface defines the magnitude of the vector $\mathbf{K}$ in some given direction. Furthermore, there exist transformations (called gauge transformations) that change the dispersion surface without affecting the physical properties of the system. This is the case when the phase of the field cannot be observed. For example, addition to $\mathbf{A}$ in (1.6) of the gradient of an arbitrary scalar function of $\mathbf{X}$ leaves the wave-packet trajectories and the interference patterns unchanged. The phase of ψ is modified, but that phase is not observable for electrons (or, more generally, for fermions).

In (1.6), the potential vector $\mathbf{A}(\mathbf{X})$ plays the role of a medium. It is useful to consider a few other examples of media. In a crystal, a charged particle is submitted to spatially periodic potentials. Solutions of the Schrödinger equation lead to a dispersion equation that is almost the same as that of a free particle with a mass $m^* \neq m$. This effective mass m^* depends on the constants of the lattice and can therefore vary with $\mathbf{X}$. Intense rf fields also have the effect of changing the apparent masses of charged particles. For optical waves, a medium is usually a collection of free charges (e.g., a cold plasma) or a collection of polarizable atoms. Artificial dielectrics, such as meander lines, also constitute acceptable media. It is not our intention to discuss microscopic models here, or to derive the dispersion equation from first principles. The function $H(\mathbf{K})$ is assumed to be known within the indeterminacies pointed out before, perhaps from interferometric measurements.

Let us now review the problems that we shall discuss in subsequent sections. In Section 1.8, the Descartes analogy between the spatial trajectories of optical rays in spatially isotropic media and nonrelativistic massive particles is discussed. Formally, this analogy amounts to stating the proportionality between the respective **k**-vectors, the frequencies being fixed. This analogy is extended to continuous media and to relativistic particles. When the charged particle is submitted to a magnetic field, the dispersion equation is no longer spatially isotropic, as (1.6) shows, the potential vector **a** introducing a preferred direction. A somewhat similar situation is encountered in optics when the medium is moving, as will be shown in Chapter 3. In Section 1.9, experiments involving dynamical concepts are discussed: radiation forces and mass-carrying momenta. Wave effects are discussed in Section 1.10. We compare first the scalar Helmholtz equation applicable to time-harmonic optical waves and the Klein–Gordon equation applicable to massive particles. Approximate forms that are applicable, respectively, to paraxial optical beams and to paraxial electron waves are subsequently obtained.

To conclude this section, let us note that the coherent interaction of matter waves with matter, relevant to our general discussion, is sometimes found in unlikely places. For instance, ultrarelativistic particles traversing a crystal are expected to interact with only one atom at a time, on the ground that the de Broglie wavelength is very small compared with the lattice constant. In fact, the interaction is coherent over macroscopic distances because the relevant quantity is the change in momentum, which is small, rather than the momentum itself. The radiation of ultrarelativistic charged particles in matter is, in fact, very similar to the radiation of optical waves, because the rest mass of ultrarelativistic particles constitutes only a small part of the total mass.

1.8 Rays in Isotropic and Anisotropic Media

Let us consider first the Descartes analogy, which compares trajectories in isotropic spaces. This analogy seems simple at first, but its interpretation is, in fact, difficult because time and space are entangled. In the next section, we clarify the assumptions that are essential to the analogy. Rather deep concepts are needed for a full understanding.

Let us associate to any ray a vector $J\mathbf{k}$, called the ray canonical momentum, which has the direction of the ray. J denotes a quantity that remains a constant along any given ray. The motivation for the notation $J\mathbf{k}$ will appear later. (Similarly, we shall denote the canonical energy $J\omega$.) The magnitude of $J\mathbf{k}$ is assumed to depend only on the medium in which the ray propagates. It is further assumed that the intrinsic properties of the ray (e.g., its color) remain the same as the incidence angle i is varied. The Descartes–Snell law of refraction, which states the constancy of $\sin(i)/\sin(i')$ at a plane interface as i varies, clearly follows from the invariance of the tangential component of the vector $J\mathbf{k}$ (see Fig. 1-26b). To make the invariance of the tangential component of $J\mathbf{k}$ plausible, Descartes noted a similar situation in mechanics. Consider a ball traversing a breakable sheet or rubber band (Fig. 1-26a). Because the rubber band does not exert any force on the ball in the horizontal direction, the horizontal component of the ball momentum $J\mathbf{k} = m\mathbf{u}$ is invariant. On the other hand, the energy absorbed by the rubber band before it is ruptured is independent of the angle of incidence of the ball as we can see from a quasi-static analysis. Assuming isotropy of space, the energy of the ball depends only on the magnitude of the momentum $J\mathbf{k}$. It follows that the ratio $|J\mathbf{k}'|/|J\mathbf{k}|$ of the magnitudes of the ball momenta before and after traversing the sheet is

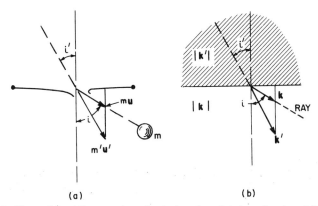

(a)   (b)

Fig. 1-26 Illustration of Descartes' mechanical analogy for the refraction of light rays. (a) A ball traverses a sheet that reduces its momentum $m\mathbf{u}$ by a factor independent of the incidence angle. The tangential component of $m\mathbf{u}$ is invariant because the sheet does not exert a force on the ball in that direction. (b) A light ray is refracted away from the normal by going to a less dense medium. The tangential component of a vector $\mathbf{k}$, whose magnitude is independent of the incidence angle, is invariant. $\sin(i)/\sin(i')$ is a constant in both cases. (From Arnaud,[38] by permission of the American Optical Society.)

independent of the angle of incidence. From the geometric construction in Fig. 1-26, the law $\sin(i)/\sin(i') = $ const follows.

Let us now generalize these results to continuous media. Considering that only *proportionality* between the canonical momenta is required for the analogy to hold, and taking as an empirical fact that the canonical momentum of a nonrelativistic particle is $m\mathbf{u}$, where m denotes the mass and $\mathbf{u}$ the velocity, the mechanical analogy just described calls for the correspondence

$$u \rightleftharpoons n \tag{1.7}$$

where n denotes the medium refractive index, defined as $|J\mathbf{k}|_{\text{med}}/|J\mathbf{k}|_{\text{vac}}$. From Newton's nonrelativistic dynamics, the trajectory in space–time $\mathbf{x}(t)$ of a particle in a gravitational potential $V(\mathbf{x})$ obeys the differential equation

$$\frac{d^2\mathbf{x}}{dt^2} = -\nabla V(\mathbf{x}) \tag{1.8a}$$

where ∇ denotes the gradient operator, with components $\partial/\partial x$, $\partial/\partial y$, and $\partial/\partial z$. If the total energy of the particle is taken equal to zero, the magnitude u of its velocity is given by $\frac{1}{2} mu^2 = -mV(\mathbf{x})$. Thus, replacing u by n according to (1.7), (1.8a) can be written[9,39]

$$\frac{d^2\mathbf{x}}{dt^2} = \nabla\left[\tfrac{1}{2} n^2(\mathbf{x}) \right] \tag{1.8b}$$

Equation (1.8b) together with the relation $ds/dt \equiv n$, where $ds \equiv |d\mathbf{x}|$ denotes the elementary ray length, is the equation for light rays (of a given color) in continuous media. This analogy is exemplified in Fig. 1-27, where we compare the trajectory of a ball in a shallow gutter and the trajectory of a light ray in a two-dimensional graded-index fiber. If V (respectively, k^2 or n^2) is quadratic in x, for example, the rays are (exactly) sinusoids in both systems. The quantity denoted t in (1.8b), which has the dimension of length, does not, in general, represent the time of flight of optical pulses. What we are comparing here are trajectories in space. Time-dependent concepts have been used in the discussion because the dispersive properties of massive particles happen to be known from Newton's dynamics. The dispersive property of the refractive medium, however, remains undefined; that is, the dispersion of the refractive medium may or may not be the same as that of massive particles. Most likely, it will be different. The pulse velocities would then be different too. When the dispersive properties of stratified media are the same as those of particles, the horizontal component of the group velocity is a constant of motion, and the transit

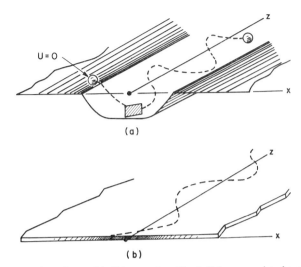

(a)

(b)

Fig. 1-27 (a) Motion of a steel ball in a gutter. The ball is assumed to have zero velocity (zero energy, by definition) when located on the top. The ball falls to the bottom of the gutter where it is deflected by a plate. Its subsequent motion is defined by the gravitational potential $V(x)$, which corresponds to the profile of the gutter if the gutter is sufficiently shallow. The axial velocity is a constant of motion. (b) Planar optical waveguide whose permittivity varies slowly with x. The ray trajectories (in space only) are the same as in (a) if the local wave vector squared $k^2(x)$ is chosen proportional to $V(x)$. (From Arnaud,[38] by permission of the American Optical Society.)

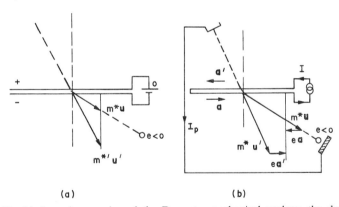

(a)   (b)

Fig. 1-28 (a) A modern version of the Descartes mechanical analogy, the sheet being replaced by a pair of parallel grids. $m^* = m - eV$ denotes the "moving mass." The ratio of the sines of the angles to the normal is a constant, as a result of isotropy. (b) The charged particle traverses two sheets of current, creating a discontinuity in potential vector $\mathbf{a}$. The tangential component of the canonical momentum $J\mathbf{k} \equiv m^*\mathbf{u} + e\mathbf{a}$ is invariant. The momentum transferred to the sheets is opposite to the change of $m^*\mathbf{u}$, as can be seen by considering a steady flow of charges (current) from the emitter to the collector. (From Arnaud,[38] by permission of the American Optical Society.)

time of a pulse depends only on the slope at the origin (Breit and Tuve theorem).[40]

Let us now consider a modern version of the Descartes analogy, applicable to relativistic particles. Consider a charged particle with charge e traversing two closely spaced grids with a difference of potential V between them, as shown in Fig. 1-28. Let $\mathbf{X}(\tau)$ represent the particle trajectory, where $\mathbf{X} \equiv \{\mathbf{x}, it\}$ and τ the proper time with $d\tau = i|d\mathbf{X}|$ if we set $c = 1$. We have, by definition,

$$\left[\frac{dx_1}{d\tau} \right]^2 + \left[\frac{dx_2}{d\tau} \right]^2 + \left[\frac{dx_3}{d\tau} \right]^2 - \left[\frac{dt}{d\tau} \right]^2 + 1 = 0 \qquad (1.9)$$

The canonical momentum of the particle, whose tangential component is invariant at a plane interface, can, in the present case, be taken equal to the mass-carrying momentum

$$J\mathbf{k} = m \frac{d\mathbf{x}}{d\tau} \qquad (1.10)$$

Upon traversing the grids, a constant energy eV needs to be added to the mass-energy $m(dt/d\tau)$ to maintain the canonical energy $J\omega$ a constant. Thus

$$J\omega = m \frac{dt'}{d\tau} = m \frac{dt}{d\tau} + eV \qquad (1.11)$$

The ratio of the magnitudes of the canonical momenta is, using (1.9),

$$\frac{|J\mathbf{k}|}{|J\mathbf{k'}|} = \frac{|d\mathbf{x}/d\tau|}{|d\mathbf{x'}/d\tau|} = \left[\frac{(dt/d\tau)^2 - 1}{(dt'/d\tau)^2 - 1} \right]^{1/2} = \left[\frac{(J\omega - eV)^2/m^2 - 1}{(J\omega)^2/m^2 - 1} \right]^{1/2}$$

$$(1.12)$$

This ratio being independent of the incidence angle, $\sin(i)/\sin(i')$ is a constant, as is the case for nonrelativistic particles. Two essential postulates were made in the above discussion: medium isotropy and invariance of the canonical momentum in directions of translational invariance of the medium. In a wave theory, the latter follows from the proportionality of the canonical momentum to the wave vector. Thus the law of refraction follows quite generally from isotropy and translational invariance of the medium. A generalized form of the law of refraction is needed if we omit the assumption of isotropy. The essential concepts of ray propagation appear, in fact, most clearly when the medium lacks isotropy. Let us consider a charged particle traversing two closely spaced grids carrying equal and opposite current densities, as shown in Fig. 1-28b.

These sheets of current create a discontinuity in the potential vector **a**, which is otherwise uniform. The refraction of the charged particle follows from the invariance of the tangential component of the canonical momentum

$$J\mathbf{k} = m\,\frac{d\mathbf{x}}{d\tau} + e\mathbf{a} \qquad (1.13a)$$

In the present case, the mass-carrying energy

$$m\,\frac{dt}{d\tau} = J\omega \qquad (1.13b)$$

is a constant, and the ratio $\sin(i)/\sin(i')$ varies with i. This is because the potential vector creates an anisotropy in space. Note that even in the absence of a magnetic field, it is permissible to add to $m(d\mathbf{x}/d\tau)$ a term of the form $e\mathbf{a} = e\nabla f(\mathbf{x})$, where f is an arbitrary scalar function of $\mathbf{x}$. This has no consequence, however, on the ray trajectories or on the interference patterns. In what follows, we shall assume, for definiteness, that **a** is the volume integral of the current density divided by the distance to the observation point (Coulomb gauge).

1.9 Experiments in Dynamics

Let us clarify further the difference between the canonical momentum $J\mathbf{k}$ and the mass-carrying momentum $m(d\mathbf{x}/d\tau)$ that were introduced in the previous section. This difference is essential to understand the Hamiltonian formalism. A confusion between these two momenta, however, is sometimes made for light waves. For example, it has been asserted that if a light pulse goes through a dielectric slab free of loss and of Fresnel reflection (see Fig. 1-29), the slab is displaced toward the source of light. This conclusion violates the law of mass–energy equivalence.[41] The error is to use the canonical momentum $(J\mathbf{k})$ of the light pulse in a problem in which the mass-carrying momentum $m(d\mathbf{x}/d\tau) = E(d\mathbf{x}/dt)$, where E denotes the pulse energy, should be used. The magnitude of the canonical momentum increases as the light pulse enters from vacuum into a medium with $n > 1$. The magnitude of the mass-carrying momentum, on the contrary, decreases since the group velocity $|d\mathbf{x}/dt|$ in a lossless medium is always less than c.

The displacement of the dielectric slab follows from the invariance of the sum of the mass-carrying momenta of the optical pulse and of the slab, with respect to an *inertial* reference system. The dielectric slab displace-

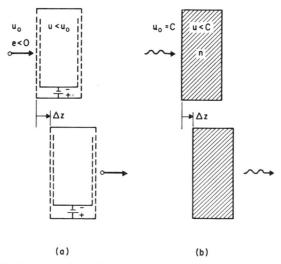

(a)   (b)

Fig. 1-29 Displacement of a slab traversed by a particle. Losses and reflection are neglected. (a) An electron traverses a potential box. Because u decreases as the electron enters the box for the voltage shown, the box displacement is forward. (b) In the optical case, the slab displacement is always forward ($u < c$). In such arrangements, only the mass-carrying momenta enter. (From Arnaud,[38] by permission of the American Optical Society.)

ment is easily found to be always in the *forward* direction and equal to $Em_0^{-1}(u^{-1} - 1)$, where m_0 denotes the mass per unit length of the slab, and u the group velocity in the slab ($c = 1$). The slab displacement can alternatively be evaluated, using classical electrodynamics, from the force exerted by the magnetic field of the incident wave on the polarization currents induced by the electric field.[42,43] A difficulty was noted for magnetizable bodies. This difficulty has been resolved recently by the introduction of a force of relativistic origin called the magnetodynamic force.[42] The displacement of the slab would be exceedingly difficult to measure. An experimental technique using the low-loss glass fibers that are now fabricated, however, may prove practicable.[43] A mechanical system analogous to the dielectric slab is the potential box shown in Fig. 1-29a. If the potential in the box is negative (for an electron, with $e < 0$), the box is displaced in the forward direction, as in the optical experiment.

If the light pulse enters under oblique incidence in the refractive medium, a momentum is transferred to the medium in the tangential direction because the tangential component of the mass-carrying momentum, unlike the tangential component of the canonical momentum, is not a constant. This tangential momentum is not present in the case of the potential box because the mass-carrying momentum happens to equal the canonical momentum. Generally speaking, the canonical momentum can differ from the mass-carrying momentum only if a third body (a medium) is present to absorb the difference in momenta. Such a medium is absent for a free particle, a cold plasma, and a smooth uniform waveguide. This is the physical connection that exists among these three seemingly unrelated systems.

Unlike mass-carrying momenta, canonical momenta are defined with respect to (noninertial) reference systems attached to the medium, or to the source of the potential vector for the case of charged particles. Because the reference system is not inertial, conservation of the total momentum is not expected to hold, except in directions of translational invariance. To understand this point, let us consider a simple mechanical system; with respect to earth (which cannot be considered inertial if it interacts with an object no matter how small the earth's acceleration might be), the momentum of a bouncing ball is not invariant since the ball can rebound off the ground. However, the horizontal component of the ball momentum is invariant if the ground is flat and smooth.

In the presence of a potential vector, the two momenta (canonical and mass carrying) are quite different. To clarify this point, let us go back to the situation described earlier of a charged particle moving through two parallel current sheets. The tangential component of the particle canonical momentum is invariant. The momentum transferred to the current sheets, however, is opposite to the change of *mass-carrying* momentum of the particle. This mass-carrying momentum has a component in the plane of the sheets that is generally different from zero.

With respect to the inertial system, we are dealing with two objects: the charged particle on the one hand, and the current carrier on the other hand. When the reference system is attached to the current sheets, the situation is different. The current sheets should no longer be viewed as physical objects, but rather as specified distributions of current. There is now an interaction electromagnetic momentum that needs to be taken into account. It is obtained by integrating $\mathbf{D} \times \mathbf{B}$, where $\mathbf{D}$ denotes the electric induction of the charge and $\mathbf{B}$ the magnetic field of the current sheets, over the volume enclosed by the sheets ($\mathbf{B}$ being equal to zero outside the sheets). It can be proved that this electromagnetic momentum is precisely the term $e\mathbf{a}$ in (1.13a). Thus, in this (noninertial) coordinate system, the sum of the tangential particle *and* interaction electromagnetic momenta is invariant.

Let us now discuss in some detail the problem of radiation force. Consider a light beam carrying a power P absorbed in a liquid, where the phase velocity is $\omega/|\mathbf{k}|$ (see Fig. 1-30). Experiments made by Jones and

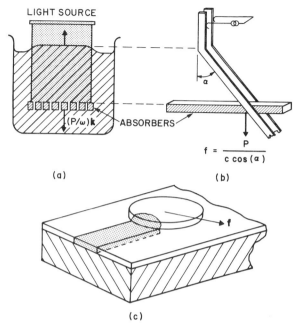

Fig. 1-30 Experiments in dynamics. A light beam emitting a power P exerts on an absorber immersed in a fluid a force $(P/\omega)\mathbf{k}$ (Jones and Richards' experiment). Perforations allow the absorber to move vertically in the fluid. The liquid is forced upward through the perforations and makes the surface of the liquid bulge out, in the steady state. (b) Equivalent model where the absorber can slide between the two wires. These wires (representing the liquid with index $n = 1/\cos \alpha$) make an angle α with the vertical. (c) Proposed experiment to verify that the force on an absorber need not have the direction of the beam if the medium lacks isotropy. This disk is a lossy dielectric picking up the optical power from the film through tunneling. (From Arnaud,[38] by permission of the American Optical Society.)

Richards[44] have shown that the force exerted by the light beam on the absorber is correctly given in magnitude by

$$\mathbf{f} = \frac{P}{\omega} \mathbf{k} \qquad (1.14)$$

A simple derivation of (1.14) is through the Doppler effect,[45] or the adiabatic invariance of the photon number in a resonator. These arguments are not completely satisfactory because appeal must be made to the second quantization ($E = \hbar\omega$). The classical derivation, applicable to incompressible fluids in the steady state, is based on the Minkowski expression for the stress density (or Abraham's, which is the same for isotropic media).[46] For scalar waves, the proof follows from the expression

$(\partial \bar{\mathcal{L}} / \partial \mathbf{K}) \mathbf{K}$ of the canonical stress-energy density, where $\bar{\mathcal{L}}$ denotes the average Lagrangian density. The so-called wave action $\partial \bar{\mathcal{L}} / \partial \mathbf{K}$ can be shown to be adiabatically conserved in lossless media.[47] The quantity denoted J is the flux of $\partial \bar{\mathcal{L}} / \partial \mathbf{k}$, the spatial component of the wave action, through the cross section of the beam. Thus J is adiabatically invariant in lossless media as stated earlier.

The balance of momenta clearly shows that in such experiments (Fig. 1-30a), in the steady state, the surface of the liquid bulges out (toward the less dense medium), the surface of the liquid being submitted to a net force[45]

$$f = -\frac{P}{c}(n-1) \tag{1.15}$$

balanced by surface tension. This bulging out is expected to take place after a time of the order of the time it takes for a sound wave to go from the absorber to the liquid surface. (See, however, Robinson.[48])

Most liquids being isotropic, only the magnitudes of $\mathbf{f}$ and $\mathbf{k}$ can be compared. It is therefore interesting to consider the arrangement in Fig. 1-30c, where the light propagates in a thin anisotropic film, perhaps a few micrometers thick. An absorbing disk that can slide on top of the film is submitted, according to (1.14), to a force $\mathbf{f} = (P/\omega)\mathbf{k}$. This force, in general, does not have the direction of the beam. This experiment has not been performed yet. It may be easier to perform than the original experiment of Jones and Richard because the radiometric forces that cause great experimental difficulties in the case of an absorber located in a gas (Crooke radiometer) or a liquid are absent in this new suggested experiment.

1.10 Wave Propagation

We wish to compare in this section the Schrödinger equation, applicable to nonrelativistic particles, and the Fock equation, applicable to paraxial beams. The latter will be discussed in more detail in Chapter 3.

The dispersion equation of an isotropic medium at some angular frequency ω is, ignoring for simplicity the y-coordinate,

$$k_x^2 + k_z^2 = k^2(x) \tag{1.16}$$

The (circular) dispersion curve is shown in Fig. 1-31a. Upon substitution of $i\mathbf{k} \to \nabla$, the scalar Helmholtz equation is obtained:

$$\left[\frac{\partial^2}{\partial x^2} + \frac{\partial^2}{\partial z^2} + k^2(x) \right]\psi = 0 \tag{1.17}$$

Consider now the dispersion equation (1.6) for massive charged particles at an angular frequency $\omega = \overline{m}$. Separating the space and time components of (1.6) and setting $\mathbf{a} = \mathbf{0}$, we obtain

$$k_x^2 + k_z^2 - [\overline{m} - \overline{e}V(x)]^2 + \overline{m}^2 = 0 \tag{1.18}$$

where $\overline{e}$ and $\overline{m}$ stand for $e/\hbar$ and $m/\hbar$, respectively. Upon substitution of $i\mathbf{k} \to \nabla$ in (1.18), the Klein–Gordon equation is obtained:

$$\left[\frac{\partial^2}{\partial x^2} + \frac{\partial^2}{\partial z^2} + \overline{e}^2 V^2(x) - 2\overline{m}\overline{e}V(x) \right]\psi = 0 \tag{1.19}$$

which is equivalent to (1.17) if

$$k^2(x) = \overline{e}^2 V^2(x) - 2\overline{m}\overline{e}V(x) \tag{1.20}$$

For the nonrelativistic case, the term $\overline{e}^2 V^2$ can be neglected. Because the total energy is equal to zero, we have $\overline{e}V + \frac{1}{2}\,\overline{m}u^2 = 0$. Thus (1.20) requires the equality of k and $\overline{m}u$ for nonrelativistic particles, as we have indicated before.

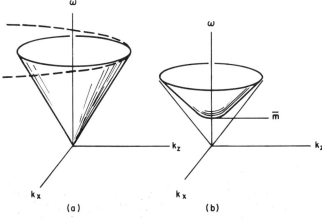

Fig. 1-31 Dispersion surfaces for optical waves in isotropic nondispersive media and massive particles (mass m) in free space. (a) The parabolic approximation used in beam optics consists in replacing the circle ($k_x^2 + k_z^2 - k^2 = 0$) at some ω by a parabola. This parabola is shown as a dashed line. (b) The nonrelativistic approximation in mechanics consists in replacing the hyperboloid by a paraboloid in the neighborhood of $\omega = \overline{m}$. The law of diffraction in free space and the law of spreading of pulses in dispersive media have the same general form. They depend solely on the curvatures of the dispersion surface in the k_x, k_z planes and k, ω planes, respectively. (From Arnaud,[38] by permission of the American Optical Society.)

Let us now introduce paraxial approximations on both (1.16) and (1.18). If we assume that k_x in (1.16) is small compared with k_z, we can write

$$k_z \approx k - \frac{1}{2k} k_x^2 \qquad (1.21)$$

With the substitution $i\mathbf{k} \to \nabla$, we obtain from (1.21) a scalar parabolic wave equation

$$- i \frac{\partial \psi}{\partial z} = \left[k(x, z) + \frac{1}{2k_0(z)} \frac{\partial^2}{\partial x^2} \right] \psi \qquad (1.22)$$

where, in the second term in the brackets, we have made the approximation $k(x, z) \approx k(0, z) \equiv k_0(z)$, on the ground that k does not vary very much with x. To make contact with the optical problem, let us indicate that in order that the reciprocity theorem be applicable in a natural way to (1.22), ψ must be defined as $[k_0(z)]^{1/2}E$, where E denotes a transverse component of the electric field of the beam. For the case where k_0 is independent of z, the wave equation (1.22) can be compared to the Schrödinger equation

$$- i \frac{\partial \psi}{\partial t} = \left[-\bar{e}U(x, z) + \frac{1}{2\bar{m}} \frac{\partial^2}{\partial x^2} \right] \psi \qquad (1.23)$$

The correspondence between (1.22) and (1.23) is $t \to z$, $\bar{m} \to k_0$, and $-\bar{e}U(x, z) \to k(x, z)$, as stated earlier. If k_0 were a function of z, a similar correspondence would be established merely by redefining the axial coordinate.

We now discuss an even closer physical similarity between paraxial optical waves and paraxial electron waves. Let us consider a charged massive particle propagating in a direction close to the z-axis. Because k_x is small compared to k_z, (1.18) can be written (in the gauge such that $\omega = \bar{m}$)

$$k_z \approx (\bar{e}^2 V^2 - 2\bar{m}\bar{e}V)^{1/2} - \tfrac{1}{2}(\bar{e}^2 V^2 - 2\bar{m}\bar{e}V)^{-1/2} k_x^2 \qquad (1.24)$$

With the substitution $i\mathbf{k} \to \nabla$, a parabolic wave equation is obtained. Assuming for simplicity that the axial motion is nonrelativistic, the wave equation is

$$- i \frac{\partial \psi}{\partial z} = \left\{ [-2\bar{m}\bar{e}V(x, z)]^{1/2} + \frac{1}{2} [-2\bar{m}\bar{e}V_0(z)]^{-1/2} \frac{\partial^2}{\partial x^2} \right\} \psi \qquad (1.25)$$

where, in the second term in the braces, we have made the approximation $V(x, z) \approx V(0, z) \equiv V_0(z)$. Equation (1.25) describes the evolution of the wave function ψ of paraxial particles. Its similarity with (1.22) is obvious.

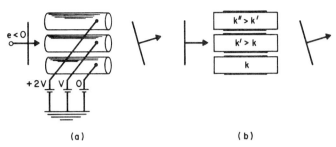

Fig. 1-32   Deflection of an electron or optical wave by wave optics gratings consisting of metal tubes (a) and electrooptic crystals (b), respectively. In both the mechanical and optical systems, the wave packet is deflected upward, even though no classical force is exerted on the particle when the voltages are turned on and off while the wave packet is traveling in the tubes or crystals (second Aharonov and Bohm effect). (From Arnaud,[38] by permission of the American Optical Society.)

Equation (1.25) also has the form of the conventional Schrödinger equation (1.23), because $dz/dt \equiv u_0 = (-2\overline{m}\overline{e}V_0)^{1/2}\overline{m}^{-1}$ is the axial velocity, and $-2\overline{e}(VV_0)^{1/2} \approx -2\overline{e}V_0 - \overline{e}U$, where $U \equiv V - V_0 \ll V_0$. The term $-2\overline{e}V_0$ corresponds to an unimportant phase factor on ψ that can be omitted. Our result simply means that for nonrelativistic axial motions the Schrödinger equation (1.23) is applicable in a frame of reference moving at the mean velocity u_0 of the electron. If the axial motion is relativistic, only slight changes are needed.

   Let us illustrate the similarity in diffraction effects between optical waves and electron waves by considering the effect of grating-type devices. The discussion will be qualitative, but it could easily be made quantitative with the previously derived expressions. The system shown in Fig. 1-32a is a sequence of metal tubes raised at increasing voltages. If a plane electron wave enters the tubes from the left, the electron is deflected upward because the phase shift introduced by the voltage is (stepwise) linear in x. If a dc voltage is applied to the tubes, this phase shift results from a change in the wavelength of the electron, the frequency being a constant because of the time invariance of the system. If the voltage is applied while the electron wave packet is traveling through the tubes, the wavelength remains a constant because of the axial invariance of the system, but the frequency is changed. The phase shift, and therefore the deflection, is the same in both cases if the voltage is applied during a time L/u, where L denotes the length of the tubes in the first experiment and u the velocity of the electron.[49] It is interesting that in the pulsed arrangement, no classical force is exerted on the electron. Yet the electron is deflected away from its original direction. This effect is similar to the second Aharonov and Bohm (AB) effect discussed, for example, in Boyer.[49] However, by considering a series of tubes rather than just two, our example makes it clear that a deflection of the average electron path can be obtained.

The discussion for the optical deflector shown in Fig. 1-32b is identical with the foregoing. The only possible difference between the two systems is that the dispersion of the electrooptic crystals used in the optical arrangement may not be the same as the dispersion of free space for matter waves. Thus, in the pulsed experiment, the pulse lag may be somewhat different.

Numerous results relevant to beam optics can be found in textbooks of quantum mechanics. It is hoped that the above discussion will bring some clarity to the physical significance of the analogy that exists between these two fields of knowledge.

References

1. A. I. Sabra, "Theories of Light; from Descartes to Newton." Oldbourne, London, 1967.
2. R. S. Longhurst, "Geometrical and Physical Optics." Longman, London, 1973.
3. M. Born and E. Wolf, "Principles of Optics." Pergamon, Oxford, 1965.
4. G. Goubau and F. Schwering, *IRE Trans. Antennas Propagat.* **AP9**, 248 (1961).
5. P. K. Tien, J. P. Gordon, and J. R. Whinnery, *Proc. IEEE* **53**, 129 (1965).
6. E. Merzbacher, "Quantum Mechanics." Wiley, New York, 1970.
7. N. S. Kapany and J. J. Burke, "Optical Waveguides." Academic Press, New York, 1972.
8. R. Ulrich and R. J. Martin, *Appl. Opt.* **10**, 2077 (1971). J. Nishizawa and A. Otsuka, *Appl. Phys. Lett.* **21**, 48 (1972).
9. R. K. Luneburg, "The Mathematical Theory of Optics." Univ. of California Press, Berkeley, California, 1964. G. Toraldo Di Francia, *Opt. Acta* **1**, 157 (1955).
10. E. A. J. Marcatili and R. A. Schmeltzer, *Bell Syst. Tech. J.* **43**, 1783 (1964).
11. H. G. Booker and W. Walkinshaw, *in* "Meteorological Factors in Radio-Wave Propagation," p. 80. Physical Soc., London, 1946.
12. S. Kawakami and J. Nishizawa, *IEEE Trans. Microwave Theory Tech.* **MTT16**, 814 (1968).
13. D. W. Berreman, *Bell Syst. Tech. J.* **43**, 1469 (1964); **44**, 2117 (1965).
14. K. C. Kao and G. A. Hockham, *Proc. IEE London* **113**, 1151 (1966). W. G. French, A. D. Pearson, G. W. Tasker, and J. B. MacChesney, *Appl. Phys. Lett.* **23**, 338 (1973). M. D. Rigterink, *Bell Labs Record* **53**, 341 (1975). A. Werts, *Onde Elec.* **45**, 967 (1966).
15. R. D. Maurer, *Proc. IEEE* **61**, 452 (1973).
16. J. Stone, *Appl. Phys. Lett.* **20**, 239 (1972).
17. W. A. Gambling, D. N. Payne, and H. Matsumura, *Electron. Lett.* **8**, 568 (1972).
18. R. Kompfner, *Appl. Opt.* **11**, 2412 (1972). S. E. Miller, E. A. J. Marcatili, and T. Li, *Proc. IEEE* **61**, 1703 (1973).
19. H. F. Taylor, *J. Appl. Phys.* **44**, 3257 (1973).
20. P. Kaiser, E. A. J. Marcatili, and S. E. Miller, *Bell Syst. Tech. J.* **52**, 265 (1973). E. A. J. Marcatili, *Bell Syst. Tech. J.* **53**, 645 (1974). J. A. Arnaud, *Bell Syst. Tech. J.* **53**, 675, 1379 (1974).
21. H. Kogelnik, *in Proc. Symp. Quasi-Opt.*, p. 333. Polytech. Inst. of Brooklyn Press, New York, 1964.

22. L. A. Vainshtein, "Open Resonators and Open Waveguides." Golem Press, Boulder, Colorado, 1969.
23. A. G. Fox and T. Li, *Proc. IRE* **48**, 1904 (1960).
24. A. E. Siegman, *Appl. Opt.* **13**, 353 (1974).
25. G. D. Boyd and J. P. Gordon, *Bell Syst. Tech. J.* **40**, 489 (1961).
26. P. W. Smith, *Proc. IEEE* **58**, 1342 (1970); **60**, 422 (1972).
27. N. G. Vakhimov, *Radio Eng. Elec. Phys.* **10**, 1439 (1965).
28. T. Li, *Bell Syst. Tech. J.* **44**, 917 (1965).
29. P. Connes, *Rev. Opt.* **35**, 37 (1956).
30. R. V. Pole, *J. Opt. Soc. Amer.* **55**, 254 (1965).
31. J. A. Arnaud, *IEEE J. Quantum Electron.* **QE-4**, 893 (1968); *Appl. Opt.* **8**, 189, 1909 (1969).
32. C. V. Heer, *Phys. Rev.* **134A**, 799 (1964).
33. L. C. Tillotson, *Science* **170**, 31 (1970).
34. V. A. Fock, *Bull Acad. Sci. URSS, Ser. Phys.* **14**, 70 (1950) [English transl.: V. A. Fock, "Electromagnetic Diffraction and Propagation Problems," Chapter 14. Pergamon, Oxford, 1965].
35. A. Ashkin and J. M. Dziedzic, *Appl. Phys. Lett.* **19**, 283 (1971).
36. J. Brown, *Proc. IEE London* **113**, 1, 27 (1966).
37. J. R. Pierce, "Almost All About Waves." MIT Press, Cambridge, Massachusetts, 1974.
38. J. A. Arnaud, *J. Opt. Soc. Amer.* **65**, 174 (1975).
39. P. Grivet, "Electron Optics." Pergamon, Oxford, 1965.
40. J. M. Kelso, "Radio Ray Propagation in the Ionosphere," p. 220. McGraw-Hill, New York, 1964.
41. O. Costa de Beauregard, *C. R. Acad. Sci. Paris* **B274**, 164 (1972); *Nuovo Cimento Lett.* **10**, 852 (1974).
42 H. A. Haus, *Physica* **43**, 77 (1969).
43. J. A. Arnaud, *Opt. Commun.* **7**, 313 (1973).
44. R. V. Jones and J. C. S. Richard, *Proc. Roy. Soc. (London)* **A221**, 480 (1954).
45. J. A. Arnaud, *Electron. Lett.* **8**, 541 (1972); *Amer. J. Phys.* **42**, 71 (1974); J. N. Dodd, *J. Phys. B* **8**, 157 (1975).
46. S. R. de Groot and L. G. Suttorp, "Foundations of Electrodynamics," p. 80. Amer. Elsevier, New York, 1972.
47. G. B. Whitham, "Linear and Nonlinear Waves." Wiley (Interscience), New York, 1974.
48. F. N. K. Robinson, *Phys. Rep. Phys. Lett. C (Netherlands)* **16**, 313 (1975).
49. T. H. Boyer, *Phys. Rev. D* **8**, 1667 (1973).

Gaussian Beams

This chapter is essentially self-contained and is aimed at giving essential results concerning gaussian beams and optical resonators on a simple theoretical basis. A few advanced results applicable, for instance, to the helical fiber, are also derived.

We are concerned with beams whose irradiance is maximum on axis and decreases, in decibels, as the square of the distance from axis. Such beams are called gaussian beams. Special consideration is given to gaussian beams because their irradiance profile remains invariant in free space and through unaberrated lenses, except perhaps for a scale factor. The laws of transformation of gaussian beams are simple, formally identical to the laws of transformation of ray pencils.[1-3] Because many field configurations encountered in practice (e.g., the fundamental mode of propagation in step-index glass fibers, or the field radiated by tapered microwave antennas) are close to gaussian, results based on a gaussian approximation are sometimes sufficiently accurate. Finally, gaussian beams have the smallest transverse extent–angular divergence product of all optical beams. Consequently, gaussian beams suffer less from diffraction losses at the edges of optical elements, which are, of necessity, of finite size. For that reason, most lasers incorporating circular apertures tend to oscillate in a mode whose field configuration is close to gaussian. Relations are first derived in this chapter for two-dimensional systems with transverse coordinate x and axial coordinate z. The optical systems and the fields are assumed invariant in the y-direction. Systems with rotational symmetry are often encountered in practice. Once the two-dimensional solution $\psi(x)$ is obtained, it is straightforward to obtain the solution $\Psi(r, z)$ for systems with rotational symmetry because, within our approximations, the relation $\Psi(r, z) = \psi(x, z)\psi(y, z)$ holds, where $r \equiv (x^2 + y^2)^{1/2}$.

2.1 The Wave Equation

The wave equation obeyed by paraxial optical beams can be obtained in a heuristic way by the requirement that it agree with the laws of geometrical optics in the limit of short wavelengths. In isotropic lossless media, the wave vector **k**, with components k_x, k_z, has a constant real magnitude

$$k \equiv \left(k_x^2 + k_z^2\right)^{1/2} \tag{2.1}$$

which is usually a function of x and z. For paraxial waves $k_x \ll k_z$, we have, solving Eq. (2.1) for k_z,

$$k_z \approx k - \tfrac{1}{2}k^{-1}k_x^2 \tag{2.2}$$

This approximation is illustrated in Fig. 2-1.

Let us consider a wave function of the form

$$\psi = \psi_0 \exp[i(k_x x + k_z z)] \tag{2.3}$$

where **k** and ψ_0 vary slowly with x, z. Neglecting the variation of ψ_0, we have

$$\frac{\partial \psi}{\partial x} = ik_x \psi, \qquad \frac{\partial \psi}{\partial z} = ik_z \psi \tag{2.4}$$

If we make in (2.2) the substitutions

$$ik_x \rightarrow \frac{\partial}{\partial x}, \qquad ik_z \rightarrow \frac{\partial}{\partial z} \tag{2.5}$$

suggested by (2.4), we obtain a wave equation of the parabolic type[4]

$$-i\frac{\partial \psi}{\partial z} = k(x, z)\psi + \frac{1}{2}k_0^{-1}\frac{\partial^2 \psi}{\partial x^2} \tag{2.6}$$

We have made the approximation that, in the second term on the rhs of (2.6), k can be replaced by its value on axis $k_0(z) \equiv k(0, z)$.

The "wave function" ψ in (2.6) is defined as $k_0^{1/2}E$, where E denotes a transverse component of the electric field. This relation follows from the power conservation requirement. Indeed, if the medium is lossless, k is

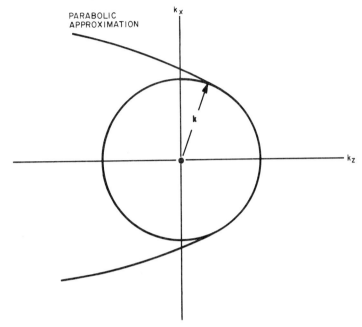

Fig. 2-1 The curve of wave vectors for isotropic media is circular. The wave equation is based on a parabolic approximation of this curve.

real. It is not difficult to show that, for any solution ψ of (2.6) that vanishes faster than $|x|^{-1}$ as $|x| \to \infty$, we have

$$\frac{d}{dz} \int_{-\infty}^{+\infty} \psi^*\psi \, dx = 0 \tag{2.7}$$

where the asterisk denotes the complex-conjugate value. Thus $\psi^*\psi$ can be interpreted as the z-component of the power density. The electromagnetic power density, on the other hand, is the vector $\mathbf{E} \times \mathbf{H}$. Its z-component $(\epsilon/\mu_0)^{1/2}E^*E$ is proportional to $k_0 E^* E$, for paraxial waves in a dielectric ($\mu_0 = $ const). Agreement between this expression and $\psi\psi^*$ requires that ψ be defined as $k_0^{1/2}E$. A more general relation, applicable to complex k, is given in Chapter 3.

2.2 The Gaussian Solution[2]

Let us now consider a square-law medium, whose wavenumber has the following dependence on x:

$$k(x, z) = k_0[1 - \tfrac{1}{2}\Omega^2(z)x^2] \tag{2.8}$$

where k_0 is now a constant and we assume that $\Omega x \ll 1$. In free space, we have of course $\Omega^2 = 0$. $\Omega^2 > 0$ corresponds to a focusing medium and $\Omega^2 < 0$ to a defocusing medium. In the latter case, Ω is an imaginary number. By letting Ω^2 be a function of z, we allow the focusing strength to vary in arbitrary manner along the z-axis. Equation (2.8) is therefore applicable to a sequence of thin lenses. Substituting (2.8) in (2.6), the wave equation becomes

$$- i \frac{\partial \psi}{\partial z} = k_0 (1 - \tfrac{1}{2} \Omega^2 x^2) \psi + \frac{1}{2} k_0^{-1} \frac{\partial^2 \psi}{\partial x^2} \qquad (2.9)$$

A solution of (2.9) of particular interest is the gaussian-beam solution

$$\psi(x, z) = (-\pi)^{-1/4} q(z)^{-1/2} \exp\left[i \tfrac{1}{2} k_0 \dot{q}(z) q(z)^{-1} x^2 \right] \exp(ik_0 z) \qquad (2.10)$$

where the overdot denotes differentiation with respect to z, and $q(z)$ is a solution of the paraxial ray equation

$$\ddot{q}(z) + \Omega^2(z) q(z) = 0 \qquad (2.11)$$

The proof is straightforward. Taking the logarithm of (2.10) and differentiating with respect to z, we have

$$\log \psi = - \tfrac{1}{2} \log(q) + i \tfrac{1}{2} k_0 \dot{q} q^{-1} x^2 + ik_0 z + \text{const}$$

$$\psi^{-1} \frac{\partial \psi}{\partial z} = - \tfrac{1}{2} \dot{q} q^{-1} + i \tfrac{1}{2} k_0 x^2 (-\dot{q}^2 q^{-2} + \ddot{q} q^{-1}) + ik_0 \qquad (2.12)$$

Differentiating now (2.10) with respect to x, we obtain

$$\psi^{-1} \frac{\partial \psi}{\partial x} = ik_0 \dot{q} q^{-1} x \qquad (2.13)$$

Differentiating (2.13) with respect to x again

$$- \psi^{-2} \left(\frac{\partial \psi}{\partial x} \right)^2 + \psi^{-1} \frac{\partial^2 \psi}{\partial x^2} = ik_0 \dot{q} q^{-1} \qquad (2.14)$$

and using (2.13), we find

$$\psi^{-1} \frac{\partial^2 \psi}{\partial x^2} = ik_0 \dot{q} q^{-1} - \left(k_0 \dot{q} q^{-1} x \right)^2 \qquad (2.15)$$

Substituting $\psi^{-1} \partial \psi / \partial z$ and $\psi^{-1} \partial^2 \psi / \partial x^2$ from (2.12) and (2.15) into (2.9), we obtain after simplifications

$$\tfrac{1}{2} k_0 x^2 (\ddot{q} q^{-1} + \Omega^2) = 0 \qquad (2.16)$$

This equation is satisfied if $q(z)$ obeys (2.11). This completes the proof. The prefactor $(-\pi)^{-1/4}$ in (2.10) normalizes the integral of $\psi^*\psi$ to unity.

When $q(z)$ is a real solution of (2.11), $\psi(x, z)$ represents the field of a ray pencil bounded by the ray $q(z)$. The physical significance of (2.10) is then clear: The second exponential term expresses the geometrical optics phase shift $(k_0 z)$ along the z-axis. The first exponential term expresses, within the paraxial approximation, the departure of the (spherical) wavefront from the tangent plane. The prefactor $q(z)^{-1/2}$, on the other hand, is a consequence of the law of conservation of power, because the product of the cross section $q(z)$ of the ray pencil and the power density $\psi\psi^*$ must remain a constant.

Equation (2.10) describes gaussian beams when complex initial values are given to $q(z)$. Ω^2 being real for lossless media, both the real and imaginary parts $q_r(z)$ and $q_i(z)$, respectively, of $q(z)$ obey the paraxial ray equation and behave like ordinary rays

$$\ddot{q}_r + \Omega^2 q_r = 0, \qquad \ddot{q}_i + \Omega^2 q_i = 0 \tag{2.17}$$

Two solutions, such as q_r and q_i, of the ray equation satisfy the invariance condition

$$\frac{d}{dz} (q_r \dot{q}_i - \dot{q}_r q_i) = 0 \tag{2.18}$$

as one readily verifies by carring out the differentiation in (2.18) and using (2.17). The quantity $q_r \dot{q}_i - q_i \dot{q}_r$ is called a Lagrange ray invariant. It is useful to represent rays in "phase space," with coordinates $q, \dot{q}$. As z varies, the point representing a ray describes a trajectory in that space. If the trajectory in real space is periodic, for instance, the trajectory in phase space is a closed curve. Equation (2.18) shows that the area of the triangle defined by $(q_r, \dot{q}_r)$, $(q_i, \dot{q}_i)$, and the origin in phase space is invariant. More generally, the area covered in phase space by any continuous set of rays is invariant. This representation is particularly useful when the focusing strength Ω^2 varies slowly with z.

Clearly, the solution $\psi(x, z)$ given in (2.10) is unaffected, except for an unimportant amplitude factor, if the ray $q(z)$ is multiplied by some constant α (because of linearity, $\alpha q(z)$ is a ray if $q(z)$ is a ray). Therefore, it is always possible to normalize the complex ray $q(z)$ by the z-invariant condition

$$q_r \dot{q}_i - \dot{q}_r q_i = k_0^{-1} \tag{2.19}$$

or, equivalently,

$$q^* \dot{q} - \dot{q}^* q = 2i k_0^{-1} \tag{2.20}$$

where the asterisks denote complex-conjugate values. This condition simplifies the relation between the beam half-width and the complex ray q, which is discussed in the next section.

2.3 Beam Half-Width, Wavefront Radius of Curvature, and On-Axis Phase

The significance of the gaussian solution (2.10) of the wave equation is best understood in terms of three quantities that have direct physical significance: the beam half-width ξ ($1/e$ point of the irradiance), the wavefront radius of curvature ρ, and the phase of the on-axis field θ.

Let us consider first the on-axis field, setting $x = 0$ in (2.10). To within a constant factor, we have

$$\psi(0, z) = q^{-1/2} \exp(ik_0 z) \tag{2.21}$$

The phase θ of this on-axis field is therefore

$$\theta = k_0 z - \tfrac{1}{2} \operatorname{phase}(q) \tag{2.22}$$

Equation (2.22) shows that the phase of the complex ray $q(z)$ is directly related to the phase of the on-axis field.

Let us set $q \equiv q_r + iq_i$ and separate the real and imaginary parts of $\mu \equiv \dot{q} q^{-1}$ in (2.10). We have

$$\mu \equiv \dot{q} q^{-1} = (\dot{q}_r + i\dot{q}_i)(q_r + iq_i)^{-1}$$

$$= [(\dot{q}_r q_r + \dot{q}_i q_i) + i(q_r \dot{q}_i - \dot{q}_r q_i)](q_r^2 + q_i^2)^{-1} \tag{2.23}$$

After a few rearrangements, we find the field pattern at z

$$\frac{\psi(x, z)}{\psi(0, z)} = \exp(i\tfrac{1}{2} k_0 \dot{q} q^{-1} x^2) = \exp\left[-\frac{1}{2}\left(\frac{x}{\xi}\right)^2\right] \exp(i\tfrac{1}{2} k_0 \rho^{-1} x^2) \tag{2.24}$$

where

$$\xi^2 \equiv qq^*[k_0(q_r \dot{q}_i - \dot{q}_r q_i)] = qq^* \tag{2.25}$$

and

$$\rho^{-1} \equiv \frac{1}{2} \frac{d\xi^2}{dz} \xi^{-2} = \frac{(d\xi/dz)}{\xi} \tag{2.26}$$

if the normalization condition (2.19) is used. The (real) quantity ξ is called the beam half-width. This is the distance from axis where the beam irradiance $\psi\psi^*$ is reduced by a factor $e = 2.718\ldots$. Thus, the modulus of the complex ray q is precisely equal to the beam half-width ξ. The curve $\xi(z)$ is called the beam profile. The second exponential term in (2.24) shows that the radius of curvature of the wavefront is

$$\rho = \left(\frac{d\xi}{dz} \right)^{-1} \xi \qquad (2.27)$$

The radius of curvature of the wavefront ρ is defined only within the paraxial approximation, $d\xi/dz \ll 1$. Equation (2.27) shows that, within this approximation, the wavefronts are circular and perpendicular to the beam profile $\xi(z)$. The quantity

$$\mu \equiv \dot{q}q^{-1} = \rho^{-1} + \frac{i}{k_0\xi^2} \qquad (2.28)$$

is called the complex wavefront curvature. It reduces to the real wavefront curvature of a ray pencil in the geometrical optics limit, $k_0 \to \infty$.

In conclusion, the three real quantities ξ, ρ, and θ that have a direct physical significance can be associated in a simple way to the complex ray $q(z)$ defining the field of a gaussian beam: The beam half-width is $\xi = (qq^*)^{1/2}$; the wavefront radius of curvature is $\rho = (d\xi/dz)^{-1}\xi$; and the phase of the on-axis field is $\theta = k_0z - \frac{1}{2}\,\text{phase}(q)$. Once the ray

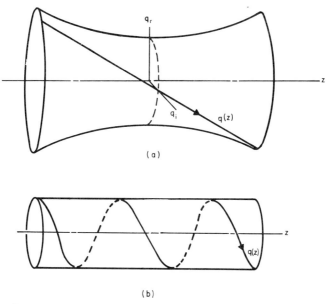

Fig. 2-2 Complex-ray representation of two-dimensional gaussian beams (a) in free space, (b) in uniform square-law fibers.

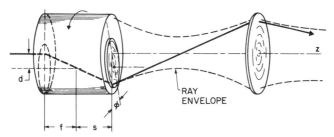

Fig. 2-3 Schematic diagram of demonstration system showing, on the left, a rotating device incorporating two offset lenses. The skew laser beam axis simulates gaussian-beam propagation through a test lens, shown on the right. (From Arnaud,[5] by permission of the American Physical Society.)

equations (2.17) have been solved, all the desired information concerning the gaussian beam can be obtained.

It is sometimes convenient to represent the complex ray $q(z)$ by a real skew ray in the q_r, q_i, z rectangular coordinate system as shown in Fig. 2-2. In this representation, ξ is the distance from the z-axis to the skew ray. Therefore, if we let the skew ray rotate about the z-axis, the beam profile is generated. The quantity $2(k_0 z - \theta)$, on the other hand, is the angular position of the skew ray in the q_r, q_i plane.

This observation can be used as a teaching aid to demonstrate the laws of propagation of gaussian beams.[5] The method requires the use of a He–Ne laser and a pair of lenses. The characteristics of the laser beam are unimportant as long as the beam has small angular divergence and does not depart too much from a geometrical optics ray. The purpose of the lenses is to transform the incident ray, which initially coincides with the z-axis, into a skew ray representing the complex ray $q(z)$. If the lens system is made to rotate, the skew ray also rotates about the z-axis. Its envelope simulates a gaussian beam. This arrangement is shown in Fig. 2-3, and the simulated gaussian beam is shown in Fig. 2-4. Figure 2-5 gives a simple construction for obtaining the parameters of the simulated gaussian beam from the lens positions and focal length in Fig. 2-3. The proof for that construction is lengthy and of little general interest. It will therefore be omitted.

In the next sections, explicit expressions are given for the beam half-width ξ, the wavefront radius of curvature ρ, and the phase of the on-axis field θ for the case of free space ($\Omega^2 = 0$) and thin lenses [$\Omega^2 = f^{-1} \delta(z)$, where f denotes the lens focal length and $\delta(z)$ the Dirac symbolic function].

It is of interest to note a formal analogy between the complex ray $q(z)$ representing a gaussian beam and the voltage $V(z)$ along an electric line.[6,7]

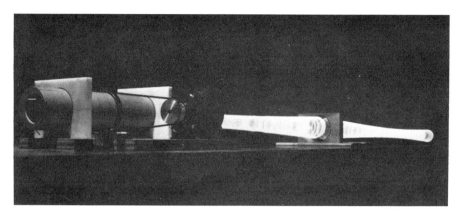

Fig. 2-4 Photograph of an experiment simulating the propagation of gaussian beams. The simulated beam is made to converge with the help of a lens. It subsequently diverges as a result of the (simulated) diffraction effects. (From Arnaud,[5] by permission of the American Physical Society.)

Consider the electric system shown in Fig. 2-6a with susceptance per unit length $C(z)\omega = k_0\Omega^2(z)$ and reactance per unit length $L\omega = 1/k_0$. The equations for the voltage $V(z) = q(z)$ and the current $I(z) = -ik_x(z)$ are, for time-harmonic sources,

$$\frac{dV}{dz} = iL\omega I \tag{2.29a}$$

$$\frac{dI}{dz} = iC\omega V \tag{2.29b}$$

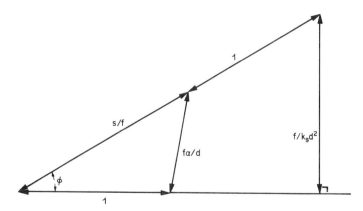

Fig. 2-5 Graphical method defining the parameters of the simulated gaussian beam. s, f, ϕ are defined in Fig. 2-3. (From Arnaud,[5] by permission of the American Physical Society.) *Note*: $k_s \equiv k_0$ and $\alpha \equiv 1/(k_0\xi_0)$.

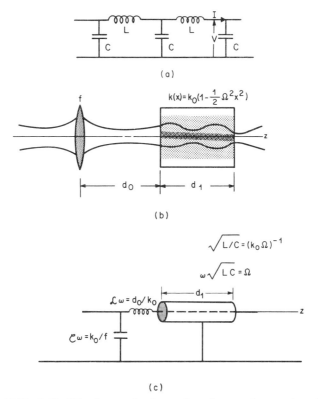

Fig. 2-6 (a) The half-width of a gaussian beam obeys the same laws as the voltage along an electric line. (b) Optical system. (c) Equivalent electrical system.

This system of equations is equivalent to

$$\frac{dq}{dz} = k_x/k_0 \tag{2.30a}$$

$$\frac{dk_x}{dz} = -k_0\Omega^2(z)q \tag{2.30b}$$

Eliminating k_x from (2.30), we obtain

$$\frac{d^2q}{dz^2} + \Omega^2 q = 0 \tag{2.31}$$

Comparing (2.31) to (2.11), we conclude that the beam radius $\xi(z) = |q(z)|$ obeys the same law as the voltage $|V(z)|$ along an electric line with susceptance per unit length $k_0\Omega^2(z)$ and reactance per unit length $1/k_0$. In particular, if Ω^2 is a constant, the beam profile is the same as that of the voltage along an unmatched coaxial line. A thin lens with focal length f

corresponds, in the electrical analog, to a lumped susceptance $\mathcal{C}\omega = k_0/f$ across the line. It causes a discontinuity in wavefront curvature, but the beam radius $\xi = |q| \rightleftharpoons |V|$ is unaffected. A section of uniform lenslike medium of length d corresponds to a series inductance $\mathcal{L}\omega = d/k_0$. The electric circuit equivalent to the optical system in Fig. 2-6b, incorporating a lens and a lenslike medium, is shown in Fig. 2-6c. Because of the equivalence just discussed, beam tracing can be accomplished using the Smith chart familiar to microwave engineers.[8,9] The procedure based on the tracing of two paraxial rays, however, has a more direct physical significance.

2.4 Propagation in Free Space

In free space we have $\Omega^2 = 0$. The paraxial ray equation (2.11) thus reduces to

$$\ddot{q}(z) = 0 \tag{2.32}$$

which has the solution

$$q(z) = \xi_0 + \frac{iz}{k_0 \xi_0} \tag{2.33}$$

if we use the normalization condition (2.19) and assume for simplicity that ξ reaches its minimum value ξ_0 at the plane $z = 0$. The skew ray representing q is now a straight line, as shown in Fig. 2-2a.

The beam half-width ξ at any plane z is, from (2.25) and (2.33),

$$\xi^2(z) = qq^* = \xi_0^2 + \left(\frac{z}{k_0 \xi_0} \right)^2 \tag{2.34}$$

We define a normalized beam half-width

$$\bar{\xi} \equiv \frac{\xi}{\xi_0} = (1 + \bar{z}^2)^{1/2} \tag{2.35}$$

where

$$\bar{z} \equiv \frac{z}{k_0 \xi_0^2} \tag{2.36}$$

The beam profile $\xi(z)$ is therefore an hyperbola. This conclusion can also be reached using the skew-ray representation previously discussed. The skew ray being a straight line, it generates, as it rotates about the z-axis, an hyperboloid of revolution that exhibits the profile of the beam.

It is interesting to consider the behavior of the beam at infinity $z \to +\infty$. The angular divergence is

$$\Delta\alpha = \lim_{z \to +\infty} \left(\frac{\xi}{z} \right) = (k_0 \xi_0)^{-1} \tag{2.37}$$

Setting $\Delta q \equiv \xi_0$ and $\lambda \equiv \lambda/2\pi \equiv k_0^{-1}$, we obtain the relation

$$\Delta\alpha \, \Delta q = \lambda \tag{2.38}$$

which is reminiscent of the Heisenberg uncertainty relation

$$\Delta p \, \Delta q = \hbar \tag{2.39}$$

applicable to matter waves. A detailed discussion of this analogy was given in Chapter 1. Physically (2.38) means that if we attempt to focus an optical beam to a small spot, the beam divergence must be large.

By substituting (2.34) in (2.26), the wavefront radius of curvature is

$$\rho^{-1} = \frac{1}{2} \left(\frac{d\xi^2}{dz} \right) \xi^{-2} \tag{2.40}$$

or

$$\rho = z \left[1 + \left(\frac{k_0 \xi_0^2}{z} \right)^2 \right] \tag{2.41}$$

$$\bar{\rho} \equiv \frac{\rho}{k_0 \xi_0^2} = \bar{z} + \frac{1}{\bar{z}} \tag{2.42}$$

These relations show that the curvature is zero at $z = 0$ (beam waist) and at $z = \infty$. The curvature is maximum when $\bar{z} = 1$, that is, when

$$z = k_0 \xi_0^2 \tag{2.43}$$

The points $z = \pm k_0 \xi_0^2$ are separated by the so-called confocal spacing.

The on-axis phase shift is obtained by substituting (2.33) in (2.22)

$$\theta = k_0 z - \tfrac{1}{2} \, \text{phase}(q)$$

$$= k_0 z - \frac{1}{2} \tan^{-1} \frac{q_i}{q_r}$$

$$= k_0 z - \frac{1}{2} \tan^{-1} \frac{z}{k_0 \xi_0^2} \tag{2.44}$$

or

$$\bar{\theta} \equiv \theta - k_0 z = -\tfrac{1}{2} \tan^{-1}(\bar{z}) \qquad (2.45)$$

Thus the variation of $\bar{\theta}$ from $z = -\infty$ to $+\infty$ is concentrated near the beam waist. In the limit where $\xi_0 \to 0$, this variation corresponds to the so-called anomalous phase shift of ray pencils. The variation of the normalized half-width, wavefront radius of curvature, and diffraction phase shift are shown in Fig. 2.7 as functions of the normalized axial coordinate $\bar{z}$.

Some of the properties of optical resonators can be understood on the basis of the previous discussion. Consider a wavefront of the gaussian beam in Fig. 2.8 and suppose that the surface of a mirror is made to coincide with this wavefront. By virtue of reciprocity, the beam is reflected back onto itself. If the surface of a second mirror is made to coincide with a second wavefront, the beam is reflected back to the first mirror, and so on. The transverse field distribution of the beam is unaffected by these bounces, assuming that the mirrors have uniform reflectivity. Thus, upon reflection, the field distribution is unaffected except for a constant factor, r, where r denotes the mirror reflectivity, a complex constant. Figure 2-8 shows various possible locations of the mirrors along the beam in (a), (b), and (c). In most lasers, a symmetrical configuration, such as the one shown

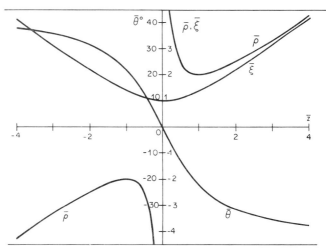

Fig. 2-7 Propagation of gaussian beams in free space. Variation of the normalized beam half-width $\bar{\xi} \equiv \xi/\xi_0$, normalized wavefront radius of curvature $\bar{\rho} \equiv \rho/k_0\xi_0^2$, and normalized diffraction phase shift $\bar{\theta} \equiv \theta - k_0 z$, as a function of the normalized axial distance $\bar{z} \equiv z/k_0\xi_0^2$. ξ, ρ, θ, and z denote the actual half-width, wavefront radius of curvature, phase shift, and distance from the beam waist, respectively.

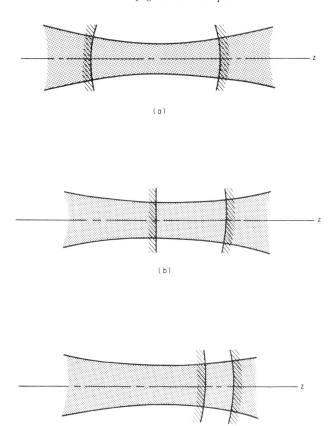

(a)

(b)

(c)

Fig. 2-8 Optical resonators incorporating two circular mirrors. (a) Symmetrical configuration. (b) Planoconcave resonator. (c) Convex–concave resonator.

in Fig. 2-8a, is used. One mirror may be flat, however, while the other
mirror is concave, as shown in Fig. 2-8b. In that case, the radius of
curvature R of the curved mirror must exceed the mirror spacing d. No
solution can be found with $R < d$. Figure 2-8c shows a configuration
where one mirror is convex, while the other mirror is concave. The
fundamental mode of resonance of a system of two spherical mirrors
facing each other can be found (provided a solution exists) by matching
the wavefronts of a gaussian beam with a properly chosen value of ξ_0 to
the mirror surfaces. A more systematic procedure applicable to any
sequence of lenses and mirrors, based on the concept of round-trip ray
matrix, is discussed in Section 2.13.

2.5 Propagation through Uniform Lenslike Media

In lenslike media with $\Omega^2 > 0$, the general solution of the ray equation (2.11) is

$$q(z) = q_0 \sin(\Omega z + \phi) \tag{2.46}$$

where q_0 and ϕ are constants of integration. If q_0 and ϕ in (2.46) are real numbers, the rays are sinusoids with period $2\pi/\Omega$ and maximum excursion q_0. Within the approximation made in this chapter, all the rays originating from a point source converge to a common point, a distance $2\pi/\Omega$ from the object point. Furthermore, all the rays have the same optical length.

A solution of the ray equation, equivalent to (2.46) but more convenient for the study of beam propagation, is

$$q(z) = q^+ e^{i\Omega z} + q^- e^{-i\Omega z} \tag{2.47}$$

where q^+ and q^- are two complex constants of integration.

Let us first assume $q^- = 0$. The beam half-width ξ is then a constant

$$\xi^2 \equiv \bar{\xi}_0^2 = qq^* = q^+ q^{+*} \tag{2.48}$$

or, using the condition (2.20) with $\dot{q} = i\Omega q^+$,

$$\bar{\xi}_0 = (k_0\Omega)^{-1/2} \tag{2.49}$$

The stronger the focusing strength Ω, the smaller is the beam half-width $\bar{\xi}_0$. For a typical graded-index glass fiber with a period of oscillation of the ray equal to 6.2 mm, we have $\Omega = 10^{-3}\ \mu\text{m}^{-1}$. If the wavelength is $\lambda = 0.62\ \mu\text{m}$, $k_0 = 10\ \mu\text{m}^{-1}$, we obtain from (2.49) $\bar{\xi}_0 = 10\ \mu\text{m}$.

The modal solution resembles that of metallic waveguides in the sense that the beam pattern is independent of z. The phase of the on-axis field is, from (2.22),

$$\theta = (k_0 - \tfrac{1}{2}\Omega)z \equiv k_z z \tag{2.50}$$

Thus the axial wavenumber k_z is smaller than the free-space wavenumber k_0, as is also the case for metallic waveguides. In the skew-ray representation, the ray follows the helicoidal path shown in Fig. 2-2b.

An optical resonator can be formed with two plane mirrors perpendicular to the guide axes, spaced a distance d apart. The resonant frequencies are obtained by stating that an integral number of half-wavelengths fits between the two mirrors, that is, from (2.50)

$$(k_0 - \tfrac{1}{2}\Omega)d = l\pi, \qquad l = 1,2,\ldots \tag{2.51}$$

This relation defines the resonant frequencies (from k_0) for the axial modes $l = 1, 2, \ldots, \Omega$ and d being known constants.

Let us now consider a more general solution of (2.11) with both q^+ and q^- in (2.47) different from zero. Let the beam be launched in the lenslike medium at the plane $z = 0$ with a half-width ξ_0 and a plane wavefront. The initial conditions are, from (2.33),

$$q(0) = q^+ + q^- = \xi_0, \qquad \dot{q}(0) = i\Omega(q^+ - q^-) = \frac{i}{k_0\xi_0} \qquad (2.52)$$

Solving (2.52) for q^+ and q^- and substituting in (2.47), we obtain

$$q(z) = \tfrac{1}{2}\left[\xi_0 + (k_0\Omega\xi_0)^{-1}\right]e^{i\Omega z} + \tfrac{1}{2}\left[\xi_0 - (k_0\Omega\xi_0)^{-1}\right]e^{-i\Omega z}$$

$$= \xi_0\cos(\Omega z) + i(k_0\Omega\xi_0)^{-1}\sin(\Omega z) \qquad (2.53)$$

The beam half-width is therefore[10-12]

$$\xi^2 = qq^* = \xi_0^2\cos^2(\Omega z) + (k_0\Omega\xi_0)^{-2}\sin^2(\Omega z) \qquad (2.54)$$

For $\Omega z = 0, \pi, 2\pi, \ldots$, we have $\xi = \xi_0$, the input beam half-width. For $\Omega z = \pi/2, 3\pi/2, \ldots$, the beam half-width ξ_m is given by

$$\xi_m^2 = (k_0\Omega\xi_0)^{-2} \qquad (2.55)$$

We observe that

$$\xi_m\xi_0 = (k_0\Omega)^{-1} \qquad (2.56)$$

Thus the product of the maximum and minimum half-widths is a constant. The behavior of the beam for that case is illustrated in Fig. 1-9a. Note that the beamwidth pulsates at twice the ray-oscillation frequency.

2.6   Refraction of a Gaussian Beam by a Lens

We have shown that a gaussian beam is completely defined by a complex ray $q(z)$. The trajectories of the real and imaginary parts of $q(z)$, $q_r(z)$ and $q_i(z)$, respectively, are obtained by conventional ray-tracing methods.

Let us illustrate this technique by evaluating the half-width of an incident gaussian beam at the image focal plane $z = f$ of a lens with focal length f located at $z = 0$ (see Fig. 2-9). The incident beam is assumed to have a half-width ξ_0. By a proper choice of the phase, which is in the

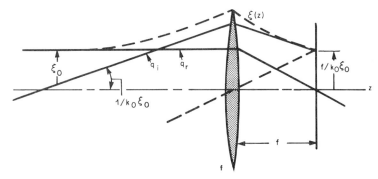

Fig. 2-9 The method of beam tracing based on the tracing of two real rays. This construction shows that the beam width at the focal plane of a lens is independent of the location of the incident beam waist. (From Arnaud,[16] by permission of North-Holland Publ. Co.)

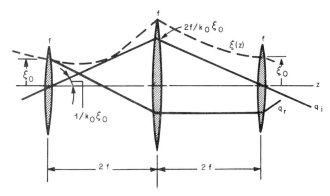

Fig. 2-10 Example of beam tracing through three identical confocal lenses. The construction shows that the beam width is the same at the output and input lenses. (From Arnaud,[16] by permission of North-Holland Publ. Co.)

present case immaterial, we can always set $q_r(z) = \xi_0$. Thus $q_r(z)$ is a line parallel to the z-axis, and $q_i(z)$ is a straight line crossing the z-axis at the beam waist plane, with a slope $1/k_0\xi_0$ [see (2.33)]. Because the ray $q_r(z)$ is parallel to the z-axis, it intersects the axis at the focal plane, after refraction, that is, $q_r(f) = 0$. Thus only the position of the ray $q_i(z)$ need be considered in the evaluation of $\xi(f)$. To obtain $q_i(f)$, consider an auxiliary ray parallel to the ray $q_i(z)$ going through the lens center. The position of this ray at $z = f$ is clearly $(1/k_0\xi_0)f$. Because this auxiliary ray meets the ray $q_i(z)$ at the lens focal plane, the answer to our problem is

$$\xi(f) = \frac{f}{k_0\xi_0} \tag{2.57}$$

It is remarkable that $\xi(f)$ does not depend on the location of the incident beam waist. This can be understood if we know from diffraction theory that the field at the focal plane of a lens is a replica of the far-field radiation pattern. The latter obviously does not depend on the location of the beam waist.

Let us work out another example. Consider three identical lenses with focal lengths f and equal separations $d = 2f$, as shown in Fig. 2-10. Let the incident beam have its waist just before the first lens, with a radius ξ_0. The construction in Fig. 2-10 shows that the exit beam has its waist located at the third lens, with the same half-width ξ_0. The beam half-width at the middle lens is easily seen to be

$$\xi = \frac{2f}{k_0 \xi_0} \tag{2.58}$$

To go further in the discussion of the transformation of gaussian beams, it is useful to introduce ray matrices. This is done in the next section.

2.7 Ray Matrices

In vacuum, geometrical optics rays follow straight lines. Let $q(z)$ denote the distance between a ray and the system axis z, a quantity that we assume to be real for the moment. We assume as before that the wavenumber on axis is a constant k_0. $q(z)$ is a linear function of z [see Fig. 2-11a]

$$q(z) = q(0) + \dot{q}(0)z \tag{2.59}$$

The slope $\dot{q} \equiv dq/dz$ of the ray is assumed to remain small compared with unity. In the present case

$$\dot{q}(z) = \dot{q}(0) \tag{2.60}$$

is a constant. Equation (2.59) can be written in matrix form

$$\begin{bmatrix} q(z) \\ \dot{q}(z) \end{bmatrix} = \begin{bmatrix} 1 & z \\ 0 & 1 \end{bmatrix} \begin{bmatrix} q(0) \\ \dot{q}(0) \end{bmatrix} \equiv \mathbf{M}(z) \begin{bmatrix} q(0) \\ \dot{q}(0) \end{bmatrix} \tag{2.61}$$

The 2×2 matrix $\mathbf{M}(z)$ is called the *ray matrix* of the section of free space of length z.

Consider next a thin lens with focal length f [Fig. 2-11b]. According to the definition of the focal length of a lens, a ray parallel to the z-axis, incident on the lens at a distance q from axis, is deflected to an axial point

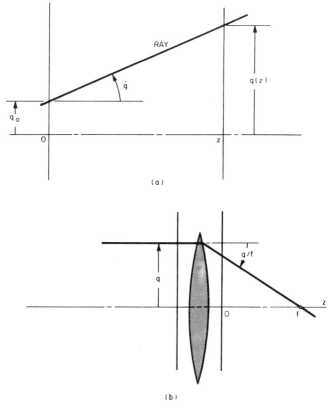

Fig. 2-11 Definition of ray matrices. (a) Section of free space. (b) Refraction by a thin lens. The ray matrix of a composite system is obtained by multiplication in reverse order of the ray matrices of the individual sections such as those shown in (a) and (b).

located at a distance f from the lens. The slope of this ray, after refraction by the lens, is therefore $-q/f$. Thus the effect of a thin lens is to deflect rays incident at q by a small angle $-q/f$. The ray transformation between a reference plane located just before the lens (primed quantities) and a reference plane located just after the lens (unprimed quantities) is consequently

$$\dot{q} - \dot{q}' = \frac{-q'}{f} \tag{2.62}$$

We also have

$$q = q' \tag{2.63}$$

because the position of a ray does not change significantly as it goes
through a thin lens. In matrix notation, (2.62) and (2.63) are

$$\begin{bmatrix} q \\ \dot{q} \end{bmatrix} = \begin{bmatrix} 1 & 0 \\ -1/f & 1 \end{bmatrix} \begin{bmatrix} q' \\ \dot{q}' \end{bmatrix} \equiv \mathbf{M}(f) \begin{bmatrix} q' \\ \dot{q}' \end{bmatrix} \tag{2.64}$$

Note that the above expression is also applicable to divergent lenses, f
being, in that case, a negative quantity.

The transformation of rays through arbitrary sequences of lenses is
obtained by multiplying (in reverse order) the matrices associated with the
lenses and the sections of free space separating the lenses. As an example
of application of this method, consider the system formed by a lens with
focal length f_1, a section of free space of length d, and a lens with focal
length f_2 (see Fig. 2-12). By application of (2.61) and (2.64), the ray matrix
of the complete system is

$$\mathbf{M} = \mathbf{M}(f_2)\mathbf{M}(d)\mathbf{M}(f_1) = \begin{bmatrix} 1 & 0 \\ -1/f_2 & 1 \end{bmatrix} \begin{bmatrix} 1 & d \\ 0 & 1 \end{bmatrix} \begin{bmatrix} 1 & 0 \\ -1/f_1 & 1 \end{bmatrix}$$

$$= \begin{bmatrix} 1 - d/f_1 & d \\ -1/f_1 - 1/f_2 + d/f_1 f_2 & 1 - d/f_2 \end{bmatrix} \tag{2.65}$$

It is interesting that the determinant of $\mathbf{M}$ is unity. This property holds
true for any ray matrix because it holds true for $\mathbf{M}(d)$ and $\mathbf{M}(f)$ and the

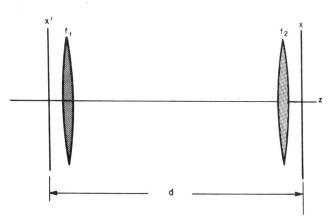

Fig. 2-12 Optical system incorporating a thin lens with focal length f_1, a section of free
space of length d, and a thin lens with focal length f_2.

determinant of a product of matrices is equal to the product of the determinants of the individual matrices.

Let us recall another general property of paraxial rays obtained before in (2.18). Consider two rays whose positions and slopes are denoted q_α, $\dot{q}_\alpha$ and q_β, $\dot{q}_\beta$, respectively. The quantity

$$q_\alpha(z)\dot{q}_\beta(z) - \dot{q}_\alpha(z)q_\beta(z) \tag{2.66}$$

known as the *Lagrange ray invariant*, assumes the same value at any transverse plane z. Let us prove it again for the case of thin lenses and sections of free space. For a thin lens with focal length f we have, from (2.64),

$$q_\alpha = q_\alpha', \qquad q_\beta = q_\beta', \qquad \dot{q}_\alpha = \dot{q}_\alpha' - \frac{q_\alpha'}{f} \qquad \dot{q}_\beta = \dot{q}_\beta' - \frac{q_\beta'}{f} \tag{2.67}$$

Therefore

$$q_\alpha\dot{q}_\beta - \dot{q}_\alpha q_\beta = q_\alpha'\left(\dot{q}_\beta' - \frac{q_\beta'}{f}\right) - \left(\dot{q}_\alpha' - \frac{q_\alpha'}{f}\right)q_\beta' = q_\alpha'\dot{q}_\beta' - \dot{q}_\alpha'q_\beta' \tag{2.68}$$

A similar proof holds, by duality $q \rightleftharpoons \dot{q}$, for a section of free space. It therefore holds true in general.

Let us now show that a lenslike medium can be considered a sequence of closely spaced thin lenses, or inversely, a thin lens can be considered a section of lenslike medium. Consider the wavenumber law

$$k(x) = k_0\left(1 - \tfrac{1}{2}\Omega^2 x^2\right) \tag{2.69}$$

Ω^2 being a real constant, the focusing strength.

To obtain the ray equation in such a medium, we may approximate it by closely spaced thin lenses. A thin lens with focal length f introduces a phase shift

$$S(x) = \frac{-k_0 x^2}{2f} \tag{2.70}$$

as we can see by noting that incident plane waves are transformed by the lens into converging waves with curvature $-1/f$. The optical thickness given by (2.70) expresses the departure of a sphere of radius $-f$ from the tangent plane. The phase shift at $x = 0$ is not truly zero as (2.70) suggests. However, the constant term that should be present on the right-hand side of (2.70) is unimportant and can be omitted.

Let us consider now a section of lenslike medium of length dz. By comparing (2.69) and (2.70), we find that this section is equivalent, with respect to both its phase shift and geometrical thickness, to a section of free space of length dz and a thin lens whose focal length f is given by the relation

$$\left(1 - \tfrac{1}{2}\Omega^2 x^2\right) dz = dz - \frac{1}{2}\frac{x^2}{f} \tag{2.71}$$

or

$$f^{-1} = \Omega^2 \, dz \tag{2.72}$$

According to (2.61), (2.64), and (2.72), the ray matrix of this short section is

$$M_{dz,f} = \begin{bmatrix} 1 & 0 \\ -\Omega^2 \, dz & 1 \end{bmatrix}\begin{bmatrix} 1 & dz \\ 0 & 1 \end{bmatrix} \approx \begin{bmatrix} 1 & 0 \\ 0 & 1 \end{bmatrix} + \begin{bmatrix} 0 & 1 \\ -\Omega^2 & 0 \end{bmatrix} dz \tag{2.73}$$

to first order in dz. Thus the ray equations are

$$q(z + dz) = q(z) + \dot{q}(z)\, dz \tag{2.74a}$$

$$\dot{q}(z + dz) = \dot{q}(z) - \Omega^2 q(z)\, dz \tag{2.74b}$$

or, in the limit $dz \to 0$,

$$\frac{dq}{dz} = \dot{q} \tag{2.75a}$$

$$\frac{d\dot{q}}{dz} = -\Omega^2 q \tag{2.75b}$$

Thus a second-order linear differential equation is obtained for $q(z)$:

$$\frac{d^2 q}{dz^2} + \Omega^2 q = 0 \tag{2.76}$$

which agrees with (2.11). Whether q is real or complex, in fact, need not be specified.

Inversely, a thin lens located at $z = 0$ can be considered a lenslike medium with focusing strength $\Omega^2 = f^{-1}\delta(z)$, where $\delta(z)$ denotes the Dirac symbolic function. Indeed, evaluating the optical thickness from $z = -\epsilon$ to $z = +\epsilon$, we have

$$S(x) = k_0 \lim_{\epsilon \to 0}\left\{ \int_{-\epsilon}^{+\epsilon}[1 - \tfrac{1}{2}f^{-1}\delta(z)x^2]dz \right\} = -\frac{k_0 x^2}{2f} \tag{2.77}$$

in agreement with (2.70).

2.8 Transformation of a Gaussian Beam by an Optical System

We have seen in the previous section that an arbitrary optical system can be described by a 2×2 unimodular matrix $[\det(\mathbf{M}) = 1]$

$$\mathbf{M} = \begin{bmatrix} A & B \\ C & D \end{bmatrix} \tag{2.78}$$

relating the input ray $(q', \dot{q}')$ to the output ray $(q, \dot{q})$

$$q = Aq' + B\dot{q}', \qquad \dot{q} = Cq' + D\dot{q}' \tag{2.79}$$

According to (2.28), the beam half-width ξ and the wavefront radius of curvature ρ are related to q and $\dot{q}$ by

$$\mu \equiv \dot{q}q^{-1} = \rho^{-1} + \frac{i}{k_0\xi^2} \tag{2.80}$$

Since (2.79) provides us with the transformation of q and $\dot{q}$, the transformation of the complex curvature μ is (Kogelnik[13])

$$\mu = (C + D\mu')(A + B\mu')^{-1} \tag{2.81}$$

where

$$\mu' \equiv \dot{q}'q'^{-1} \tag{2.82}$$

It should be noted that μ does not contain any information about the phase of the on-axis field. The on-axis phase shift is obtained from $(2.22)^2$

$$\theta - \theta' = k_0L - \tfrac{1}{2}\,\mathrm{phase}(q) + \tfrac{1}{2}\,\mathrm{phase}(q')$$

$$= k_0L - \tfrac{1}{2}\,\mathrm{phase}\!\left(\frac{q}{q'}\right)$$

$$= k_0L - \tfrac{1}{2}\,\mathrm{phase}(A + B\mu')$$

$$= k_0L - \frac{1}{2}\,\cot^{-1}\!\left[k_0\xi'^2\!\left(\rho'^{-1} + \frac{A}{B} \right) \right] \tag{2.83}$$

where we have introduced the definition of μ' in term of the incident beam half-width ξ' and wavefront radius of curvature ρ', and L denotes the length of the optical system. It is worth noting that the phase shift $\theta - \theta'$ depends only on the ratio A/B and the incident beam parameters.

Let us summarize the procedure discussed in this section. The half-width ξ' of the incident beam and its wavefront radius of curvature ρ' are lumped together into a complex quantity μ' called the complex wavefront curvature. The transformation of the complex wavefront curvature $\mu' \to \mu$ through an arbitrary optical system described by its ray matrix $\mathbf{M}$ is given by a simple law in (2.81). The half-width ξ and the wavefront radius of curvature ρ of the output beam are then obtained by separating the real and imaginary parts of μ. The phase of the field is obtained independently with the help of (2.83).

2.9 Media with Nonuniform Loss or Gain

The gain of most gas lasers is maximum on the axis of the discharge tube and decreases in decibels approximately as the square of the distance from axis for sufficiently small distances. The behavior of gaussian beams propagating through such lasers is significantly affected by the non-uniformity of the gain. A related problem is the effect of gaussian apertures (i.e., of apertures whose transmissivity is maximum on axis and decreases, in decibels, as the square of the distance from axis) on incident optical beams.

The transmission of gaussian beams through systems with quadratic loss or gain profiles is obtained simply by giving complex values to the focusing strength Ω^2 or the mirror curvatures.[2,7,14] The complex ray $q(z)$ remains a solution of the paraxial ray equations; these ray equations, however, now involve the complex function $\Omega^2(z)$.

Consider as an example a medium with complex wavenumber

$$k(x) = k_0(1 + i\alpha^2 x^2) \tag{2.84}$$

where k_0 is a real constant and α^2 is a real positive. Comparing (2.84) to (2.69), we find that $\Omega^2 \equiv -2i\alpha^2$. Equation (2.84) describes a system whose loss is zero on axis but increases off axis, and which has no real focusing strength. Yet beams whose patterns are independent of z can be found. They are represented by a complex ray $q(z)$ obtained by replacing Ω^2 by $-2i\alpha^2$, or Ω by $\alpha(1 - i)$ in previous expressions. From (2.50) we obtain

$$k_z = k_0 - \tfrac{1}{2}\Omega = k_0 - \tfrac{1}{2}\alpha(1 - i)$$

whose imaginary part is

$$k_{zi} = \tfrac{1}{2}\alpha \tag{2.85}$$

Thus the loss suffered by the fundamental mode is

$$\text{loss} = 0.5\alpha \text{ neper/unit length} \approx 4.3\alpha \text{ dB/unit length} \tag{2.86}$$

indicating an exponential decay of the power as a function of z, with a loss proportional to α. The general expression for the field is obtained by substituting $\Omega = \alpha(1 - i)$ into (2.10) and (2.11). The normalization condition (2.20), however, may not be used, because relation (2.18) does not hold when Ω^2 is complex.

The example given above is sufficient to show how the case of complex Ω^2 can be handled. Because it is more difficult to separate the real and imaginary parts in (2.10) to obtain quantities that have direct physical significance, this separation should be done only when the final result is needed. Intermediate steps are best handled in the complex ray representation.

2.10 Coupling between Gaussian Beams

The problem of evaluating the coupling between two antennas arises frequently. The theory given in this section provides an easy way of estimating this coupling when the field distributions are close to gaussian. This theory is also useful in evaluating the excitation of modes of various orders in optical resonators.

Let $\psi_\alpha(x, z)$ and $\psi_\beta(x, z)$ denote the fields of two optical beams propagating in a direction close to the z-axis. The coupling between these beams is defined by the integral[15]

$$c_{\alpha\beta} \equiv \langle \psi_\alpha^*, \psi_\beta \rangle \equiv \int_{-\infty}^{+\infty} \psi_\alpha^*(x, z)\psi_\beta(x, z) \, dx \qquad (2.87)$$

where the asterisks denote complex-conjugate values. Note, incidentally, that $\langle \psi_\alpha^*, \psi_\beta \rangle$ represents a Hermitian product in function space. Let us substitute in (2.87) the expressions obtained before in (2.10) for the field of gaussian beams and integrate. We obtain

$$c_{\alpha\beta} = (-\pi)^{-1/2} q_\alpha^{*\,-1/2} q_\beta^{-1/2} \int_{-\infty}^{+\infty} \exp\left[i\tfrac{1}{2} k_0(\dot{q}_\alpha^* q_\alpha^{*\,-1} - \dot{q}_\beta q_\beta^{-1})x^2 \right] dx$$

$$(2.88)$$

We now make use of the mathematical identity

$$\int_{-\infty}^{+\infty} \exp(-ax^2) \, dx = (\pi/a)^{1/2} \qquad (2.89)$$

Thus (2.88) is

$$c_{\alpha\beta} = \langle q_\alpha^*, q_\beta \rangle^{-1/2} \qquad (2.90)$$

where we have set

$$\langle q_\alpha^*, q_\beta \rangle \equiv (k_0/2i)(q_\alpha^* \dot{q}_\beta - \dot{q}_\alpha^* q_\beta) \qquad (2.91)$$

If the normalization condition (2.20) is used, we have $\langle q_\alpha^*, q_\alpha \rangle = \langle q_\beta^*, q_\beta \rangle = 1$. The notation $\langle q_\alpha^*, q_\beta \rangle$ is justified by the fact that this quantity has all the properties of a Hermitian product. In particular, $\langle q_\beta^*, q_\alpha \rangle = \langle q_\alpha^*, q_\beta \rangle^*$. For lossless media, $\langle q_\alpha^*, q_\beta \rangle$ is a Lagrange ray invariant. Then $c_{\alpha\beta}$ is independent of z. It is not difficult to verify this result directly by differentiating (2.87) with respect to z and using the wave equation (2.6) and its complex conjugate, k being assumed real. Thus the coupling between two optical beams is independent of the plane where the integration is performed. Note that the fields ψ_α, ψ_β in (2.10) are normalized by the conditions

$$\langle \psi_\alpha^*, \psi_\alpha \rangle = 1 \qquad (2.92a)$$

$$\langle \psi_\beta^*, \psi_\beta \rangle = 1 \qquad (2.92b)$$

that is, we have $c_{\alpha\alpha} = c_{\beta\beta} = 1$.

Before we give the expression for $c_{\alpha\beta}$ in terms of quantities that have a direct physical significance, let us observe that (2.90) can be obtained in a heuristic way by requiring that the correct geometrical optics limit be obtained. Consider two ray pencils α and β whose centers are located on the z-axis (see Fig. 2-13). Let the positions and slopes of two representative (real) rays be denoted q_α, $\dot{q}_\alpha$ and q_β, $\dot{q}_\beta$, respectively. The coupling between the ray pencils is significant only at the points where the phase of $\psi_\alpha^* \psi_\beta$ is stationary, that is, on the ray that belongs to both ray pencils; this ray coincides with the z-axis in the present case. The *intensity* of the coupling is expected to be proportional to the field intensities, which are proportional

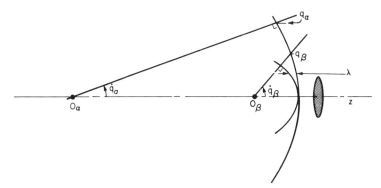

Fig. 2-13 Coupling between ray pencils. The expression for the coupling between ray pencils is suggestive of the result applicable to gaussian beams.

to $q_\alpha^{-1/2}$ and $q_\beta^{-1/2}$, and to the arc length over which the wavefronts of the two ray pencils do not depart from each other by more than, say, a wavelength. Thus the coupling is proportional to

$$q_\alpha^{-1/2}q_\beta^{-1/2}\left(\dot{q}_\alpha q_\alpha^{-1} - \dot{q}_\beta q_\beta^{-1}\right)^{-1/2} = \left(q_\alpha\dot{q}_\beta - \dot{q}_\alpha q_\beta\right)^{-1/2} \qquad (2.93)$$

because the wavefront curvatures are, respectively, $\dot{q}_\alpha q_\alpha^{-1}$ and $\dot{q}_\beta q_\beta^{-1}$, within the paraxial approximation. In the case of gaussian beams, represented by complex rays q_α and q_β, respectively, the natural extension of (2.93) is

$$\left(q_\alpha^*\dot{q}_\beta - \dot{q}_\alpha^* q_\beta\right)^{-1/2} \qquad (2.94)$$

a result that we have just proved by direct integration.

Equation (2.90) can be written

$$c_{\alpha\beta} = \langle q_\alpha^*, q_\beta\rangle^{-1/2} = \left[\tfrac{1}{2}ik_0 q_\alpha^* q_\beta\left(\mu_\alpha^* - \mu_\beta\right)\right]^{-1/2} \qquad (2.95)$$

where $\mu_\alpha \equiv \dot{q}_\alpha q_\alpha^{-1}$ and $\mu_\beta \equiv \dot{q}_\beta q_\beta^{-1}$. Introducing the beam half-widths

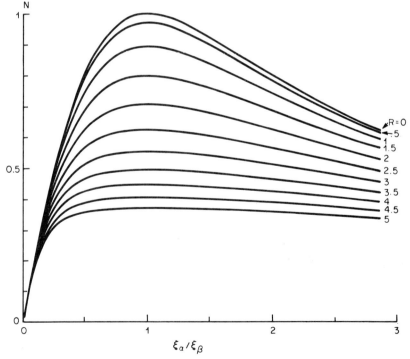

Fig. 2-14   Power coupling as a function of the mismatch ξ_α/ξ_β (or ξ_β/ξ_α) between two gaussian beams. The parameter R is the wavefront mismatch.

$\xi^2_{\alpha,\beta} = q_{\alpha,\beta}q^*_{\alpha,\beta}$ and the wavefront curvature radii $\rho_{\alpha,\beta}$ by (2.28), we obtain for the power coupling, after a few rearrangements,

$$N_{\alpha\beta} \equiv c_{\alpha\beta}c^*_{\alpha\beta} = 2\left[\left(\frac{\xi_\alpha}{\xi_\beta} + \frac{\xi_\beta}{\xi_\alpha}\right)^2 + k_0^2\xi_\alpha^2\xi_\beta^2\left(\rho_\alpha^{-1} - \rho_\beta^{-1}\right)^2\right]^{-1/2} \qquad (2.96)$$

This power coupling, which is unity when $\xi_\alpha = \xi_\beta$ and $\rho_\alpha = \rho_\beta$, is plotted in Fig. 2-14 as a function of the mismatch parameter ξ_α/ξ_β for various values of $R \equiv \bar{\rho}_\alpha^{-1} - \bar{\rho}_\beta^{-1}$, where $\bar{\rho}_{\alpha,\beta} \equiv \rho_{\alpha,\beta}/k_0\xi_\alpha\xi_\beta$.

Consider now two identical antennas α and β whose field distribution can be approximated by gaussian curves with equal half-widths ξ_0, and let us evaluate the coupling between these two antennas as a function of their separation d. Rather than using the explicit form (2.96), it is more convenient to use the complex-ray formulation, (2.90). Let the transmitting antenna α be located at $z = 0$ and the receiving antenna β at $z = d$. The beam radiated by the transmitting antenna is represented at $z = d$ by a complex ray

$$q_\alpha = \xi_0 + \frac{id}{k_0\xi_0} \qquad (2.97a)$$

$$\dot{q}_\alpha = \frac{i}{k_0\xi_0} \qquad (2.97b)$$

At the same plane, the beam *accepted* by the receiving antenna is represented by

$$q_\beta = \xi_0 \qquad (2.98a)$$

$$\dot{q}_\beta = \frac{i}{k_0\xi_0} \qquad (2.98b)$$

Substituting in (2.90), the normalized coupling is found to be

$$c_{\alpha\beta} = \left(1 - \frac{id}{2k_0\xi_0^2}\right)^{-1/2} \qquad (2.99)$$

and the power coupling, that is, the power collected by antenna β when a power of unity is fed into antenna α, is

$$N \equiv c_{\alpha\beta}c^*_{\alpha\beta} = \left[1 + \left(\frac{d}{2k_0\xi_0^2}\right)^2\right]^{-1/2}$$

$$\approx \frac{2k_0\xi_0^2}{d}, \qquad \text{if} \quad d \to \infty \qquad (2.100)$$

If the antennas have circular symmetry, the power coupling N^2 is obtained by squaring the previous expression

$$N^2 \equiv \left(c_{\alpha\beta} c_{\alpha\beta}^* \right)^2 = \left[1 + \left(\frac{d}{2k_0 \xi_0^2} \right)^2 \right]^{-1}$$

$$\approx \left(\frac{2k_0 \xi_0^2}{d} \right)^2, \qquad \text{if} \quad d \to \infty \qquad (2.101)$$

For comparison with formulas in antenna theory, let us note the following. Two antennas with field distributions

$$\psi(x) = \left(\pi \xi_0^2 \right)^{-1/4} \exp\left(-\frac{1}{2} \frac{x^2}{\xi_0^2} \right) \qquad (2.102)$$

and

$$\psi'(x) = \begin{cases} \left(4\pi \xi_0^2 \right)^{-1/4} & \text{for} \quad |x| \leqslant \pi^{1/2} \xi_0 \\ 0 & \text{for} \quad |x| > \pi^{1/2} \xi_0 \end{cases} \qquad (2.103)$$

respectively, radiate power of unity, that is, $\langle \psi^*, \psi \rangle = \langle \psi'^*, \psi' \rangle = 1$, and have the same on-axis far field. The far field is indeed proportional to

$$\langle 1, \psi \rangle = \langle 1, \psi' \rangle = \left(4\pi \xi_0^2 \right)^{1/4} \qquad (2.104)$$

These two antennas have, therefore, the same "gain." Thus an antenna with the gaussian field in (2.102) has an effective half-width $a = \pi^{1/2} \xi_0$. If two such antennas face each other at a large distance d, the power coupling is, according to (2.100),

$$N = \frac{2k_0 \xi_0^2}{d} = \frac{4a^2}{\lambda d} \qquad (2.105)$$

For antennas that have circular symmetry, we find similarly that the normalized fields

$$\psi(r) = \left(\pi \xi_0^2 \right)^{-1/2} \exp\left(-\frac{1}{2} \frac{r^2}{\xi_0^2} \right) \qquad (2.106)$$

and

$$\psi'(r) = \begin{cases} \left(4\pi \xi_0^2 \right)^{-1/2} & \text{for} \quad r \leqslant 2\xi_0 \\ 0 & \text{for} \quad r > 2\xi_0 \end{cases} \qquad (2.107)$$

have the same "gain." The effective radius of an antenna with the gaussian radial field in (2.106) is therefore $a \equiv 2\xi_0$. If two such antennas face each other at a large distance d, the power coupling is, from (2.101),

$$N^2 = 4k_0^2\xi_0^4/d^2 \equiv (\pi a^2)^2/(\lambda d)^2 \tag{2.108}$$

in agreement with the well-known Friis formula $N^2 = \text{area} \times \text{area}/(\lambda d)^2$.

The coupling between the most general Hermite–Gauss beams in three dimensions can be obtained without integration with the help of the mode-generating system concept, explained in Section 2.16. The result has the form of the product of a function of Gauss and a Hermite polynomial in four complex variables that are generalized complex ray invariants. This very general formulation will not be given here (see Arnaud[17]).

2.11 Bent Fibers

One of the major problems in guiding optical waves is the problem of alignment. Bends are introduced when the guide is installed in the trench, or at a later time because of changes in the gradient of temperature of the earth, or because of the motion of the ground. It is therefore essential to estimate the sensitivity of light-guiding systems to bends.

Let z denote the arc length along the axis of a fiber (a curved line) and $-C(z)$ the curvature of that axis. $-C(z)$ is the reciprocal of the radius of curvature. It is not difficult to show that, within the paraxial approximation, a curvature $-C(z)$ is equivalent to an increment $C(z)x$ in refractive index.[18] Let us, for instance, take as the curved z-axis a great circle on the surface of the earth with radius $1/C$. In the absence of any atmosphere, an optical ray is a straight line in the ordinary Euclidean space. Thus, if this ray is originally tangent to the earth, its distance to earth increases as $x(z) \approx \frac{1}{2}Cz^2$, assuming that $dx/dz \ll 1$. Thus, to an earth-bound observer, the ray trajectory appears to be parabolic. The same $x(z)$ curve is obtained in a rectangular xz-coordinate system when the medium wavenumber depends on x according to a law $k(x) = k_0Cx$. This can be seen by considering each section of this inhomogeneous medium, from z to $z + dz$, as a small prism. This prism increments the ray slope by the angle $C\,dz$. The solution of the equation $d(dx/dz) = C\,dz$ is indeed $x(z) = \frac{1}{2}Cz^2$.

Let us consider a straight square-law fiber with wavenumber as in (2.8)

$$k(x, z) = k_0[1 - \tfrac{1}{2}\Omega^2(z)x^2] \tag{2.109}$$

where Ω is an arbitrary function of z. As we have seen, a sequence of thin lenses is a special case of (2.109) with $\Omega(z)$ a sequence of δ-functions. If the fiber is bent with a curvature law $-C(z)$, the ray equation becomes, instead of (2.11),

$$\ddot{q}(z) + \Omega^2(z)q(z) = C(z) \tag{2.110}$$

An interesting analog of the bent fiber is the harmonic mechanical oscillator with spring constant $\Omega(t)$ submitted to a driving force $C(t)$, t denoting time.[19] The position $q(t)$ of the mass (the mass is assumed to be unity) is given by (2.110) with z changed to t.

Let us assume for a moment that the fiber is uniform ($\Omega = $ const) as is usually the case for glass fibers, and let a ray be launched along the fiber axis at $z = 0$. To evaluate the ray amplitude at some $z > 0$, it is convenient to introduce the so-called normal-mode amplitude

$$a(z) = \Omega^{1/2}q(z) + i\Omega^{-1/2}\dot{q}(z) \qquad (2.111)$$

Substituting (2.111) in (2.110), we find that the complex-valued function $a(z)$ obeys the differential equation

$$\dot{a}(z) + i\Omega a(z) = i\Omega^{-1/2}C(z) \qquad (2.112)$$

whose solution is

$$a(z) = i\Omega^{-1/2}\exp(-i\Omega z)\int_0^z C(z)\exp(i\Omega z)\,dz \qquad (2.113)$$

If we define the ray "energy" (by analogy with the mechanical problem) as $E = \tfrac{1}{2}(\dot{q}^2 + \Omega^2 q^2)$, we obtain from (2.111) and (2.113)

$$E(z) = \tfrac{1}{2}\Omega aa^* = \tfrac{1}{2}\left|\int_0^z C(z)\exp(i\Omega z)\,dz\right|^2 \qquad (2.114)$$

This expression clearly shows that only the spectral components of $C(z)$ at the angular frequency Ω contribute to the ray amplitude. In most practical cases, the curvature law $C(z)$ has a broad spatial frequency spectrum. The only important quantity is the spectral density of $C(z)$ at Ω. For curvature laws with broad frequency spectra, the guide sensitivity to bends does not depend critically on the refractive-index profile (e.g., square law or step index).[20]

Let us now go back to the general case with arbitrary $\Omega(z)$ and demonstrate a simple and general result. If a beam is launched, possibly off axis, in a bent square-law fiber, the beam shape is unaffected, and the center of the beam follows a classical ray trajectory. Let $\psi(x, z)$ denote a solution of the wave equation (2.9) applicable to a straight (but possibly nonuniform) fiber, and let $\bar{q}(z)$ denote a ray trajectory in the bent fiber, that is, an arbitrary solution of (2.110). Then the function

$$\Psi(x, z) \equiv \psi(x - \bar{q}, z)\exp\left\{ik_0\left[\dot{\bar{q}}(x - \bar{q}) + \int_0^z \bar{L}\,dz\right]\right\} \qquad (2.115)$$

where

$$\bar{L}(z) \equiv 1 + \tfrac{1}{2}\dot{\bar{q}}^2 - \tfrac{1}{2}\Omega^2\bar{q}^2 + C\bar{q} \qquad (2.116)$$

is a solution of the wave equation for the bent fiber.[2,21] The integration in (2.115) can be partially carried out. We have

$$\int_0^z \overline{L} \, dz = z + \tfrac{1}{2} \left[\bar{q}(z)\dot{\bar{q}}(z) - \bar{q}(0)\dot{\bar{q}}(0) + \int_0^z C\bar{q} \, dz \right] \quad (2.117)$$

We leave the proof of (2.115) as a problem.

The exponential term in (2.115) affects only the phase of the field. The important conclusion is that the general solution $\Psi(x, z)$ is obtained from the particular solution $\psi(x, z)$ essentially by offsetting that solution transversely by a length corresponding to a ray trajectory $\bar{q}(z)$. This result can be generalized to any optical system where the axial component k_z of the wave vector is at most quadratic in the transverse components k_x, k_y of the wave vector and at most quadratic in the transverse coordinates x, y. The z-variation of the coefficients, however, is arbitrary. This result, thus, has great generality.

2.12 Integral Transformations

A wave equation describes how the field ψ evolves as a function of z. In many problems the field is known at the plane $z = 0$, and we want to know the field at some fixed plane $z > 0$. This relation can be written in general, because of linearity, as

$$\psi(x, z) = \int_{-\infty}^{+\infty} G(x, z; x', 0)\psi(x', 0) \, dx' \quad (2.118)$$

where primes are used at the plane $z = 0$ for clarity. For fixed z, the Green function G in (2.118) is a function of x and x'. It is clear that for $\psi(x, z)$ to be a solution of the wave equation, $G(x, z; x', 0)$ must be a solution of the wave equation for any x'. This is a necessary but not sufficient condition that G must satisfy.

Note that in practice $\psi(x', 0)$ is known only at a finite number of values of x'. Thus $\psi(x', 0)$ can be represented by a vector ψ', whose components are the values of $\psi(x', 0)$ at these selected points. Similarly, $\psi(x, z)$ can be represented by a vector ψ, and (2.118) can be written in the abstract form

$$\psi = \mathbf{K}\psi' \quad (2.119)$$

where $\mathbf{K}$ denotes a matrix.

This representation can be made exact if we introduce a vector space of a special kind having an uncountably infinite number of dimensions, called the Hilbert space. We do not intend to discuss here the properties of the Hilbert space. We merely wish to point out that (2.119) is often a convenient shorthand notation for (2.118). Most of the rules of matrix algebra are applicable to (2.119). No formal justification, however, can be given here for that fact.

Let us now consider an aligned square-law medium, perhaps a sequence of lenses, whose wavenumber is given in (2.109). The Green function for that medium is [2,21]

$$G(x, z; x', 0) = \pm (2\pi)^{-1/2}(-ik_0 Q^{-1})^{1/2} \exp(ik_0 z)$$
$$\times \exp\left[ik_0 \left(\tfrac{1}{2} \dot{Q} Q^{-1} x^2 - Q^{-1} xx' + \tfrac{1}{2} Q^{-1} Q^{\dagger} x'^2 \right) \right]$$

(2.120)

where Q and $Q^{\dagger}$ are solutions of the ray equations

$$\ddot{Q} + \Omega^2(z)Q = 0, \qquad Q(0) = 0, \quad \dot{Q}(0) = 1 \qquad (2.121a)$$

$$\ddot{Q}^{\dagger} + \Omega^2(z)Q^{\dagger} = 0, \qquad Q^{\dagger}(0) = 1, \quad \dot{Q}^{\dagger}(0) = 0 \qquad (2.121b)$$

The arguments z of Q and $Q^{\dagger}$ are omitted for brevity.

Note that the prefactor in (2.120) can be written in the simpler form $\pm(i\lambda Q)^{-1/2}$, where $k_0 \equiv 2\pi/\lambda$. The form given in (2.120), however, has a deeper significance. The discussion of the sign ambiguity in (2.120) is omitted here. In order to prove (2.120), we must first prove that $G(x, z; x', 0)$ is, for any x', a solution of the wave equation (2.9). For $x' = 0$, the proof is identical to that given in Section 2.2. The general proof will be left as a problem.

Let us consider in more detail the special case of free space: $\Omega = 0$. The solutions of the ray equations (2.121) are

$$Q = z, \qquad Q^{\dagger} = 1 \qquad (2.122)$$

The Green function in (2.120) therefore becomes

$$G(x, z; x', 0) = (i\lambda z)^{-1/2} \exp\left\{ ik_0 \left[z + \frac{1}{2} \frac{(x - x')^2}{z} \right] \right\} \qquad (2.123)$$

This is a well-known result, usually obtained by making the paraxial (or Fresnel) approximation in the Kirchhoff diffraction integral. Equation (2.120) shows that the simplicity of the Green function in (2.123) is essentially preserved for arbitrary sequences of lenses as long as aperture blockage can be neglected. The results given earlier for the transformation of gaussian beams through a sequence of lenses can be obtained either from the integral transformation (2.118) with G in (2.120) or by repeated application of (2.118) with G in (2.123) for the successive sections of free space.

In three dimensions, the Green function is essentially the product of two two-dimensional Green functions, one written for x and one for y. We obtain

$$\psi(x, y, z) = (i\lambda z)^{-1} \exp(ik_0 z)$$

$$\times \int \int_{-\infty}^{+\infty} \exp\left\{ \tfrac{1}{2} ik_0 \left[(x - x')^2 + (y - y')^2 \right] \frac{1}{z} \right\}$$

$$\times \psi(x', y', 0) \, dx' \, dy' \qquad (2.124)$$

If $\psi(x', y', 0)$ has the form $\psi'(r') \exp(i\mu\varphi')$, where

$$x' \equiv r' \cos(\varphi'), \qquad y' \equiv r' \sin(\varphi')$$

the integration over φ' can be carried out. The result is[22]

$$\psi(r, \varphi) = -(-i)^\mu \exp(i\mu\varphi) \frac{ik_0}{z} \exp(ik_0 z)$$

$$\times \int_0^\infty \exp[\tfrac{1}{2} ik_0 z^{-1}(r^2 + r'^2)] J_\mu \left(\frac{k_0 r r'}{z} \right) \psi'(r') r' \, dr' \qquad (2.125)$$

where we have used the mathematical expression for the Bessel function of order μ

$$\int_0^{2\pi} \exp[i(\mu\varphi - x \cos \varphi)] \, d\varphi = 2\pi(-i)^\mu J_\mu(x) \qquad (2.126)$$

The integral transformation given above is important to discuss the transformation of the optical field through a system of circular apertures and to obtain the diffraction loss of optical resonators incorporating circular apertures.

2.13 Two-Dimensional Optical Resonators

The interest in optical resonators was stimulated in the 1960s by the invention of the laser and the need for resonators with large volumes and low losses. Single-mode operation was found desirable to optimize the coherence of the radiation. The first resonator to be considered for that application was the plane-parallel Fabry–Perot resonator. Numerical calculations have shown that the finite size of the mirrors introduces rather large diffraction losses. A system of two concave cylindrical mirrors facing each other was subsequently found to have considerably lower diffraction loss for a given degree of mode discrimination and less sensitivity to misalignments. If R denotes the radii of curvature of the mirrors and d their spacing, the wavenumbers at resonance are given by [23,24]

$$2k_{ml}d = 2l\pi + 2(m + \tfrac{1}{2}) \cos^{-1}\left(1 - \frac{d}{R}\right) \qquad (2.127)$$

assuming the mirror size unlimited. The integer l in (2.127) is the axial-mode number and the integer m the transverse-mode number, i.e., the number of nodes of the field in the transverse direction. Expression (2.127) will be derived in this chapter by specializing more general formulas. Equation (2.127) shows that a two-mirror resonator is stable, that is, k_{ml} is real, if the mirrors are concave ($R > 0$) and have a radius of curvature larger than $d/2$. Otherwise k_{ml} has an imaginary part, and the resonator is unstable.[25] The round-trip loss, in nepers, is the imaginary part of $2k_{ml}d$. The result in (2.127) was originally obtained by solving an integral equation based on the Kirchhoff–Fresnel diffraction theory. For unlimited mirror sizes and $m = 0$, (2.127) can be obtained by fitting the wavefronts of gaussian beams to the mirror surfaces, as we have seen before. This method becomes complicated for resonators that incorporate focusing elements or dielectric slabs, often considered in applications. For example, the effect of tilted quartz windows, which are commonly used in gas lasers, cannot always be neglected. Furthermore, the heat generated by laser beams introduces various degrees of focusing or defocusing along the optical path, which must sometimes by taken into consideration.

When the mirrors of a resonator have a finite size or when there are sharp-edge apertures along the path, the expressions for the resonant fields become complicated. Analytic expressions are, in fact, known only for the confocal case[24] $R = d$ and, approximately, for the plane-parallel case.[26] If, however, the mirrors have a gaussian reflectivity, (2.127) remains formally applicable, provided the mirror curvature is replaced by a complex curvature.[14] Although gaussian apertures are more difficult to manufacture than sharp-edge apertures, they have the advantage over the latter that the fundamental mode of oscillation that they help select is, in principle, exactly gaussian in shape. In this section, a theory of optical resonators incorporating focusing elements and gaussian apertures is given.

Let us choose a reference plane along the path of the resonator, evaluate the round-trip ray matrix, and find by algebraic methods what beam reproduces itself after a round trip. Because we restrict ourselves, as before, to the approximation of Gauss, this approach can be made simple and general. It also constitutes a good starting point for more advanced theories involving edge diffraction and aberrations. Ring-type resonators, which are conceptually simpler than resonators with folded optical axis (though less commonly used), will be investigated first. A ring-type optical resonator is a section of optical waveguide closed on itself; the output plane of the section of waveguide coincides with the input plane. We shall ignore, for the time being, how this can be done in practice.

Equations (2.79), which relate the position and slope q, $\dot{q}$ of a (perhaps complex) ray at the output plane of an optical system to the values q', $\dot{q}'$ assumed by the same ray at the input plane, are

$$q = Aq' + B\dot{q}' \tag{2.128a}$$

$$\dot{q} = Cq' + D\dot{q}' \tag{2.128b}$$

with $AD - BC = 1$. If the optical system incorporates gaussian apertures, A, B, C, and D are complex quantities; otherwise, they are real.

The resonance condition is that, after a round trip, the field of a beam launched from some reference plane reproduces itself exactly. Actually, we only require that the field reproduces itself to within a constant complex factor. The field experiences a phase shift that, as we have seen before, is the sum of two terms: the geometrical optics phase shift $k_0 L$, where L denotes the round-trip path length, and a phase shift that results from the transverse limitation of the beam. Because a very large number of wavelengths usually exists along the ring path, only a minute change in frequency is needed to change the geometrical optics phase shift by an angle large enough to compensate for the phase shift introduced by the transverse limitation of the beam. If the resonator has a uniform gain, resulting for instance from the presence of a homogeneous active medium (laser) or a uniform loss resulting from a mirror reflectivity r with $|r| < 1$, L has an imaginary part, negative and positive, respectively. More precisely, the gain (in nepers) required to sustain an oscillation is equal to the imaginary part of $-k_0 L$.

We have seen that the field of a gaussian beam is completely defined by a complex ray $q(z)$. Let us see if, by a proper choice of $q(z)$ at the reference plane, it is possible to have the gaussian beam repeat itself after a round trip, except perhaps for a complex factor. In other words, let us see whether solutions can be found for the pair of equations

$$q = \bar{\lambda}q' \tag{2.129a}$$

$$\dot{q} = \bar{\lambda}\dot{q}' \tag{2.129b}$$

where $\bar{\lambda}$ denotes a constant, not to be confused with the wavelength. Substituting (2.129) in (2.128), a system of two homogeneous equations is obtained:

$$(A - \bar{\lambda})q' + B\dot{q}' = 0 \tag{2.130a}$$

$$Cq' + (D - \bar{\lambda})\dot{q}' = 0 \tag{2.130b}$$

These equations have nontrivial solutions only if

$$\det \begin{bmatrix} A - \bar{\lambda} & C \\ B & D - \bar{\lambda} \end{bmatrix} = 0 \qquad (2.131)$$

The constant factor $\bar{\lambda}$ (called an eigenvalue) must therefore be

$$\bar{\lambda} = \frac{A + D}{2} \pm \left[\left(\frac{A + D}{2} \right)^2 - 1 \right]^{1/2} \qquad (2.132)$$

where we have made use of the relation $AD - BC = 1$. Equation (2.132) can be rewritten[2]

$$\bar{\lambda} = \exp(\pm i\theta) \qquad (2.133)$$

if we set

$$\frac{A + D}{2} \equiv \cos(\theta) \qquad (2.134)$$

Notice now that, if q' and $\dot{q}'$ reproduce themselves, except for a constant $\bar{\lambda}$, the field (2.10) of the gaussian beam that they define also reproduces itself identically, except for a multiplicative factor $\bar{\lambda}^{-1/2}$. The resonant frequencies and losses of the system are therefore given by the condition

$$k_0 L = 2l\pi \pm \frac{1}{2} \cos^{-1} \left(\frac{A + D}{2} \right) \qquad (2.135)$$

where l denotes an integer. Let us now evaluate the beam complex curvature at the reference plane. We have, from (2.130a) and (2.133),

$$\mu' = \mu \equiv \dot{q}q^{-1} = \frac{D - A}{2B} \pm i \frac{\sin(\theta)}{B} \qquad (2.136)$$

The beam radius ξ at the reference plane is given, according to (2.28) and (2.136), by

$$\xi^{-2} = \text{Im}(k_0 \dot{q}q^{-1}) = k_0 \, \text{Im} \left(\frac{D - A}{2B} \pm i \frac{\sin(\theta)}{B} \right) \qquad (2.137)$$

and the wavefront curvature radius by

$$\rho^{-1} = \text{Re} \left(\frac{D - A}{2B} \pm i \frac{\sin(\theta)}{B} \right) \qquad (2.138)$$

Simple conclusions can be drawn from these expressions when the resonator does not incorporate apertures, that is, when A, B, C, and D are real. In that case, the first question is to decide whether θ in (2.134) is real. Clearly, θ is real if

$$-2 < A + D < +2 \tag{2.139}$$

Equation (2.139) is the *condition of stability* of the resonator. If either

$$A + D > 2 \tag{2.140}$$

or

$$A + D < -2 \tag{2.141}$$

the resonator is said to be *unstable*. In the limit case, i.e., when $A + D$ is equal to either $+2$ or -2, the resonator is *mode degenerate*. The two values given for $\bar{\lambda}$ in (2.133) coalesce in that case, hence the name "degenerate."

Let us assume that the resonator is stable and lossless, condition (2.139) being satisfied. Equations (2.137) and (2.138) reduce, respectively, to

$$\xi^{-2} = \pm k_0 B^{-1} \sin(\theta) \tag{2.142}$$

and

$$\rho^{-1} = \frac{D - A}{2B} \tag{2.143}$$

The condition that ξ be real, that is, ξ^{-2} positive, is clearly satisfied for one and only one sign in (2.142). Whether we should select the upper or lower sign depends on the sign of θ and B, i.e., on the particular resonator considered.

Before giving a few examples of applications of the previous results, let us see how the case of resonators with folded axis can be reduced to the general case just discussed. Note first that a curved mirror with radius of curvature R can be replaced by a plane mirror and a lens in front of it with focal length $f = R$. It is easy to check that the focusing properties are the same in both cases, to first order. We can therefore assume that the end mirrors of the resonator are plane and take the reference plane at one of these mirrors. In a round trip, two optical systems, which are the mirror images of each other, are encountered. Let us denote

$$\mathbf{M} \equiv \begin{bmatrix} a & b \\ c & d \end{bmatrix} \tag{2.144}$$

the ray matrix of the system comprised between mirror 1 and mirror 2 (see Fig. 2-15b). To move backward from mirror 2 to mirror 1, we take the reciprocal of $\mathbf{M}$ and change the sign of the terms of the trailing diagonal to take into account the change in sign of the slopes. We get in that way the round-trip ray matrix $\mathbf{M}_{rt}$ of the resonator

$$\mathbf{M}_{rt} = \begin{bmatrix} A & B \\ C & D \end{bmatrix} = \begin{bmatrix} d & b \\ c & a \end{bmatrix}\begin{bmatrix} a & b \\ c & d \end{bmatrix} = \begin{bmatrix} ad + bc & 2db \\ 2ac & ad + bc \end{bmatrix} \quad (2.145)$$

Notice that in a resonator with folded axis, A is always equal to D at the (plane) end mirrors. According to (2.143), this implies that the wavefront is plane and, more generally, coincides with the surface of the end mirrors. If d denotes the mirror separation, we have $L = 2d$, and the resonance condition (2.135) becomes

$$2k_0 d = \tfrac{1}{2}\cos^{-1}(ad + bc) + 2l\pi$$

$$= \tfrac{1}{2}\cos^{-1}(2ad - 1) + 2l\pi$$

$$= \cos^{-1}(ad)^{1/2} + 2l\pi \quad (2.146)$$

where we have used the condition $ad - bc = 1$ and introduced trig-

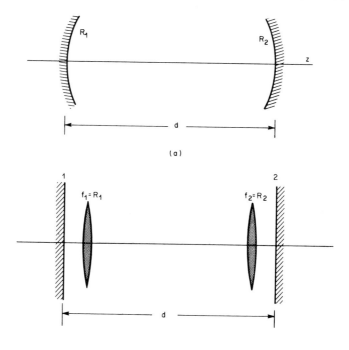

(a)

(b)

Fig. 2-15 (a) Optical resonator incorporating two circular mirrors. (b) Equivalent resonator incorporating lenses.

onometric identities. Note also that the stability condition (2.139) becomes

$$0 < ad < 1 \qquad (2.147)$$

Consider as an example a resonator formed by two spherical mirrors with radii R_1 and R_2 facing each other at a distance d (see Fig. 2-15a). R_1 and R_2 are taken as positive if the concave sides of the mirrors are facing inward.

Replacing the spherical mirrors by lenses with focal lengths $f_1 = R_1$ and $f_2 = R_2$ and plane mirrors, and using (2.65), the ray matrix from the lhs mirror to the rhs mirror is

$$\mathbf{M} = \begin{bmatrix} 1 - \dfrac{d}{R_1} & d \\[2mm] -\dfrac{1}{R_1} - \dfrac{1}{R_2} + \dfrac{d}{R_1 R_2} & 1 - \dfrac{d}{R_2} \end{bmatrix} \equiv \begin{bmatrix} a & b \\ c & d \end{bmatrix} \qquad (2.148)$$

We obtain for the resonant frequencies

$$2k_0 d = \theta + 2l\pi = \cos^{-1}\left\{ \left[\left(1 - \frac{d}{R_1}\right)\left(1 - \frac{d}{R_2}\right) \right]^{1/2} \right\} + 2l\pi \qquad (2.149)$$

For $R_1 = R_2$, (2.149) agrees with the fundamental solution $m = 0$ of (2.127). The result for any m will be derived in Section 2.18. The stability condition (2.139) is, in the present case,

$$0 < \left(1 - \frac{d}{R_1}\right)\left(1 - \frac{d}{R_2}\right) < 1 \qquad (2.150)$$

The region of stability is represented in Fig. 2-16 by a clear area, in the coordinate system $1 - d/R_1$, $1 - d/R_2$. Assuming now for simplicity that $R_1 = R_2 \equiv R$, we find from (2.142) and (2.145) that the beam half-width at the mirrors is given by

$$k_0 \xi^2 = \frac{B}{\sin \theta} = db(ad - a^2 d^2)^{-1/2} = b(1 - a^2)^{-1/2}$$

$$= d\left[1 - \left(1 - \frac{d}{R}\right)^2 \right]^{-1/2} \qquad (2.151)$$

This expression is not applicable in case of degeneracy. In particular, it is not necessarily applicable to confocal resonators with $d = R$. A (non-unique) solution for the symmetrical confocal resonator is obtained by setting $d = R$ in (2.151). We obtain

$$k_0 \xi^2 = d \qquad (2.152)$$

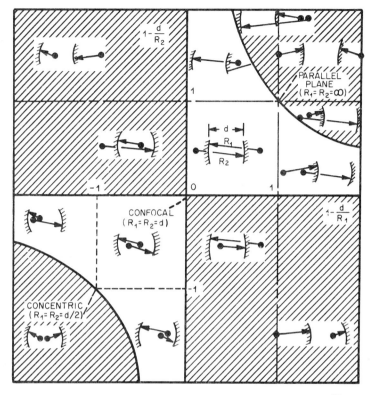

Fig. 2-16 Stability diagram for optical resonators. (From Kogelnik and Li,[24] by permission of the Optical Society of America.)

This is in agreement with (2.34) and (2.41). Indeed, setting $z = d/2 = k_0 \xi_0^2$ in (2.34), we have

$$\xi = \sqrt{2}\, \xi_0 \tag{2.153}$$

that is, the beam half-width at half the confocal distance z from the beam waist is $\sqrt{2}$ times the beam waist half-width ξ_0. By (2.41) we verify that at that distance the wavefront radius of curvature $\rho = R$ is equal to $2z = d$. Thus $\xi^2 = 2\xi_0^2 = d/k_0$. As R decreases to $d/2$ or increases to ∞, our formulas indicate that the spot size ξ at the mirrors tends to infinity. Before this can happen, however, the approximation of Gauss ceases to be applicable.

Let us give a numerical example for the nonconfocal case

$$1 - \frac{d}{R_1} = 1 - \frac{d}{R_2} = \frac{1}{\sqrt{2}} \tag{2.154}$$

Equation (2.150) shows that this resonator is stable. The beam half-width at the mirrors is, according to (2.151),

$$\xi = 2^{1/4} \left(\frac{d}{k_0} \right)^{1/2} \tag{2.155}$$

For $d = 1$ m and $\lambda = 1$ μm, we have $\xi = 0.48$ mm.

2.14 Resonators with Gaussian Apertures

Let us now assume that the end mirrors, instead of being perfectly reflecting, have a field reflectivity

$$r = \exp\left[-\frac{1}{2} \left(\frac{x}{a} \right)^2 \right] \tag{2.156}$$

where a denotes some effective radius. Because the action of a mirror with radius of curvature R on an incident field amounts to multiplying this field by a factor

$$\exp\left(-\frac{ik_0 x^2}{R} \right) \tag{2.157}$$

a curved mirror with the gaussian reflectivity in (2.156) can be assumed to have a complex radius of curvature R' given by

$$\frac{1}{R'} = \frac{1}{R} - \frac{i}{2k_0 a^2} \tag{2.158}$$

The round-trip loss of a resonator incorporating such mirrors is obtained by replacing R^{-1} in (2.149) by R'^{-1} from (2.158). The round-trip transmission is therefore

$$\mathfrak{I} = \exp[-2\,\text{Im}(2k_0 d)]$$

$$= \exp\left(-2\,\text{Im}\left\{ \cos^{-1}\left[1 - d\left(R^{-1} - \frac{i}{2k_0 a^2} \right) \right] \right\} \right) \tag{2.159}$$

Curves of constant loss are ellipses with equation

$$\left[\frac{1 - d/R}{\cosh(y)} \right]^2 + \left[\frac{d/2k_0 a^2}{\sinh(y)} \right]^2 = 1 \tag{2.160}$$

where we have defined $\exp(2y) \equiv \mathfrak{I}^{-1}$. These curves are shown in Fig.

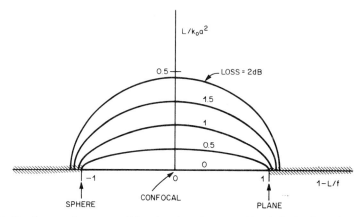

Fig. 2-17 Curves of constant diffraction loss. For two identical circular mirrors with gaussian reflectivity the vertical axis is $d/2k_0 a^2$ and the horizontal axis is $1 - d/R$. The parameter is the loss per transit. (From Arnaud,[31] by permission of the IEEE.)

2-17 for various values of the loss per transit. The segment $-1 < 1 - d/R < 1$ corresponds to stable lossless resonators. It should be noted that $\cos^{-1}(\;)$ in (2.159) has an undefined sign. The proper sign in (2.159) is obtained by specifying that the power flowing in the resonator is finite. The upper part ($a^2 > 0$) of the curves corresponds to a loss. The lower part ($a^2 < 0$), not shown, corresponds to a gain because, in the latter case, the mirror reflectivity increases as a function of the distance from axis, and, therefore, exceeds unity (see Section 2.18).

2.15 Mode Discrimination in Resonators with Rotational Symmetry

Within the approximation of Gauss, the wave equation is separable, and the field $\psi(x, y)$ is the product $\psi(x)\psi(y)$ of two two-dimensional solutions. For resonators with rotational symmetry, expression (2.149) for the resonant frequencies becomes

$$2k_0 d = 2(m + n + 1) \cos^{-1}\left\{\left[\left(1 - \frac{d}{R_1}\right)\left(1 - \frac{d}{R_2}\right)\right]^{1/2}\right\} + 2l\pi$$

$$(2.161)$$

The integers m and n in (2.161) denote the transverse-mode numbers. The derivation of the complete equation (2.161) will be given in Section 2.18.

For the fundamental mode $m = n = 0$, the diffraction phase shift and the diffraction loss that follow from (2.161) are just twice as large as for the two-dimensional case discussed in the previous section. The ratio of the diffraction loss for the first spurious modes $m = 1, n = 0$ or $m = 0, n = 1$ to that of the fundamental mode $m = n = 0$ is clearly independent of the mirror curvature and is equal to 2. The loss per transit of the fundamental mode $m = n = 0$ from (2.161) can be written, in decibels,

$$\mathcal{L}_{dB} = 8.68 \, \text{Im}[\cos^{-1}(g + iF^{-1})] \qquad (2.162)$$

where

$$g \equiv 1 - \frac{d}{R} \quad \text{and} \quad F \equiv \frac{2k_0 a^2}{d} = \frac{4\pi a^2}{\lambda d}$$

After a few algebraic transformations, we find that (2.162) can be written

$$\mathcal{L}_{dB} = 8.68 \, \log_e\left[(1 + b^2)^{1/2} + b\right] \qquad (2.163)$$

where

$$b \equiv \left\{\left[\tfrac{1}{4}(1 - g^2 - F^{-2})^2 + F^{-2}\right]^{1/2} - \tfrac{1}{2}(1 - g^2 - F^{-2})\right\}^{1/2} \qquad (2.164)$$

This loss is plotted in Fig. 2-18 as a function of $a^2/\lambda d$ for various values of g. The case $g = 1$ corresponds to the plane-parallel Fabry–Perot, $g = 0$ to the confocal resonator, and $g > 1$ to unstable resonators.

It is interesting to compare the diffraction loss of resonators incorporating mirrors with gaussian reflectivity to that of resonators incorporating mirrors whose reflectivity is unity up to a radius a and zero outside that radius. The latter problem is equivalent to having apertures of radii a between the plane mirrors and the lenses in Fig. 2-15b. The transformation of the field between planes 1 and 2 is given in (2.125). We need only add the phase shift introduced by the lenses with $f_1 = f_2 = R$ at both ends and restrict the integration to $r' < a$. Because of symmetry, it is sufficient to specify that the transformed field $\psi(r)$ coincides with the initial field $\psi'(r')$. Thus we have to solve the integral equation

$$\bar{\lambda}\psi(r) = -(-i)^\mu \frac{ik_0}{d} \int_0^a \exp\left[\frac{1}{2} ik_0 d^{-1}\left(1 - \frac{d}{R}\right)(r^2 + r'^2)\right]$$

$$\times J_\mu\left(\frac{k_0 r r'}{d}\right) \psi(r') r' \, dr' \qquad (2.165)$$

The solution $\psi(r)$ is called the eigenfunction and the constant $\bar{\lambda}$ the eigenvalue by analogy with the matrix equation $\bar{\lambda}\psi = \mathbf{K}\psi$.

The solution $\psi(r), \bar{\lambda}$ of this integral equation is known in terms of tabulated functions only for the confocal case $d = R$. For that special case, $\psi(r)$ is expressible in terms of generalized prolate spheroidal wave

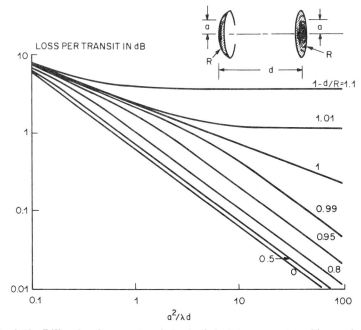

Fig. 2-18 Diffraction loss per transit in decibels for a resonator with gaussian mirror reflectivity as a function of the parameter $a^2/\lambda d$ for various values of $1 - d/R$, R being the radius of curvature of the mirrors.

functions.[24] In general, $d \neq R$, and we have to resort to numerical integration. Numerical values for the loss per transit $|\bar{\lambda}|$ were first obtained by Goubau for the equivalent problem of a periodic sequence of apertured lenses.[22] Detailed numerical results applicable to various resonator geometries were given by Fox and Li[27] and Li[28] in a series of papers. Further discussion can be found in di Francia,[29] Fox and Li,[30] Arnaud,[31] and experimental observation of laser modes in Rigrod.[32] $|\bar{\lambda}|$ is given in Fig. 2-19 as a function of $a^2/\lambda d$ for various values of $1 - d/R$, for the fundamental mode in (a) and for the first spurious mode in (b). Note that the transit loss in decibels is related to $|\bar{\lambda}|$ by $\mathcal{L}_{dB} = -20 \log_{10}(|\bar{\lambda}|)$. By comparing the curves in Figs. 2-18 and 2-19, we note that the diffraction loss of resonators with gaussian reflectivity is much larger than that of resonators with uniform reflectivity, if the radii a are the same in both cases. (There is, of course, some arbitrariness in the definition of a for mirrors with gaussian reflectivity. This arbitrariness, however, cannot account for the difference in slope of the curves.) The diffraction losses are similar for the two types of resonators only for very low Fresnel number, corresponding to very large diffraction losses. A qualitative explanation for

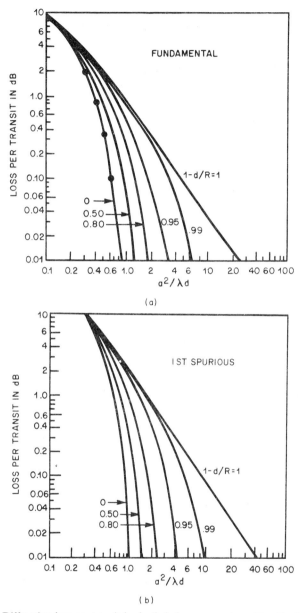

Fig. 2-19   Diffraction loss per transit in decibels for a resonator with reflectivity unity for $r < a$ and zero for $r > a$, as a function of the parameter $a^2/\lambda d$ for various values of $1 - d/R$. (a) Fundamental mode. (The dots are from the 1958 numerical results of Goubau.[22]) (b) First spurious mode. (From Li,[28] by permission of American Telegraph and Telephone Co.)

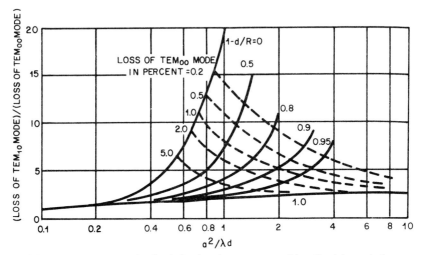

Fig. 2-20 Transverse-mode discrimination for resonator with reflectivity unit for $r < a$ and zero for $r > a$. Ratio of the losses per transit of the two lower-order modes. The dotted curves are contours of constant loss for the fundamental mode. (From Li,[28] by permission of American Telegraph and Telephone Co.)

the difference observed at moderate and large Fresnel numbers is that, in a resonator with gaussian reflectivity, the field always suffers a reflection loss, even if it is concentrated near the axis.

The main problem in designing a resonator for lasers is to have a large loss ratio between the spurious transverse modes and the fundamental mode. A confocal resonator with uniform reflectivity provides a ratio of the order of 10, as shown in Fig. 2-20, for a 1% loss on the fundamental mode. Such a resonator is therefore superior to resonators with gaussian reflectivity profiles, for which the ratio is only 2. The difference in loss ratio between the two types of resonators is less pronounced for the plane-parallel case ($R = \infty$) and is almost nonexistent for unstable resonators ($R < d/2$ or $R < 0$).

2.16 Hermite–Gauss Modes in Two Dimensions

Essential results concerning the propagation of gaussian beams in lenslike media were derived in previous sections. The results concerning the higher-order modes of propagation were stated without proofs. In this section, we investigate the propagation of Hermite–Gauss modes in general media, which are possibly anisotropic. The approximation of Gauss is maintained, however, which postulates that the point-eikonal (to be defined later) is a function at most quadratic of the transverse coordinates. The indiffused crystal waveguide is an example of anisotropic medium with quadratic variation of the point-eikonal.

It is a remarkable fact that a unified presentation can be given for these rather complicated optical systems, based on the following concept. The modes of propagation are the fields radiated by multipoles (monopoles for the fundamental gaussian beam, dipoles for the first-order modes, etc.). Mathematically, these fields are the coefficients of the expansion of the Green function in power series of the source coordinates.[33,17,31] For clarity, this formulation is given first for two-dimensional systems. The only generalization made at this point, compared to the results in previous sections, consists therefore in the introduction of the higher-order modes.

The approximation of Gauss consists in assuming that k_z is at most quadratic in k_x and x. We assume, for simplicity, that terms linear in x and k_x are absent. Thus the dependence of the axial wavenumber k_z on x, z, and k_x has the form

$$k_z = k_z(k_x, x, z) = k_0(z) - \tfrac{1}{2} k_0(z)\Omega^2(z)x^2 - \tfrac{1}{2} h(z)k_x^2 \quad (2.166)$$

For an isotropic medium with wavenumber $k(x, z)$, we have

$$k_z^2 = \left[k^2(x, z) - k_x^2\right]^{1/2} \approx k(x, z) - \tfrac{1}{2} k^{-1}(0, z)k_x^2 \quad (2.167)$$

Thus, for isotropic media, $h(z)$ in (2.166) is equal to $k_0^{-1}(z)$. We shall not make that assumption, however, and leave $h(z)$ an arbitrary function of z. If the optical fiber is uniform (z-invariant), (2.167) assumes the special form

$$k_z(k_x, x) = k_0 - \tfrac{1}{2} k_0\Omega^2 x^2 - \tfrac{1}{2} k_0^{-1} k_x^2 \quad (2.168)$$

where k_0 and Ω are constants. k_0 is the wavenumber of plane waves of the medium on axis ($x = 0$), and $2\pi/\Omega$ represents the ray period.

Let us now go back to the more general form (2.166). The wave equation is obtained by substituting ∇ for $i\mathbf{k}$ in (2.166):

$$\left[i \frac{\partial}{\partial z} + k_0(z) - \frac{1}{2} k_0(z)\Omega^2(z)x^2 + \frac{1}{2} h(z) \frac{\partial^2}{\partial x^2} \right]\psi = 0 \quad (2.169)$$

where the wave function ψ is defined as $k_0^{1/2}E$, where E denotes a transverse component of the electric field.

The Green Function

Let us show that the following function of x, z, and α

$$G(x, z; \alpha) \equiv V_g(z)^{1/2}\exp\left[iS_g(x, z; \alpha)\right] \quad (2.170)$$

is, for any α, a solution of the wave equation (2.169). The point-eikonal S_g is assumed to have the form

$$S_g(x, z; \alpha) = d_g(z) + \tfrac{1}{2} U_g(z)x^2 - V_g(z)x\alpha + \tfrac{1}{2} W_g(z)\alpha^2 \quad (2.171)$$

and $k_x = \partial S_g/\partial x$, $k_z = \partial S_g/\partial z$ is a solution of the eikonal equation (2.166). Let us first establish the differential equations obeyed by $d_g(z)$, $U_g(z)$, $V_g(z)$, and $W_g(z)$ by substituting S_g from (2.171) into (2.166). We obtain

$$\dot{d}_g = k_0 \qquad (2.172a)$$

$$\dot{U}_g + k_0\Omega^2 + hU_g^2 = 0 \qquad (2.172b)$$

$$\dot{V}_g + hU_gV_g = 0 \qquad (2.172c)$$

$$\dot{W}_g + hV_g^2 = 0 \qquad (2.172d)$$

where overdots denote differentiation with respect to z.

We shall now prove that

$$\left(i\,\frac{\partial}{\partial z} + k_0 - \frac{1}{2}\, k_0\Omega^2x^2 + \frac{1}{2}\, h\,\frac{\partial^2}{\partial x^2} \right)[V_g^{1/2}\,\exp(iS_g)] = 0 \quad (2.173)$$

The first term in (2.173) is

$$i\,\frac{\partial G}{\partial z} = \left[\tfrac{1}{2} i\dot{V}_gV_g^{-1} - \left(\dot{d}_g + \tfrac{1}{2}\dot{U}_gx^2 - \dot{V}_gx\alpha + \tfrac{1}{2}\dot{W}_g\alpha^2 \right) \right]G \quad (2.174)$$

The last term in (2.173) is

$$\frac{1}{2}\, h\,\frac{\partial^2 G}{\partial x^2} = \tfrac{1}{2}h\left[iU_g - (U_gx - V_g\alpha)^2 \right]G \qquad (2.175)$$

By using (2.172), it readily follows that (2.173) indeed holds for any α.

The Modes

Because $G(x, z; \alpha)$ is a solution of the wave equation (2.169) for any α, each coefficient $\psi_m(x, z)$ of the expansion of G in powers of α is also a solution of the wave equation. Let us evaluate these coefficients. Setting

$$v \equiv \left(\frac{-iV_g^2}{2W_g} \right)^{1/2} x, \qquad w \equiv -\left(\frac{-iW_g}{2} \right)^{1/2} \alpha \qquad (2.176)$$

the function G in (2.170) and (2.171) can be written

$$G(x, z; \alpha) = V_g^{1/2}\,\exp\left[i\left(d_g + \tfrac{1}{2}U_gx^2\right) \right]\exp(2vw - w^2) \quad (2.177)$$

Using the Taylor series expansion of $\exp(u)$ in power series of u, we obtain

$$\exp(2vw - w^2) = \sum_{r=0}^{\infty} \sum_{s=0}^{\infty} \frac{(2v)^r}{r!} (-)^s \frac{w^{2s+r}}{s!}$$

$$= \sum_{m=0}^{\infty} (m!)^{-1} w^m \sum_{s=0}^{[m/2]} m!(-)^s [s!(m - 2s)!]^{-1} (2v)^{m-2s}$$

$$\equiv \sum_{m=0}^{\infty} (m!)^{-1} w^m H_m(v) \tag{2.178}$$

r being set equal to $m - 2s$ in the first expression. $[m/2]$ denotes the largest integer smaller or equal to $m/2$. $H_m(v)$ is called a Hermite polynomial. The first polynomials are

$$H_0(v) = 1, \qquad H_1(v) = 2v, \qquad H_2(v) = 4v^2 - 2 \tag{2.179}$$

If we use this mathematical result, we can rewrite G in the form

$$G(x, z; \alpha) = \sum_{m=0}^{\infty} \alpha^m \frac{-(-i/2)^{m/2}}{m!} \psi_m(x, z) \tag{2.180}$$

where

$$\psi_m(x, z) = V_g^{1/2} W_g^{m/2} \exp\left[i\left(d_g + \tfrac{1}{2} U_g x^2\right)\right] H_m\left[\left(\frac{-iV_g^2}{2W_g}\right)^{1/2} x\right] \tag{2.181}$$

Note that the mode fields depend only on U_g and V_g^2/W_g, except for a function of z. The values of U_g, V_g, and W_g at some plane, say $z = 0$, can be selected at our convenience. Whatever these initial conditions are, $\psi_m(x, z)$ is a solution of the wave equation, provided $U_g(z)$, $V_g(z)$, and $W_g(z)$ evolve for $z > 0$ as required by the differential equations (2.172). To understand how our result (2.181) can be used in practice, let us consider stationary states (modes) in uniform media. They are obtained by specializing (2.181).

Uniform Fibers

Let the fiber be uniform along z. k_0, Ω, and h are now constants. Stationary states (ordinary modes) by definition have the form

$$\psi_m(x, z) = \psi_m(x) \exp(ik_z z) \tag{2.182}$$

For the fundamental mode of propagation $m = 0$ in (2.181), the field is stationary in the sense of (2.182) if U_g is independent of z, that is, if

$\dot{U}_g = 0$. Thus setting $\dot{U}_g = 0$ in (2.172b), U_g is found to be

$$U_g = i\left(\frac{k_0}{h}\right)^{1/2}\Omega \qquad (2.183)$$

Integration of (2.172a) and (2.172b) is straightforward because k_0, h, and U_g are constants in that example. We obtain

$$d_g = k_0 z \qquad (2.184)$$

$$V_g = \exp(-hU_g z) \qquad (2.185)$$

If results (2.183)–(2.185) are substituted in (2.181), we obtain the field of the fundamental mode (omitting a constant factor)

$$\psi_0(x, z) = \exp\left\{i\left[k_0 - \tfrac{1}{2}(k_0 h)^{1/2}\Omega\right]z\right\}\exp\left[-\frac{1}{2}\left(\frac{k_0}{h}\right)^{1/2}\Omega x^2\right] \qquad (2.186)$$

Thus the axial wavenumber is

$$k_z = k_0 - \tfrac{1}{2}(k_0 h)^{1/2}\Omega \qquad (2.187)$$

For isotropic media, $h = k_0^{-1}$, and (2.187) reduces to

$$k_z = k_0 - \tfrac{1}{2}\Omega \qquad (2.188)$$

a well-known result of the quantum theory of harmonic oscillators. The beam half-width, defined as the $1/e$ point of the irradiance $\psi_0\psi_0^*$, is

$$\bar{\xi}_0 = \left(\frac{k_0}{h}\right)^{-1/4}\Omega^{-1/2} \qquad (2.189)$$

and, for isotropic media ($h = k_0^{-1}$),

$$\bar{\xi}_0 = (k_0\Omega)^{-1/2} \qquad (2.190)$$

Having obtained the stationary fundamental solution, let us turn our attention to the higher-order modes. Integration of (2.172d) with V_g in (2.185) gives

$$W_g = -h\int\exp(-2hU_g z)\,dz = \tfrac{1}{2}U_g^{-1}\exp(-2hU_g z) \qquad (2.191)$$

The integration constant must clearly be chosen equal to zero in order that V_g^2/W_g in the argument of H_m in (2.181) be independent of z. For that choice, (2.181) becomes, to within a constant factor,

$$\psi_m(x, z) = \exp\left\{i\left[k_0 - \frac{(m + \tfrac{1}{2})h}{\bar{\xi}_0^2}\right]z\right\}\exp\left[-\frac{1}{2}\frac{x^2}{\bar{\xi}_0^2}\right]H_m\left(\frac{x}{\bar{\xi}_0}\right) \qquad (2.192)$$

where the beam half-width $\bar{\xi}_0$ has the value in (2.189). This is the final result for two-dimensional systems. For isotropic media, the term $h/\bar{\xi}_0^2$ in (2.192) is equal to Ω.

Indiffused Waveguides

This result can be applied to indiffused uniaxial crystal waveguides.[34] The uniaxial crystal is cut with the optical axis parallel to the surface and parallel to the direction of propagation z (see Fig. 2-21). The curve of wave vector for the extraordinary wave, as we shall see in Chapter 3, is an ellipse with equation

$$\left(\frac{k_x}{k'} \right)^2 + \left(\frac{k_z}{k} \right)^2 = 1 \qquad (2.193)$$

where k and k' are constants

$$k \equiv \omega(\epsilon\mu_0)^{1/2}, \qquad k' = \omega(\epsilon'\mu_0)^{1/2} \qquad (2.194)$$

Thus, to second order in k_x,

$$k_z \approx k - \frac{1}{2} \frac{k}{k'^2} k_x^2 \qquad (2.195)$$

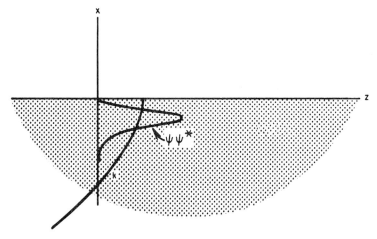

Fig. 2-21 Modes of indiffused uniaxial waveguides. The optical c axis is parallel to the z axis. The curve of wave normal for the extraordinary wave (electric field in the plane of the figure) is approximated by a parabola whose curvature is different from that applicable to an isotropic medium. Because the parameters vary approximately as the square of the distance from the air–crystal boundary and because the field almost vanishes at the boundary, the irradiance profile $\psi\psi^*$ shown is almost that of the first antisymmetrical Hermite–Gauss mode.

The parameter h is, in the present case, equal to k/k'^2. The process of indiffusion induces a small decay of k as a function of the distance from the surface. In some cases, $k(x)$ can be approximated by a truncated square law

$$k(x) = \begin{cases} k_0 - \frac{1}{2} k_0 \Omega^2 x^2 & x < 0 \\ \omega/c & x > 0 \end{cases} \tag{2.196}$$

The variation of k' with x can be neglected. Because of the large discontinuity in refractive index at the crystal–air boundary, the field can be assumed to vanish at the boundary (see Chapter 5), and the modes almost coincide with the antisymmetrical modes of square-law media. The fundamental mode $\Psi_0(x)$ of the indiffused waveguide is, therefore, almost identical to the $\psi_1(x)$ mode in (2.192). We have the irradiance pattern $[H_1(x) = 2x]$

$$\Psi_0(x)^2 = |\psi_1|^2 = \left(\frac{x}{\bar{\xi}_0} \right)^2 \exp\left(-\frac{x^2}{\bar{\xi}_0^2} \right) \tag{2.197}$$

where

$$\bar{\xi}_0 = (k'\Omega)^{-1/2} \tag{2.198}$$

In optics, k' is not very different from k, and the effect of anisotropy on the beam radius is small.

The theory presented in this section has perhaps greater practical significance for artificial dielectrics, which often exhibit large anisotropies, as we shall see in Section 2.22. Our theory, based on the concept of complex point-eikonals, agrees with the theory presented by Ash and Mason[38] for acoustical surface waves propagating on anisotropic but homogeneous substrates. Ash and Mason's theory is based on plane-wave expansions of the field. It would not be easily generalized to inhomogeneous media.

2.17 Hermite–Gauss Modes in Three Dimensions; The Helical Fiber

The results given in the previous section can be greatly generalized. In this section we shall use the notation $x_1 \equiv x$, $x_2 \equiv y$. The vector $\mathbf{x}$ has rectangular components x_1, x_2. The subscripts g are omitted for brevity.

The Green function for gaussian systems was discovered in 1928 by Van Vleck[35] in the realm of wave mechanics. The Green function has the form

$$G(\mathbf{x}, z; \mathbf{x}', z') = \pm (2\pi)^{-N/2} \left[\det \frac{-\partial^2 iS}{\partial x_i \, \partial x_j'} \right]^{1/2} \exp(iS) \tag{2.199}$$

where $S(\mathbf{x}, z; \mathbf{x}', z')$ denotes the point-eikonal, that is, the phase shift along the ray that goes from $\mathbf{x}'$, z' to $\mathbf{x}$, z. N denotes the number of transverse dimensions. In the present optical problem, $N = 2$. The sign ambiguity in (2.199) can be ignored for the moment. The physical interpretation of (2.199) is simple when S is real, as is the case in conventional geometrical optics. The term $\exp(iS)$ expresses the phase shift along the ray, according to the definition of S. It can be shown that the prefactor in (2.199) follows from power conservation requirements. Expression (2.199) is general; it is applicable even when the medium has losses, in which case S is complex valued. The wave equation must, however, be symmetrically ordered; that is, a term such as $\mathbf{x}\partial/\partial\mathbf{x}$ in the wave equation operator should be replaced by $\frac{1}{2}[\mathbf{x}(\partial/\partial\mathbf{x}) + (\partial/\partial\mathbf{x})\mathbf{x}]$. Note that expression (2.170) is a special case of (2.199). The derivation of (2.199) is rather lengthy. We shall therefore omit the proof and refer the reader to Van Vleck's paper.[35]

For $S(\mathbf{x}, \boldsymbol{\alpha})$, a quadratic function of $\mathbf{x}$ and $\boldsymbol{\alpha}$, the Green function (2.199) has the form

$$G(\mathbf{x}, z; \boldsymbol{\alpha}) = (\det \mathbf{V})^{1/2} \exp[i(\theta_0 + \tfrac{1}{2}\mathbf{x}\mathbf{U}\mathbf{x} - \mathbf{x}\mathbf{V}\boldsymbol{\alpha} + \tfrac{1}{2}\boldsymbol{\alpha}\mathbf{W}\boldsymbol{\alpha})] \quad (2.200)$$

to within a constant factor that we shall not need. The modes are defined as the coefficients of the expansion of G in power series of $\boldsymbol{\alpha}$, that is, of both α_1 and α_2. Two procedures can be used to expand the term

$$\exp[i(-\mathbf{x}\mathbf{V}\boldsymbol{\alpha} + \tfrac{1}{2}\boldsymbol{\alpha}\mathbf{W}\boldsymbol{\alpha})] \quad (2.201)$$

in (2.200) in power series of α_1 and α_2. It is possible to write the term $\boldsymbol{\alpha}\mathbf{W}\boldsymbol{\alpha}$ as a sum of squares, with the help of a coordinate transformation in the $\boldsymbol{\alpha}$ plane, since the real and imaginary parts of $\mathbf{W}$ can be diagonalized simultaneously. The new coordinate system is real, but it may not be rectangular. Furthermore, it varies with z, and may differ from the coordinate system that diagonalizes $\mathbf{U}$. This is not very convenient. Thus we find it preferable to make use of the Hermite–Grad polynomials in two variables, defined as follows from their generating function (with some changes of notation from Arnaud[17]):

$$\exp(\boldsymbol{\xi}\boldsymbol{\zeta} + \boldsymbol{\zeta}\boldsymbol{\phi}\boldsymbol{\zeta}) \equiv \sum_{m_1, m_2 = 0}^{\infty} \zeta_1^{m_1}\zeta_2^{m_2} He_{m_1 m_2}(\boldsymbol{\xi}; \boldsymbol{\phi}) \quad (2.202)$$

To obtain an explicit expression for $He_{m_1 m_2}(\boldsymbol{\xi}; \boldsymbol{\phi})$, we proceed as for the one-dimensional case. The argument of the exponential function in (2.202) is written explicitly, and the five exponentials are expanded in power series of their respective arguments (remember that $\phi_{12} = \phi_{21}$). We obtain

$$\exp(\boldsymbol{\xi}\boldsymbol{\zeta} + \boldsymbol{\zeta}\boldsymbol{\phi}\boldsymbol{\zeta}) = \exp(\xi_1\zeta_1)\exp(\xi_2\zeta_2)\exp(\phi_{11}\zeta_1^2)\exp(\phi_{22}\zeta_2^2)\exp(2\phi_{12}\zeta_1\zeta_2)$$

$$= \sum_{r, s, \alpha, \beta, \gamma = 0}^{\infty} \xi_1^{(r)}\zeta_1^r \xi_2^{(s)}\zeta_2^s \phi_{11}^{(\alpha)}\zeta_1^{2\alpha}\phi_{22}^{(\beta)}\zeta_2^{2\beta}(2\phi_{12})^{(\gamma)}\zeta_1^\gamma\zeta_2^\gamma \quad (2.203a)$$

where we have used the notation $a^{(\alpha)} \equiv a^\alpha/\alpha!$ (The integer α should not be confused with the components of the vector $\boldsymbol{\alpha}$.) Setting

$$m_1 = r + 2\alpha + \gamma, \qquad m_2 = s + 2\beta + \gamma \qquad (2.203b)$$

we find that (2.203) can be written in the form (2.202), with

$$He_{m_1 m_2}(\boldsymbol{\xi};\boldsymbol{\phi}) = \sum_{\alpha,\,\beta,\,\gamma\,=\,0} \phi_{11}^{(\alpha)}\phi_{22}^{(\beta)}(2\phi_{12})^{(\gamma)}\xi_1^{(m_1-2\alpha-\gamma)}\xi_2^{(m_2-2\beta-\gamma)} \qquad (2.204)$$

The summation in (2.204) terminates when one of the exponents vanishes.

We shall make use of the following relation: For any diagonal matrix $\mathbf{D}$ with elements λ_1, λ_2, we have

$$He_{m_1 m_2}(\mathbf{D}\boldsymbol{\xi};\mathbf{D}\boldsymbol{\phi}\mathbf{D}) = \lambda_1^{m_1}\lambda_2^{m_2}He_{m_1 m_2}(\boldsymbol{\xi};\boldsymbol{\phi}) \qquad (2.205)$$

This relation follows either from definition (2.202) or from the explicit expression in (2.204).

We now make use of these mathematical results, setting

$$\boldsymbol{\phi} = \frac{-i\mathbf{W}}{2}, \qquad \boldsymbol{\zeta} = i\boldsymbol{\alpha}, \qquad \boldsymbol{\xi} = -\tilde{\mathbf{V}}\mathbf{x} \qquad (2.206)$$

and obtain the mode fields

$$\psi_{m_1 m_2}(\mathbf{x}, z) = (\det \mathbf{V})^{1/2}\exp[i(\theta_0 + \tfrac{1}{2}\mathbf{x}\mathbf{U}\mathbf{x})] \times He_{m_1 m_2}(-\tilde{\mathbf{V}}\mathbf{x}; -i\mathbf{W}/2) \qquad (2.207)$$

where the tilde denotes transposition. If linear terms are included in the expression of S

$$S = \theta_0 + \mathbf{u}\mathbf{x} + \mathbf{w}\boldsymbol{\alpha} + \tfrac{1}{2}\mathbf{x}\mathbf{U}\mathbf{x} - \mathbf{x}\mathbf{V}\boldsymbol{\alpha} + \tfrac{1}{2}\boldsymbol{\alpha}\mathbf{W}\boldsymbol{\alpha} \qquad (2.208a)$$

the following expression is obtained for the modes:

$$\psi_{m_1 m_2}(\mathbf{x}, z) = (\det \mathbf{V})^{1/2}\exp[i(\theta_0 + \mathbf{u}\mathbf{x} + \tfrac{1}{2}\mathbf{x}\mathbf{U}\mathbf{x})]$$

$$\times He_{m_1 m_2}(\mathbf{w} - \tilde{\mathbf{V}}\mathbf{x}; -i\mathbf{W}/2) \qquad (2.208b)$$

which is the most general form consistent with the approximation of Gauss.

Systems with Circular Symmetry. The Laguerre–Gauss Modes

Optical systems with rotational symmetry, such as conventional lenses or optical fibers, are of particular interest. The solutions of the parabolic wave equation can be written as products of two two-dimensional solutions:

$$\psi_{mn}(x, y) = \psi_m(x)\psi_n(y) \qquad (2.209a)$$

where $\psi_m(x)$, $\psi_n(y)$ are, for instance, the Hermite–Gauss modes defined in (2.181). The rectangular coordinate form in (2.209a) is not the most convenient because it does not exhibit the circular symmetry of the system. Thus, we shall instead derive the beam mode solution for circularly symmetric systems from the general form in (2.207).

The parameters $\mathbf{U}$, $\mathbf{V}$, $\mathbf{W}$ in (2.207) obey the following matrix equations:

$$\dot{\mathbf{U}} + \mathbf{U}^2 + \mathbf{\Omega}^2 = 0$$

$$\dot{\mathbf{V}} + \mathbf{UV} = 0 \tag{2.209b}$$

$$\dot{\mathbf{W}} + \tilde{\mathbf{V}}\mathbf{V} = 0$$

which generalize (2.172). For a system with circular symmetry the focusing matrix $\mathbf{\Omega}^2$ has the form

$$\mathbf{\Omega}^2(z) = \Omega^2(z)\mathbf{1} \tag{2.209c}$$

where $\Omega^2(z)$ is a scalar function of z and $\mathbf{1}$ denotes the 2×2 unit matrix. Using the above differential equations, it is easy to see that if the matrices $\mathbf{U}$, $\mathbf{V}$, $\mathbf{W}$ have, at the input plane, the form

$$\mathbf{U} = U\mathbf{1}$$

$$\mathbf{V} = V \begin{bmatrix} 1 & i \\ i & 1 \end{bmatrix} \tag{2.209d}$$

$$\mathbf{W} = 2iW \begin{bmatrix} 0 & 1 \\ 1 & 0 \end{bmatrix}$$

where U, V, and W are scalar functions of z, these forms are maintained for all z. If we substitute these forms for $\mathbf{U}$, $\mathbf{V}$, $\mathbf{W}$ in (2.207), we find that the triple sum in (2.204) reduces to a single sum over γ because $\phi_{11} = \phi_{22} = 0$. Assuming for definiteness that $m_2 < m_1$, and setting

$$Z \equiv x_1 + ix_2 \equiv re^{i\varphi} \tag{2.209e}$$

we obtain a rather simple expression for ψ:

$$\psi_{m_1 m_2}(Z, z) = V \exp\left(\frac{iUZZ^*}{2} \right)$$

$$\times \sum_{\gamma=0}^{m_2} (2W)^{(\gamma)}(-VZ)^{(m_1-\gamma)}(-iVZ^*)^{(m_2-\gamma)} \tag{2.209f}$$

where, as before, $a^{(\gamma)} \equiv a^\gamma/\gamma!$. Setting in (2.209f) $s \equiv m_2 - \gamma$, $m_2 = \alpha$, and $m_1 - m_2 = \mu$, this expression can be written, restoring the k_0 factor and omitting a constant,

$$\psi_{\mu\alpha}(r, z) = V \exp\left(\frac{ik_0 U r^2}{2} \right)\left(\frac{W}{ik_0} \right)^\alpha (VZ)^\mu$$

$$\times L_\alpha^\mu\left(\frac{ik_0 V^2 r^2}{2W} \right)$$

where

$$L_\alpha^\mu(x) \equiv \sum_{s=0}^{\alpha} \frac{(\alpha + \mu)!(-x)^s}{(\mu + s)!(\alpha - s)!s!} \tag{2.209g}$$

is a generalized Laguerre polynomial. In particular, $L_0^\mu = 1$. The above form clearly shows that μ is the azimuthal mode number and α the radial mode number. An explicit relation with one-dimensional Hermite–Gauss functions is given by Arnaud.[33]

For the special case of a uniform square-law medium with focusing constant Ω, the stationary modal solutions are obtained by setting $U = i\Omega$, $V = \exp(-i\Omega z)$, and $W = -V^2/2i\Omega$ in (2.209g). We obtain

$$\psi_{\mu\alpha}(r, z) = \exp\left(-\frac{1}{2}\,\frac{r^2}{\xi_0^2} \right)\left(\frac{r}{\xi_0} \right)^\mu L_\alpha^\mu\left(\frac{r^2}{\xi_0^2} \right)$$

$$\times \exp\{i[k_0 - (2\alpha + \mu + 1)\Omega]z\}$$

where

$$\xi_0^2 \equiv \frac{1}{k_0\Omega} \tag{2.209h}$$

Note that the axial wave number depends on α and μ only through the sum $2\alpha + \mu$. This means that many different modes have the same axial wave number. This degeneracy is peculiar to square-law media (harmonic oscillators in mechanics) and a few other very special refractive index laws. Another example is the Maxwell fish-eye medium: $n(r) = (1 + r^2)^{-1}$, whose degeneracy is a consequence of the symmetry exhibited in Fig. 1.14a.

Nonorthogonal Optical Systems

The solution (2.207) of the parabolic wave equation is applicable to the four optical systems shown in Fig. 2-22. Nonorthogonal resonators such as those shown in Fig. 2–22b and d are discussed in Section 2.20. In Fig. 2-22a, a fundamental gaussian beam with circular symmetry is shown

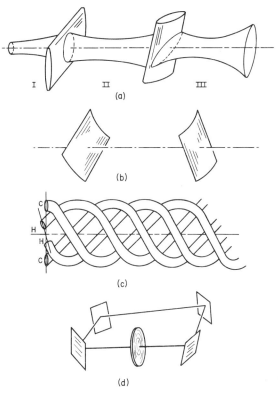

Fig. 2-22 Nonorthogonal optical systems. (a) A gaussian beam with circular symmetry (I) is transformed into a beam with simple astigmatism (II) and into a beam with general astigmatism (III) by a pair of cylindrical lenses. (b) Resonator incorporating two cylindrical mirrors whose generatrices make an angle different from $\mu\pi/2$, μ integer. (c) Helical gas lens. The four coaxial helices create a saddle-shaped distribution of refractive index that rotates along the system axis. Optical beams are kept confined by the strong-focusing effect. (d) Cavity with image rotation incorporating a lens with circular symmetry and four plane mirrors. The nonorthogonality originates from the twist of the path. Because of this twist, modes with $\exp(i\mu\varphi)$ and $\exp(-i\mu\varphi)$ azimuthal variations are not degenerate. The irradiance of all the modes is circularly symmetric. (From Arnaud,[16] by permission of North-Holland Publ. Co.)

transmitted through a pair of astigmatic (specifically, cylindrical) lenses.[36] The first lens transforms the circular incident beam into a beam with elliptical cross section, and the second lens, located some distance away from the first lens, generates a beam with so-called general astigmatism. The principal axes of the irradiance ellipse rotate in free space,[36] the total rotation angle from $z = -\infty$ to $z = +\infty$ being equal to π. We shall only briefly indicate how this rotation comes about. The only term in (2.207) relevant to the present discussion is the term $\exp(\frac{1}{2} i\mathbf{x}\mathbf{U}\mathbf{x})$. The symmetrical

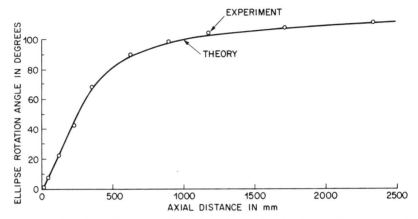

Fig. 2-23 Rotation in free space of the principal axes of the irradiance ellipse of a gaussian beam with general astigmatism. The theory (curve) is compared to experimental results (circles). (From Arnaud and Kogelnik,[36] by permission of the American Optical Society.)

matrix $\mathbf{U}$ obeys in free space the differential equation $d\mathbf{U}/dz + \mathbf{U}^2/k_0 = 0$, which is the same as (2.172b), with $\Omega = 0$ and a straightforward generalization to matrix parameters (for more details, see Chapter 4). To solve this equation, it is convenient to introduce a complex matrix ray $\mathbf{Q}(z)$ defined by $\mathbf{U} \equiv k_0(d\mathbf{Q}/dz)\mathbf{Q}^{-1}$. If we substitute this expression in the above differential equation for $\mathbf{U}$ and remember that $d(\mathbf{Q}^{-1})/dz = -\mathbf{Q}^{-1}(d\mathbf{Q}/dz)\mathbf{Q}^{-1}$, it follows that $d^2\mathbf{Q}/dz^2 = 0$, that is, $\mathbf{Q}$ is a linear function of z that we may write $\mathbf{Q} = \mathbf{Q}_0 + \mathbf{Q}_1 z$, where $\mathbf{Q}_0$ and $\mathbf{Q}_1$ denote constant complex matrices. Thus, $\mathbf{U}(z) = k_0\mathbf{Q}_1(\mathbf{Q}_0 + \mathbf{Q}_1 z)^{-1}$. The beam irradiance is defined by the imaginary part of $\mathbf{U}$ whose principal axes clearly vary with z if $\mathbf{Q}_0$ and $\mathbf{Q}_1$ do not have the same eigenvectors. This is the case for the initial conditions imposed by the two cylindrical lenses in Fig. 2-22a.

A different but equivalent approach was used by Arnaud and Kogelnik[36] to discuss gaussian beams with general astigmatism. It was observed that when an optical system is rotationally symmetric (free space is a trivial example of such a system), a solution of the wave equation remains a solution of the wave equation when the coordinates are transformed by an arbitrary rotation about the system axis. It turns out that the field of gaussian beams with general astigmatism can be obtained by introducing a complex angle of rotation on the x, y-coordinates in the expression of the field of a beam with simple astigmatism (e.g., the field of the beam between the two lenses in Fig. 2-22a). This approach is not quite as general as the one that we have discussed earlier because optical elements (such as cylindrical lenses) that lack rotational symmetry need to be dealt with independently from the rest of the optical system. Figure 2-23 shows that there is excellent agreement between the theory sketched above and the experimental results.

The Helical Fiber

Let us discuss in more detail the modal solutions of the helical lens (or fiber) shown in Fig. 2-22c. An explicit and exact solution can be obtained for that system by application of (2.207).

The focusing properties of electrostatic lenses incorporating four coaxial helices at potentials $+V$, $-V$, $+V$, and $-V$, respectively, are well known in the technology of particle accelerators. These electrostatic lenses evolved from the concept of strong focusing, according to which a periodic sequence of converging and diverging lenses with equal absolute powers has a net focusing effect, provided the period does not exceed a certain critical value. A similar technique can be applied to the guidance of optical beams.[10,37,38] In the helical gas lens, four coaxial helices are raised at alternately high and low temperatures (see Fig. 2.22c). Because of this difference in temperature, gradients of refractive index are created in the gas filling the space inside the helices. The gas thus acts as a quadrupole lens whose principal axes rotate along the system axis. Alternatively, a glass preform can be given the required refractive-index law, of the form $n = 1 - \frac{1}{2}(x^2 - y^2)$. The helical rotation that is needed for beam confinement is obtained by spinning the glass preform as the fiber is being pulled.

We assume that the medium is isotropic and has wavenumber

$$k(x_1, x_2) = k_0\left(1 - \tfrac{1}{2}\mathbf{x}\Omega^2\mathbf{x}\right) \qquad (2.210a)$$

in a coordinate system that rotates at a spatial rate τ about the guide axis. Ω (and therefore Ω^2) denotes a symmetrical 2×2 matrix. For simplicity, in this section we drop the factor k_0. It can easily be restored from dimensional considerations. Alternatively, we can assume that the free-space wavelength is 2π times the unit of length (e.g., micrometers). Because of the medium isotropy, we have, within the approximation of Gauss,

$$k_z = k(\mathbf{x}) - \tfrac{1}{2}k_{\mathbf{x}}^2 = 1 - \tfrac{1}{2}\mathbf{x}\Omega^2\mathbf{x} - \tfrac{1}{2}k_{\mathbf{x}}^2 \qquad (2.210b)$$

If we substitute $k_{\mathbf{x}} \to \partial S / \partial \mathbf{x}$, $k_z \to \partial S / \partial z$, with the expression of S in (2.200), into (2.210b) and identify terms that have the same power in $\mathbf{x}$ and α, we obtain the differential equations obeyed by $\mathbf{U}$, $\mathbf{V}$, $\mathbf{W}$ in a fixed rectangular coordinate system. These equations, given before in (2.209b), are

$$\dot{U} + U^2 + \Omega^2 = 0, \qquad \dot{V} + UV = 0, \qquad \dot{W} + \tilde{V}V = 0 \quad (2.211)$$

and $\theta_0 = z - z'$. (An even more general result is given in Chapter 4.) Let us now rewrite these equations for the case where the x_1, x_2 coordinate system rotates at a spatial rate τ about the z-axis. A rotation τz is conveniently expressed in matrix notation by the coordinate transformation

$$\mathbf{x} \to e^{-\mathbf{T}z}\mathbf{x} \tag{2.212}$$

where

$$\mathbf{T} \equiv \tau \begin{bmatrix} 0 & 1 \\ -1 & 0 \end{bmatrix} \tag{2.213}$$

as we can verify by diagonalizing $\mathbf{T}$.

To obtain the transformation of the parameters defining the point-eikonal, it suffices to specify that these scalar quantities have the same values in the fixed and rotated coordinate systems. The laws of transformation are therefore obtained by introducing (2.212) in (2.210a) and (2.200). We obtain

$$\Omega^2 \to e^{-\mathbf{T}z}\Omega^2 e^{\mathbf{T}z}, \qquad \mathbf{U} \to e^{-\mathbf{T}z}\mathbf{U}e^{\mathbf{T}z}, \qquad \mathbf{V} \to e^{-\mathbf{T}z}\mathbf{V}, \qquad \mathbf{W} \to \mathbf{W}$$

$$\tag{2.214}$$

Upon substitution of these expressions in (2.211), generalized differential equations for $\mathbf{U}$, $\mathbf{V}$, and $\mathbf{W}$ are obtained:

$$\dot{U} - \mathbf{T}U + U\mathbf{T} + U^2 + \Omega^2 = 0 \tag{2.215a}$$

$$\dot{V} - \mathbf{T}V + UV = 0 \tag{2.215b}$$

$$\dot{W} + \tilde{V}V = 0 \tag{2.215c}$$

Neglecting aberration, the wavenumber profile of a helical fiber has, in the rotating coordinate system, the form

$$k = 1 - \tfrac{1}{2}\Omega^2(x_1^2 - x_2^2) \equiv 1 - \tfrac{1}{2}\mathbf{x}\Omega^2\mathbf{x} \tag{2.216}$$

where we have defined the diagonal 2×2 matrix

$$\Omega^2 = \begin{vmatrix} \Omega^2 & 0 \\ 0 & -\Omega^2 \end{vmatrix} \tag{2.217}$$

The parameter Ω is real if the medium is lossless, but this assumption need not be made.

It is clear from (2.207) that the field configuration of the fundamental mode ($m_1 = m_2 = 0$) is invariant if the matrix $\mathbf{U}$ does not depend on z. The reader may verify that the solution of (2.215a) with $\dot{\mathbf{U}} \equiv d\mathbf{U}/dz = \mathbf{0}$ is

$$\mathbf{U} = \tau \tan(\nu) \begin{bmatrix} i(\cos \nu - \sin \nu) & 1 \\ 1 & i(\cos \nu + \sin \nu) \end{bmatrix} \tag{2.218}$$

where we have defined

$$\sin(2\nu) \equiv \left(\frac{\Omega}{\tau} \right)^2 \tag{2.219}$$

Assuming that the medium is lossless, i.e., that Ω is real, the angle ν defined in (2.219) is real when the stability condition

$$\Omega < \tau \tag{2.220}$$

is met. The helical lens is stable when ν is comprised between 0 and $\pi/4$. We shall give without derivation the complete modal solution in the rotating coordinate system, restoring the k_0 factor[38]

$$\psi_{m_1 m_2}(x_1, x_2) = \exp(ik_0 z)$$

$$\times \exp\{ i\tau z [(m_1 + \tfrac{1}{2})(\cos \nu - \sin \nu)$$

$$- (m_2 + \tfrac{1}{2})(\cos \nu + \sin \nu)] \}$$

$$\times \exp\{ -\tfrac{1}{2} \tan \nu [(\cos \nu - \sin \nu)\chi_1^2$$

$$+ (\cos \nu + \sin \nu)\chi_2^2] + i(\tan \nu)\chi_1\chi_2 \}$$

$$\times \sum_{\alpha, \beta, \gamma} (-)^\gamma \left(\frac{1}{2} \frac{\sin \nu \cos \nu}{\cos \nu - \sin \nu} \right)^{(\alpha)}$$

$$\times \left(\frac{1}{2} \frac{\sin \nu \cos \nu}{\cos \nu + \sin \nu} \right)^{(\beta)} (i \cot \nu)^{(\gamma - \alpha - \beta)}$$

$$\times [\chi_1 + i\chi_2(\cos \nu + \sin \nu)]^{(m_1 - \gamma - \alpha + \beta)}$$

$$\times [\chi_2 + i\chi_1(\cos \nu - \sin \nu)]^{(m_2 - \gamma - \beta + \alpha)} \tag{2.221}$$

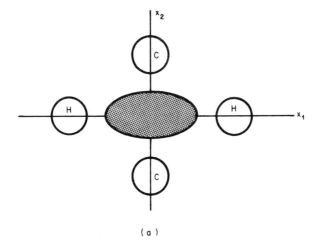

(a)

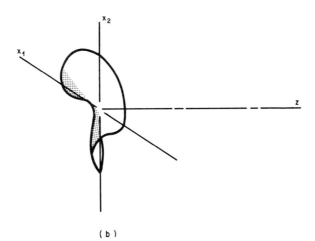

(b)

Fig. 2-24 Fundamental mode of the helical fiber. (a) Irradiance pattern. The irradiance ellipse is elongated along the direction of maximum focusing. (b) The wavefront has the shape of a saddle. (From Arnaud,[38] by permission of the American Optical Society.)

where we have defined

$$\chi_1 \equiv (k_0 \tau)^{1/2} x_1 \qquad\qquad (2.222a)$$

$$\chi_2 \equiv (k_0 \tau)^{1/2} x_2 \qquad\qquad (2.222b)$$

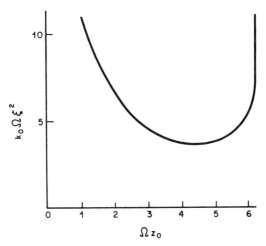

Fig. 2-25   Size of the fundamental mode of propagation in a helical fiber. ξ is half the major axis of the intensity ellipse ($1/e$ point), Ω the focusing strength, $z_0 \equiv 2\pi/\tau$ the helix period, and k_0 the wavenumber on axis. The instability boundary is at $\Omega z_0 = 2\pi$. For the optimum period, the beam radius is comparable to that obtained with a circularly symmetric fiber having the same total difference of refractive index. (From Arnaud,[38] by permission of the American Optical Society.)

As before, we have

$$\sin(2\nu) \equiv \left(\frac{\Omega}{\tau} \right)^2 \tag{2.222c}$$

$$a^{(\alpha)} \equiv \frac{a^\alpha}{\alpha!} \tag{2.222d}$$

and the summation over α, β, γ begins at $\alpha = \beta = \gamma = 0$ and terminates when one of the exponents is zero.

The irradiance and wavefront of the fundamental mode are shown in Fig. 2-24. The variation of the size of the fundamental mode with the focusing strength is shown in Fig. 2-25. The irradiance of modes of various order $|\psi_{m_1 m_2}|^2$, calculated from (2.222), is shown in Fig. 2-26. A possible advantage of the helical fiber compared with a conventional graded-index fiber is that the number of modes that can propagate in the (cladded) helical fiber can be selected by changing τ, that is, by changing the speed of rotation of the preform as the fiber is being pulled. Pulse broadening in helical fibers is very small, as in any near-square-law fiber. To minimize further pulse broadening it is, in principle, possible to optimize the coefficients of all fourth-order terms in the expression of $k^2(x, y)$. These techniques are discussed in Chapter 4 for simpler axially uniform fibers.

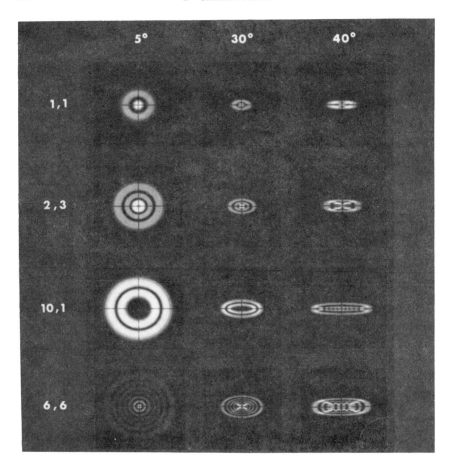

Fig. 2-26 Calculated irradiances of modes of the helical fiber. The mode numbers are shown on the left. The angle on top (5°, 30°, 40°) is a monotonically increasing function of the focusing strength. The helical fiber becomes unstable when that angle reaches 45°. (From Arnaud,[38] by permission of the American Optical Society.)

2.18 Modes in Two-Dimensional Resonators

Essential results concerning unaberrated optical resonators were given in previous sections. The theory was based on the complex-ray representation of gaussian beams. In the present section, we derive results applicable to the higher-order modes. We choose not to refer explicitly to complex rays or to complex ray matrices. For two-dimensional systems, it is simpler to proceed directly from the point-eikonal.[31] The underlying principle, however, is the same. The complete solution is obtained in a straightforward manner, which does not require prior knowledge of the solution of any differential or integral equation.

Let us consider the ring-type resonator shown in Fig. 2-27a. The path is shown circular, but it could be rectangular as well and defined by four plane mirrors. A focusing element and a gaussian aperture (split in two for later convenience) are shown for illustrative purposes. The general case is discussed in this section. An arbitrary (two-sided) reference plane $(x'; x)$ is chosen, and the point-eikonal $S(x; x')$ from x' to x for rays in some specific direction, for instance, counterclockwise, is evaluated. This can be

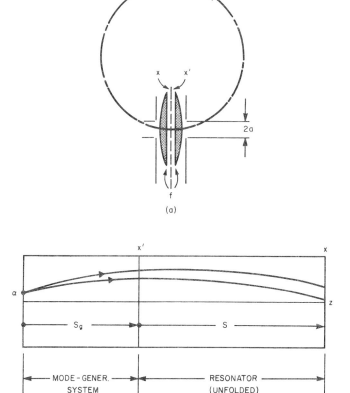

Fig. 2-27   (a) Ring-type resonator with a lens and a gaussian aperture, symmetrically located with respect to the reference plane. The reference plane $(x'; x)$ is a plane of symmetry. (b) Schematic representation of the resonator (unfolded) from x' to x and its mode-generating system (from α to x'). The modes of the resonator are obtained by expanding the Green function of the mode-generating system in power series of α. (From Arnaud,[31] by permission of the IEEE.)

done, as explained in more detail in Chapter 4, by adding the point-eikonals of the successive elements (sections of free space, lenses, gaussian apertures, . . .) encountered along the path. In that way, a round-trip point eikonal of the form

$$S(x; x') = \theta_0 + \tfrac{1}{2} U x^2 - V x x' + \tfrac{1}{2} W x'^2 \tag{2.223}$$

is obtained. θ_0, U, V, W are complex constants that fully characterize the resonator under investigation. (Alternatively, the resonator can be characterized by its round-trip ray matrix.)

For clarity, we have represented in Fig. 2.27b the optical resonator unfolded, with the planes x and x' separated. Suppose now that there is adjacent to this unfolded resonator at plane x', an optical system, with input plane α, characterized by a point-eikonal S_g

$$S_g(x'; \alpha) = \theta_{0g} + \tfrac{1}{2} U_g x'^2 - V_g x' \alpha + \tfrac{1}{2} W_g \alpha^2 \tag{2.224}$$

The point-eikonal $S'_g(x; \alpha)$ of the system from plane α to plane x is obtained by adding the point-eikonals and eliminating the intermediate variable x' with the help of the Fermat principle $\partial S_{\text{total}} / \partial x' = 0$. More details are given in Chapter 4. The result is

$$\theta'_{0g} = \theta_0 + \theta_{0g} \tag{2.225a}$$

$$U'_g = U - V^2 (W + U_g)^{-1} \tag{2.225b}$$

$$V'_g = V V_g (W + U_g)^{-1} \tag{2.225c}$$

$$W'_g = W_g - V_g^2 (W + U_g)^{-1} \tag{2.225d}$$

We have observed in Section 2.16 that the mode fields $\psi_m(x)$ depend only on two parameters, namely, U_g and V_g^2 / W_g. Thus the fields ψ_m generated by S_g reproduce themselves after a round trip in the resonator (from x' to x) to within a constant factor, if $U'_g = U_g$ and $V_g'^2 / W'_g = V_g^2 / W_g$. Setting first $U'_g = U_g$ in (2.225b), we obtain

$$(U_g - U)(U_g + W) + V^2 = 0 \tag{2.226a}$$

whose solution is

$$U_g = \tfrac{1}{2}(U - W) + \tfrac{1}{2} \Delta \tag{2.226b}$$

$$\Delta \equiv \pm \left[(U + W)^2 - 4V^2 \right]^{1/2} \tag{2.226c}$$

Evaluating now $V_g'^2/W_g' = V_g^2/W_g$ from (2.225c) and (2.226d), we obtain

$$V_g^2/W_g = \Delta \tag{2.227}$$

We also note that, from (2.225) and (2.226b), we have

$$\frac{V_g'}{V_g} = \frac{2V}{U + W + \Delta} = \frac{U + W - \Delta}{2V} \tag{2.228}$$

If we substitute the expressions obtained for U_g and V_g^2/W_g, (2.226) and (2.227), in (2.181), the field of a mode m is, to within an unimportant constant,

$$\psi_m(x) = \exp[\tfrac{1}{4} i(U - W + \Delta)x^2]$$

$$\times \sum_{s=0}^{[m/2]} \left[-(i\Delta)^{1/2}x \right]^{m-2s} [2^s s!(m - 2s)!]^{-1}$$

$$\Delta \equiv \pm\left[(U + W)^2 - 4V^2\right]^{1/2} \tag{2.229}$$

This result can be cast in a more familiar form if we set

$$\frac{1}{\rho} \equiv \frac{1}{2} \frac{U - W}{k_0}$$

$$\frac{1}{\xi^4} \equiv V^2 - \tfrac{1}{4}(U + W)^2 \equiv -\tfrac{1}{4}\Delta^2 \tag{2.230}$$

$$H_m(x) \equiv \sum_{s=0}^{[m/2]} (-)^s (2x)^{m-2s} [s!(m - 2s)!]^{-1}$$

where H_m denotes a Hermite polynomial, defined in (2.178). The result (2.229) is, with this notation,

$$\psi_m(x) = \exp\left(\frac{1}{2} \frac{ik_0 x^2}{\rho} \right) \exp\left(-\frac{1}{2} \frac{x^2}{\xi^2} \right) H_m\left(\frac{x}{\xi} \right) \tag{2.231}$$

Form (2.231) is useful when the parameters θ_0, U, W, $i\Delta$ are real, because ρ and ξ are then real. In that special case, ρ represents the radius of curvature of the wavefront and ξ the beam half-width ($1/e$ point of the irradiance) at the reference plane.

Let us now turn our attention to the factor that multiplies ψ_m as we go from plane x' to plane x. The prefactor $V_g'^{1/2}W_g^{m/2}$ in (2.181) can be written

$V_g^{m+1/2}(V_g^2/W_g)^{-m/2}$, where V_g^2/W_g is invariant, and $V_g/V_g' \equiv e^{i\theta}$ is given in (2.228). Thus, after a round trip, ψ_m becomes

$$\psi_m \exp\{i[\theta_0 - (m + \tfrac{1}{2})\theta]\} \qquad (2.232)$$

where, from (2.228), θ is given by

$$2\cos\theta = V^{-1}(U + W) \qquad (2.233)$$

The resonance condition is that the argument of the exponential factor in (2.232) is an integral multiple of 2π, that is,

$$\theta_0 - (m + \tfrac{1}{2})\cos^{-1}[\tfrac{1}{2}V^{-1}(U + W)] = 2l\pi \qquad (2.234)$$

where l denotes the axial-mode number, that is, the number of wavelengths along the optical axis. m is the transverse-mode number. It is equal to the number of nodes in the transverse direction.

Equations (2.229) or (2.230) and (2.234) are our final general results. Substitution in these equations of the values of the parameters θ_0, U, V, W, that describe the resonator provides us with the mode fields $\psi_m(x)$ at the reference plane (to within arbitrary constants) and the resonant frequencies. Note that the dependence of $(U + W)/V$ on frequency can usually be neglected compared with the very fast dependence of the geometrical optics phase shift ($\theta_0 \equiv k_0 \times$ length of optical axis) on the frequency. Therefore, (2.234) is essentially an explicit equation for the resonant frequencies.

Condition on the Field Behavior

Because the sign in front of Δ is either $+$ or $-$, it appears at first that we have obtained two sets of modes. We should, however, impose on these modes the condition that the field decays as x tends to infinity. This condition is now discussed. Equation (2.229) shows that the field amplitude decays, grows, or is a constant as x tends to infinity, depending on whether the imaginary part of $U - W + \Delta$ is positive, negative, or zero, respectively. If there are no apertures in the resonator, the parameters U, V, W are real. We are then confronted with three possibilities

$$(U + W)^2 - 4V^2 > 0, \qquad \theta \quad \text{imaginary}$$

$$(U + W)^2 - 4V^2 = 0, \qquad \theta = 0, \quad \text{modulo} \quad \pi \qquad (2.235)$$

$$(U + W)^2 - 4V^2 < 0, \qquad \theta \quad \text{real}$$

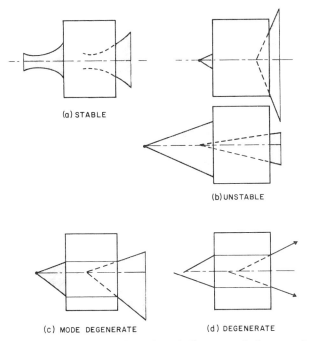

Fig. 2-28   Self-reproducing beams and rays in periodic systems (only one period is shown) or resonator (unfolded). (a) Self-reproducing gaussian beam in a stable system. (b) Self-reproducing ray pencils in unstable systems. (c) Self-reproducing rays in mode-generate resonators. (d) In degenerate resonators all rays recycle. (From Arnaud,[16] by permission of the North Holland Publ. Co.)

In the first case, $U - W + \Delta$ has no imaginary part. The field amplitude remains a constant. The resonator is called unstable. The second case is a case of mode degeneracy, which will be discussed in a subsequent section. In the last and most important case, θ is real and Δ imaginary. The resonator is then stable. Because $U - W$ is real, it is clear that, for one sign of Δ, the field is growing as $x \to \infty$ and, for the other sign, the field is decaying. Thus we have just one set of physical solutions. We note further from (2.231) that for that case, the polynomial in x has real coefficients. In other words, the polynomial does not contribute to the phase pattern of the field. The wavefronts are circular, and they are the same for all the modes. The self-reproducing beams, or rays, are shown in Fig. 2-28 for these three cases. A detailed discussion of the practical applications of unstable resonators (mainly high-gain lasers) is given in Siegman.[25] If there are gaussian apertures along the resonator path, or a laser with quadratic transverse variation of the gain, and the reference plane is a plane of symmetry (this plane always exists in resonators with folded optical axis incorporating reciprocal media), U is equal to W. Here again we find that one set of solutions grows if the other one decays and that there exists at most one set of physical solutions. If the resonator is both symmetrical and apertureless, the wavefronts are plane. The result that the wavefronts are plane is, in fact, more general than the above derivation suggests. Indeed, if a resonator is lossless, reciprocal, symmetrical, and free of internal reflection, the round-trip transformation of the wave function can be written $\psi = \mathbf{K}\psi'$, where the operator $\mathbf{K}$ is symmetrical and unitary. The wavefronts are plane because nondegenerate eigenvalues of symmetrical unitary operators are real, as one can easily show.

In the most general case where U, V, and W are arbitrary complex numbers, the wavefronts are different for different modes, and they may not be circular. It seems at first that two sets of physical solutions may exist. This is not the case, however, because one must impose the condition that the field decays as $x \to \infty$, not just at the reference plane, but also beyond the optical elements that are represented by the coefficients U or W in the expression of the point-eikonal. When these additional conditions are imposed, we find that just one set of mode, at most, is "physical." The solution obtained from the previous equations must also be checked for stability. For example, if the resonator medium has a gain that increases as a function of distance from axis (this case is the opposite of a gaussian aperture), we can find a solution that decays as a function of distance from axis. This solution is only formal, however, because it is unstable. For any small deviation of the field from that solution, the successive passes diverge.[40] Such solutions are therefore not acceptable, even though they decay in an acceptable manner as a function of distance.

If the resonator contains only reciprocal materials, the point-eikonal for clockwise waves is obtained from the point-eikonal for counterclockwise waves simply by exchanging x and x', that is, by exchanging U and W in all the expressions. Since the sum $U + W$ remains the same, it is clear that the resonant frequencies and losses, given by (2.234), are the same for clockwise and counterclockwise waves. The mode fields, however, are different, because $U - W$ changes sign.

Example of Resonator with Gaussian Aperture

To exemplify the previous general results, let us consider the resonator shown in Fig. 2-27a, which incorporates a lens with focal length $f/2$ and a gaussian aperture with effective half-width $a/\sqrt{2}$. For convenience, these elements are shown split into two elements of equal strength, namely, two thin lenses with focal length f and two gaussian apertures with half-width a. The two-sided reference plane $(x; x')$ is a plane of symmetry. The round-trip length is denoted L [L is the physical length of the path plus $(n - 1)$ times the thickness of the lens at its vertex, n being the refractive index of the lens material]. The point-eikonal, for either clockwise or counterclockwise waves, is readily obtained

$$S(x; x') = k_0 L + \tfrac{1}{2} k_0 L^{-1}(x - x')^2$$

$$- \tfrac{1}{2} k_0 f^{-1}(x^2 + x'^2) + \tfrac{1}{2} i a^{-2}(x^2 + x'^2) \qquad (2.236)$$

Thus the parameters θ_0, U, V, and W are, for that resonator,

$$\theta_0 = k_0 L, \qquad U = W = k_0 L^{-1} - k_0 f^{-1} + i a^{-2}, \qquad V = k_0 L^{-1}$$

$$(2.237)$$

The resonant frequencies [$\omega/c = \text{Re}(k_0)$] and losses are obtained by substituting these expressions in (2.234), which we rewrite here for convenience:

$$k_0 L = 2l\pi + (m + \tfrac{1}{2}) \cos^{-1}[\tfrac{1}{2} V^{-1}(U + W)] \qquad (2.238)$$

For the fundamental mode ($m = 0$), using for U, V, W the expressions in (2.237), the loss in nepers is found to be

$$\text{\textsterling} \equiv k_{0i} L \equiv \text{Im}(k_0 L) = \frac{1}{2} \text{Im}\left[\cos^{-1}\left(1 - \frac{L}{f} + \frac{iL}{k_0 a^2} \right) \right] \qquad (2.239)$$

Separating the real and imaginary parts, we find that the curves of constant loss are ellipses

$$\left[\frac{1 - (L/f)}{\cosh(2\text{\textsterling})} \right]^2 + \left[\frac{L/k_0 a^2}{\sinh(2\text{\textsterling})} \right]^2 = 1 \qquad (2.240)$$

Curves with constant round-trip loss (in decibels) are shown in Fig. 2-17. This figure shows that if the aperture width is infinite ($L/k_0 a^2 = 0$), the loss vanishes when $|1 - (L/f)|$ is less than unity (stable resonator), but is very high when $|1 - (L/f)|$ exceeds unity (unstable resonator).

The configuration shown in Fig. 2-15 is similar to the one just discussed. In a round trip, however, one encounters twice the optical system shown in Fig. 2-17. Thus the diffraction phase shifts and the losses are twice as large; that is, the term $(m + \tfrac{1}{2})$ in (2.238) should be multiplied by 2. Note also that the effective focal length of a mirror with radius R is $f = R/2$. When these substitutions are made in (2.238), and for $a = \infty$ (no aperture), the resonance condition given earlier is recovered.

2.19 Excitation of Resonators

The problem that we have considered so far can be stated in physical terms as follows. We have assumed that there is at our disposal in the resonator a spatially uniform medium with adjustable gain. The gain is increased until some oscillation sets up. What we have evaluated is the spatial distribution of the field that appears at a reference plane at the threshold of oscillation. Because the linear approximation is made, the amplitude of the field is undefined. If modes of various orders are well resolved in frequency and the linewidth of the active medium is sufficiently narrow, it is possible to have only one mode oscillating by properly choosing the length of the cavity. The requirement that the field vanishes as $x \to \infty$, which we stated before, is natural. If this were not the case, the perturbation caused by the fact that the optical elements are, in fact, of finite size could not be neglected.

The situation is somewhat different if the resonator is used as an interferometer excited by external sources. Let us assume that we have at our disposal a set of functions $\psi_m^\dagger(x)$ biorthogonal to the modes $\psi_m(x)$, that is, satisfying

$$\langle \psi_m^\dagger, \psi_n \rangle \equiv \int_{-\infty}^{+\infty} \psi_m^\dagger(x)\psi_n(x)dx = 0, \qquad \text{if} \quad m \neq n \qquad (2.241)$$

The incident field $\psi(x)$ at some reference plane x can be expanded in terms of the modal fields $\psi_m(x)$

$$\psi(x) = \sum_{m=0}^{\infty} a_m \psi_m(x) \qquad (2.242a)$$

The coefficients a_m are obtained by multiplying (2.242a) by $\psi_m^\dagger(x)$, integrating over x, and using (2.241). We obtain

$$a_m = \frac{\langle \psi_m^\dagger, \psi \rangle}{\langle \psi_m^\dagger, \psi_m \rangle} \qquad (2.242b)$$

as the amplitude of the mode m (primes on x have been dropped for simplicity). Now each mode after a round trip is multiplied by a factor $\bar{\lambda}_m \equiv \exp(i\theta_0) \exp[-i(m+\frac{1}{2})\theta]$ as we have seen. Thus the total field associated with each mode m is multiplied by

$$\sum_{s=0}^{\infty} \left(\bar{\lambda}_m\right)^s = \left(1 - \bar{\lambda}_m\right)^{-1} \qquad (2.243)$$

within the radius of convergence $|\bar{\lambda}_m| < 1$ of the series. The total field ψ_T circulating in the resonator (that is, essentially the field that reaches the detector, to within some coupling constant) is obtained by adding the contribution of each mode

$$\psi_\mathrm{T}(x) = \sum_{m=0}^{\infty} \left(1 - \bar{\lambda}_m\right)^{-1} \psi_m(x) a_m \qquad (2.244)$$

The restriction made before concerning the behavior of ψ_m at infinity does not necessarily apply here. For instance, a plane-parallel Fabry–Perot resonator is a mode-degenerate resonator ($\theta = 0$) whose "modes" are plane waves. These plane waves can be used in (2.244) in spite of the fact that the field does not vanish at infinity.

When the approximation of Gauss is used, the adjoint field $\psi_m^\dagger(x)$ is obtained simply by exchanging U and W in the expression for $\psi_m(x)$. This follows from the fact that the integral operator that relates $\psi(x)$ to $\psi'(x')$ is changed to its transpose (in the operator sense) when $S(x; x')$ is changed to $S(x'; x)$. It is known that the eigenvectors of an operator and of its transpose corresponding to distinct eigenvalues are mutually orthogonal. The result concerning $\psi_m^\dagger(x)$ can also be verified by direct integration. (Note that it is easier to perform the integration of the product of the generating functions for ψ_m and $\psi_m^\dagger$ than directly the integral of the product of the explicit expressions for ψ_m and $\psi_m^\dagger$, involving Hermite polynomials.)

Physically, the adjoint modes $\psi_m^\dagger$ are the modes for waves propagating in the clockwise rather than counterclockwise direction in the resonator, when the medium is reciprocal. If the resonator has a plane of symmetry (as is the case for the resonator in Fig. 2-27a), the adjoint modes $\psi_m^\dagger$ and the modes ψ_m coincide at that plane. Another special case of interest is when the resonator is lossless and stable. Then U, W, and $i\Delta$ are real as we have seen, and

$$\psi_m^\dagger(x) = \psi_m^*(x) \qquad (2.245)$$

where the asterisk denotes complex conjugation. In that special case, the modes ψ_m are mutually orthogonal in the sense of the Hermitian product.

The response of a resonator can be obtained without any reference to the concept of mode, by adding the successive passes of the incident field. Thus, if $\psi = \mathbf{K}\psi'$ denotes the field transformation, (2.244) must be equivalent to

$$\psi_\mathrm{T} = \sum_{s=0}^{\infty} \mathbf{K}^s \psi' = (1 - \mathbf{K})^{-1} \psi' \qquad (2.246)$$

That this is indeed the case will now be shown in the abstract operator form. We shall assume that $(1 - \mathbf{K})^{-1}$ exists and that the $\psi_m, \psi_m^\dagger$ form a complete biorthogonal set of eigenvalues of $\mathbf{K}$ and of its transpose (formal adjoint).

Formal Comparison between Modal and Multipass Expansions

Let us first recall the function space notation, already used in (2.246), where the functions $\psi(x)$, $\psi'(x)$, and $\psi_T(x)$ are represented by vectors ψ, ψ', and ψ_T, respectively.

Let us first assume, for simplicity, that the transverse coordinate x can assume only a finite number of values such as $x = 0, 1, 2, \ldots, N$. The components of the vector ψ on the $0, 1, 2, \ldots, N$ axes are, respectively, $\psi(0), \psi(1), \ldots, \psi(N)$. The above representation can be generalized to an infinite countable number of values of x. Then $\psi(x)$ is represented by a vector in a space having a countable number of dimensions. The full generalization consists of letting x be a continuous variable. We shall overlook, for the sake of clarity, the various mathematical difficulties that are involved in this representation and assume that finite matrix algebra is applicable. (The reader may find it useful to write explicitly the subsequent expressions in this section for the case where the coordinate x can assume only two values, namely, 0 and 1.) In this representation, a linear operator such as $\mathbf{K}$, applied to a function $\psi'(x)$, has the effect of changing the vector ψ' representing $\psi'(x)$ to some other vector ψ. This transformation is written $\psi = \mathbf{K}\psi'$.

A mode is a solution of the eigenvalue equation

$$\mathbf{K}\psi_m = \psi_m \bar{\lambda}_m \tag{2.247}$$

where the $\bar{\lambda}_m$, called the eigenvalues, are complex numbers, and the ψ_m represent the modal fields. Note that ψ_m is defined only up to an arbitrary multiplicative factor. The vector ψ_m has a well-defined direction in function space but arbitrary length. We can lump together all the eigenvectors ψ_m with $m = 0, 1, \ldots$ to form a modal matrix $\mathbf{W}$:

$$\mathbf{W} \equiv [\psi_0 \psi_1 \cdots] \tag{2.248a}$$

For example, if x assumes only the values 0 and 1, $\mathbf{W}$ is the 2×2 matrix

$$\mathbf{W} \equiv \begin{bmatrix} \psi_0(0) & \psi_1(0) \\ \psi_0(1) & \psi_1(1) \end{bmatrix} \tag{2.248b}$$

Going back to the general case, we write (2.247), for all m, in the form

$$\mathbf{KW} = \mathbf{WD} \tag{2.249}$$

where $\mathbf{D}$ is the spectral matrix of $\mathbf{K}$, a diagonal matrix with diagonal elements $\bar{\lambda}_0, \bar{\lambda}_1, \ldots$. Similarly, let $\mathbf{W}^\dagger$ denote the modal matrix of $\tilde{\mathbf{K}}$, where $\tilde{\mathbf{K}}$ denotes the transpose (adjoint) of the original operator $\mathbf{K}$. We have

$$\tilde{\mathbf{K}}\mathbf{W}^\dagger = \mathbf{W}^\dagger\mathbf{D} \qquad (2.250)$$

because $\mathbf{K}$ and $\tilde{\mathbf{K}}$ have the same eigenvalues. As is well known in matrix algebra, $\mathbf{W}$ and $\mathbf{W}^\dagger$ can always be normalized to satisfy the biorthogonality condition

$$\tilde{\mathbf{W}}^\dagger\mathbf{W} = 1 \qquad (2.251)$$

where 1 denotes the unit matrix, even in case of degeneracy.

The incident field ψ' can be expanded in term of the modes ψ_m in the form

$$\psi' = \sum_m a_m\psi_m \equiv \mathbf{W}\mathbf{a} \qquad (2.252)$$

where the vector $\mathbf{a}$ has components $a_0, a_1, \ldots$.

With the help of (2.251), we find

$$\mathbf{a} = \mathbf{W}^{-1}\psi' = \tilde{\mathbf{W}}^\dagger\psi' \qquad (2.253)$$

We assume that $\mathbf{W}$ is not singular.

By definition, each mode ψ_m reproduces itself after a round trip in the resonator except for a factor $\bar{\lambda}_m$. Thus the total field associated with a particular mode ψ_m is

$$\psi_{m\mathrm{T}} = \sum_{s=0}^\infty \bar{\lambda}_m^s\psi_m = \left(1 - \bar{\lambda}_m\right)^{-1}\psi_m \qquad (2.254)$$

within the radius of convergence $|\bar{\lambda}_m| < 1$ of the series. When the modes are well resolved in frequency, the cavity field has significant amplitude only for narrow ranges of frequency of the source, and we need to consider only one mode at a time. In general, the field is obtained by summing $\psi_{m\mathrm{T}}$ over all the values of m. By lumping together all the modes, Eq. (2.254) can be rewritten

$$\mathbf{W}_\mathrm{T} = \mathbf{W}(1 - \mathbf{D})^{-1} \qquad (2.255)$$

where $\mathbf{W}$ is, as before, the modal matrix of $\mathbf{K}$ defined in (2.248), $\mathbf{D}$ is the spectral matrix of $\mathbf{K}$ (or $\tilde{\mathbf{K}}$) with diagonal elements $\bar{\lambda}_0, \bar{\lambda}_1, \ldots$, and we have defined

$$\mathbf{W}_\mathrm{T} \equiv [\psi_{0\mathrm{T}}\psi_{1\mathrm{T}} \cdots] \qquad (2.256)$$

Finally, the total field is, using (2.555) and (2.253),

$$\psi_\mathrm{T} = \mathbf{W}_\mathrm{T}\mathbf{a} = \mathbf{W}(1 - \mathbf{D})^{-1}\tilde{\mathbf{W}}^\dagger\psi' \equiv \mathbf{K}_\mathrm{T}\psi' \qquad (2.257)$$

where

$$\mathbf{K_T} = \mathbf{W}(1 - \mathbf{D})^{-1}\tilde{\mathbf{W}}^{\dagger} \qquad (2.258)$$

The expression for $\mathbf{K_T}$ obtained in (2.246) by adding the field of the successive passes of the wave, on the other hand, is

$$\mathbf{K_T} = \sum_{s=0}^{\infty} \mathbf{K}^s \qquad (2.259)$$

To show that the expressions for $\mathbf{K_T}$ given in (2.258) and (2.259) are, at least formally, identical, it suffices to replace $\mathbf{K}$ in (2.259) by its expression $\mathbf{WDW}^{-1}$ from (2.249). Then we obtain

$$\mathbf{K_T} = \sum_{s=0}^{\infty} \left(\mathbf{WDW}^{-1}\right)^s = \sum_{s=0}^{\infty} \mathbf{WD}^s\mathbf{W}^{-1} = \mathbf{W}(1 - \mathbf{D})^{-1}\mathbf{W}^{-1} \quad (2.260)$$

We can use (2.251) to change $\mathbf{W}^{-1}$ to $\tilde{\mathbf{W}}^{\dagger}$. Thus, the equivalence of (2.258) and (2.259) is established. The derivation of (2.260) that we gave earlier exhibits the successive steps encountered in the modal expansion method.

Let again ψ' denote the incidence field, and let us consider now the total energy flow P_T in the resonator

$$P_T = \psi_T^*\psi_T = \psi'^*\tilde{\mathbf{K}}_T^*\mathbf{K_T}\psi' \equiv \psi'^*\mathbf{K}_P\psi' \qquad (2.261)$$

Except for a constant coupling factor, P_T is the power that reaches the detector. Asterisks denote complex-conjugate values, and, as in the rest of this book, the transposition signs on first vectors are omitted. In the second expression in (2.261), we have used the form (2.257) of ψ_T. Thus evaluation of the detected power for a given incident field requires the evaluation of the operator $\mathbf{K}_P \equiv \tilde{\mathbf{K}}_T^*\mathbf{K_T}$. If we use the multipass approach described by (2.245), $\mathbf{K}_P$ is expressed as a double infinite sum. In fact, it can be shown that this double sum reduces to a single sum when the cavity losses are independent of x, e.g., when the losses are due solely to the reflectivity r of the coupling mirror, which is assumed uniform. In such resonators, the operator $\mathbf{K}$ is the product of a complex number r and a unitary operator $\mathbf{K}_u$.

$$\mathbf{K} = r\mathbf{K}_u \qquad (2.262a)$$

where

$$\tilde{\mathbf{K}}_u^*\mathbf{K}_u = 1 \qquad (2.262b)$$

The unitarity of $\mathbf{K}_u$ follows from the fact that, once r has been factored out, the resonator is lossless. From (2.261) and (2.246), we have first

$$\mathbf{K}_P = \tilde{\mathbf{K}}_T^*\mathbf{K_T} = \sum_{s=0}^{\infty} \sum_{t=0}^{\infty} r^{*s}r^t\tilde{\mathbf{K}}_u^{*s}\mathbf{K}_u^t \qquad (2.263)$$

Using the fact that $\mathbf{K}_u$ is unitary and the identity

$$\sum_{t=0}^{\infty}\sum_{s=0}^{\infty} = \sum_{t-s=0}^{\infty}\sum_{s=0}^{\infty} + \sum_{s-t=0}^{\infty}\sum_{t=0}^{\infty} - \sum_{t=s=0}^{\infty} \qquad (2.264)$$

we obtain, after a few rearrangements,

$$\mathbf{K}_P = \tilde{\mathbf{K}}_\mathbf{T}^*\mathbf{K}_\mathbf{T} = (1 - rr^*)^{-1}\left\{\sum_{s=0}^{\infty}[(\mathbf{K})^s + (\tilde{\mathbf{K}}^*)^s] - 1\right\} \qquad (2.265)$$

This expression is simpler to evaluate than the original expression for $\mathbf{K}_P$ in (2.263).

We shall now give without proof the expression for the power in term of the modal field. For a single-mode m, the power in the resonator is

$$P_{\mathrm{T}m} = \boldsymbol{\psi}'^*\mathbf{K}_{P_m}\boldsymbol{\psi}'$$

where

$$\mathbf{K}_{P_m} = \boldsymbol{\psi}_m(1 - \bar{\lambda}_m^*)^{-1}(1 - \bar{\lambda}_m)^{-1}\tilde{\boldsymbol{\psi}}_m^* \qquad (2.266)$$

The total power, obtained by summing P_m over all m, is

$$P_\mathrm{T} = \boldsymbol{\psi}'^*\mathbf{K}_P\boldsymbol{\psi}'$$

where

$$\mathbf{K}_P = \mathbf{W}(1 - \mathbf{D}^*)^{-1}(1 - \mathbf{D})^{-1}\tilde{\mathbf{W}}^* \qquad (2.267)$$

Let us rewrite our main result (2.261) and (2.265) in a more explicit form. The transmission of the resonator is now defined as

$$T_0 = \frac{P_\mathrm{T}(1 - rr^*)}{\boldsymbol{\psi}'^*\boldsymbol{\psi}'} = \frac{\boldsymbol{\psi}'^*\left\{\displaystyle\sum_{s=0}^{\infty}[(\mathbf{K})^s + (\tilde{\mathbf{K}}^*)^s] - 1\right\}\boldsymbol{\psi}'}{\boldsymbol{\psi}'^*\boldsymbol{\psi}'} \qquad (2.268)$$

Let us introduce the coupling c_s between the incident beam $\boldsymbol{\psi}_s' = \mathbf{K}_u^s\boldsymbol{\psi}'$ after s passes in the resonator and the incident beam $\boldsymbol{\psi}'$, where, as before, $\mathbf{K}_u = \mathbf{K}/r$. The effect of the mirror reflectivity on the field transformation is not included in the definition of $\boldsymbol{\psi}_s'$. The coupling c_s is defined as

$$c_s = \frac{\boldsymbol{\psi}'^*\boldsymbol{\psi}_s'}{\boldsymbol{\psi}'^*\boldsymbol{\psi}'} \equiv \frac{\displaystyle\int_{-\infty}^{+\infty}\boldsymbol{\psi}'^*(x)\boldsymbol{\psi}_s'(x)\,dx}{\displaystyle\int_{-\infty}^{+\infty}\boldsymbol{\psi}'^*(x)\boldsymbol{\psi}'(x)\,dx} \qquad (2.269)$$

(The transformed field $\boldsymbol{\psi}_s'$ should not be confused with the modal field $\boldsymbol{\psi}_m$.

The former is distinguished from the latter by a prime.) Note that $c_0 = 1$. Equation (2.268) is rewritten[2,39]

$$T_0 = 2 \operatorname{Re}\left[\sum_{s=0}^{\infty} r^s c_s \right] - 1 = 1 + 2 \operatorname{Re}\left[\sum_{s=1}^{\infty} r^s c_s \right] \qquad (2.270)$$

It should be noted that the frequency dependence of c_s is, in first approximation, expressed by $\exp(isk_0L)$, where L denotes the round-trip path length of the resonator. Consequently the average transmission of the resonator over a free spectral range [namely, from $k_0L \approx 2l\pi$ to $2(l + 1)\pi$, where l denotes the axial-mode number] is always unity and does not depend on the excitation field.

Degenerate Resonator

In the case of lossless degenerate optical resonators, with one mirror with reflectivity r and round-trip optical length L, the field reproduces itself exactly after a round trip, except for a multiplication factor $r \exp(ik_0L)$. For such cavities, the transmission T_0 is simply

$$T_0 = 2 \operatorname{Re}[1 - r \exp(ik_0L)]^{-1} - 1 \qquad (2.271)$$

If we neglect the phase of the mirror reflectivity, we can set $r = R^{1/2}$, where R denotes the power reflectivity of the mirror. At resonance $[\exp(ik_0L) = 1]$, the transmission T_0 is

$$T_0 = \frac{1 + \sqrt{R}}{1 - \sqrt{R}} \qquad (2.272)$$

The normalization of T_0 is such that $T_0 = 1$ if $R = 0$. Thus T_0 expresses the enhancement in energy flow inside the resonator due to the resonance.

Mismatched Stable Resonator

Let us consider now a stable circularly symmetric resonator with round-trip ray matrix

$$\mathbf{M}_{\text{rt}} = \begin{bmatrix} A & B \\ C & D \end{bmatrix} \qquad (2.273a)$$

where

$$-2 < A + D < 2 \tag{2.273b}$$

The characteristic angle θ is defined in (2.134)

$$\cos(\theta) = \tfrac{1}{2}(A + D) \tag{2.274}$$

The beam radius $\bar{\xi}$ and the wavefront curvature radius $\bar{\rho}$ at the reference plane are given, respectively, by (2.142) and (2.143):

$$\bar{\xi}^{-2} = |k_0 B^{-1} \sin(\theta)| \tag{2.275}$$

$$\bar{\rho}^{-1} = \frac{D - A}{2B} \tag{2.276}$$

Let the incident beam be a gaussian beam with radius ξ and wavefront curvature radius ρ. The coupling between that beam and the same beam after s round trips in the resonator is found to be (e.g., by direct integration)

$$c_s = [\cos(s\theta) + iQ \sin(s\theta)]^{-1} \exp(isk_0 L) \tag{2.277}$$

where we have introduced a mismatch parameter Q

$$Q = \frac{1}{2} \left[\left(\frac{\xi}{\bar{\xi}} \right)^2 + \left(\frac{\bar{\xi}}{\xi} \right)^2 \right] + \frac{1}{2} k_0^2 \xi^2 \bar{\xi}^2 (\rho^{-1} - \bar{\rho}^{-1})^2 \tag{2.278}$$

This parameter is related to the power coupling N defined in (2.96)

$$Q \equiv \frac{2}{N^2} - 1 \tag{2.279}$$

When the incident beam is matched to the fundamental mode of resonance, we have $N = Q = 1$ and therefore $c_s = \{\exp[i(k_0 L - \theta)]\}^s$. This simple result follows from the fact that a matched gaussian beam reproduces itself in a round trip, except for the geometrical phase shift $k_0 L$ and the diffraction phase shift θ. The derivation of (2.277) and (2.278) can be found in Ref. 2. It is based on the evaluation of the powers of the round-trip ray matrix with the help of the Sylvester theorem. If we substitute (2.277) in the general expression (2.270), we obtain the normalized transmission

$$T_0 = 1 + 2 \operatorname{Re} \sum_{s=1}^{\infty} r^s \exp(isk_0 L)[\cos(s\theta) + iQ \sin(s\theta)]^{-1} \tag{2.280}$$

For high cavity finesses, $|r| \to 1$, and $k_0 L = \theta(1 + 2\alpha) + 2l\pi$, where l denotes the axial-mode number and α a nonnegative integer, the sum in (2.280) can be replaced by an integral over the interval $s\theta = 0$ to 2π. Taking $\exp(is\theta)$ as a complex variable, the integral is easily evaluated with the calculus of residues. Equation (2.280) becomes (with $r = R^{1/2}$ assumed real for simplicity)[2]

$$T_0 = \frac{1 + \sqrt{R}}{1 - \sqrt{R}} \frac{2(Q - 1)^\alpha}{(Q + 1)^{\alpha+1}} \tag{2.281}$$

A result identical to (2.281) can be obtained by expanding the incident gaussian beam in the Laguerre–Gauss modes of the resonator. When $Q = 1$ (incident beam matched to the resonator), we find from (2.281) that $T_0 = 0$ unless $\alpha = 0$, that is, only the fundamental mode is excited as one expects. For $\alpha = 0$, T_0 coincides with the result in (2.272) applicable to degenerate cavities. However, the frequency of resonance is now given by $k_0 L - \theta = 2l\pi$, instead of $k_0 L = 2l\pi$. The integer $\alpha = 0, 1, 2, \ldots$ is the radial mode number.

The variation of the cavity response T_0 as a function of frequency is shown in Fig. 2-29 for various values of the mismatch parameter Q. If we assume that the input mirror is plane, as shown in Fig. 2-29, and that the incident beam waist is located at that plane, the mismatch originates entirely from the difference in beam radii. The value of $\xi/\bar{\xi}$ corresponding to Q is also shown in Fig. 2-29.

As another example of application of the general results of this section, let us consider the case where the incident beam is matched to the cavity except for an offset $\bar{q}$ and a tilt $\dot{q}$ of the axis. We shall assume for simplicity that $A = D$. Then we find the coupling between the incident beam and the beam after s passes in the resonator

$$c_s = \exp(-b) \exp[b \exp(is\theta)] \exp[is(k_0 L - \theta)] \tag{2.282a}$$

where

$$b \equiv \tfrac{1}{2} k_0 |\Omega^{1/2}\bar{q} + i\Omega^{-1/2}\dot{q}|^2, \qquad \Omega \equiv (k_0 \bar{\xi}^2)^{-1} \tag{2.282b}$$

In the limit of high cavity finesses $R \to 1$ and $k_0 L = 2l\pi + (\mu + 1)\theta$, μ being a nonnegative integer, we obtain the response

$$T_0 = \frac{1 + \sqrt{R}}{1 - \sqrt{R}} \frac{b^\mu}{\mu!} \exp(-b) \tag{2.283}$$

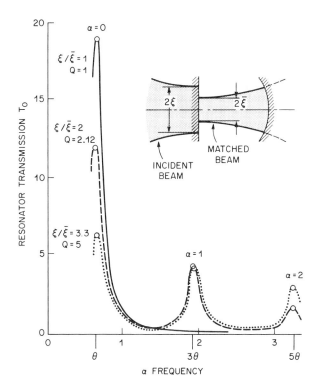

Fig. 2-29 Response of a stable resonator with circular symmetry to an incident gaussian beam versus frequency ($\phi \equiv 2k_0 d - 2\pi l$, where l is the axial mode number). When the incident beam is matched to the fundamental mode of resonance of the resonator, $\xi = \bar{\xi}$ or $Q = 1$, only one peak is observed for a given axial-mode number. For $\xi \neq \bar{\xi}$, higher-order Laguerre–Gauss modes are excited. (Radial mode number $\alpha = 0, 1, 2, \ldots$.)

This result (2.283) can alternatively be obtained by modal expansion.

If $\bar{q} = 0$, $\tilde{q} = 0$, then $b = 0$, and the response of the resonator is equal to zero unless $\mu = 0$ (fundamental mode). The results (2.281) and (2.283) are of great practical importance. Equation (2.281) gives the response of stable resonators to incident gaussian beams that are properly aligned but mismatched to the fundamental mode of resonance in radius or wavefront. Equation (2.283) gives the response of the resonator when the axis of the incident gaussian beam is offset and tilted.

For a stable resonator, the two methods—addition of the successive passes and modal expansion—are equivalent, as we have shown in a general abstract form and also for a few special cases. For a degenerate

misaligned cavity, the multipass method is to be preferred because the
modes are ill-defined. A two-dimensional misaligned degenerate resonator
can be described by a 3×3 degenerate ray matrix. To evaluate the sth
power of that matrix the confluent form of the Sylvester theorem must be
used. We shall not discuss that case here (see Arnaud[2]).

Response of Fabry–Perot Resonators

The response of plane-parallel Fabry–Perot resonators is of great practical importance because these resonators are commonly used in optics
(sometimes under the name étalons) inside laser resonators, for axial-mode
selection, and as narrow-band filters. They are also used as diplexers in the
millimeter wave region of the frequency spectrum to separate or combine
various channels.

A detailed discussion of this type of resonator is given by Arnaud *et al.*[39]

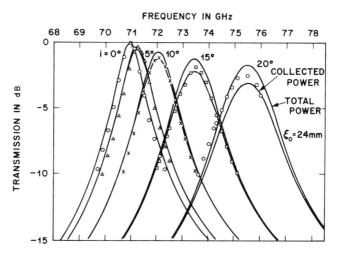

Fig. 2-30 Transmission of a plane-parallel Fabry–Perot resonator with gaussian-beam
excitation as a function of frequency. The angle $i = i'$ is the incidence angle. The radius of
the incident gaussian beam is $\xi_0 = 24$ mm. The plain curves are theoretical. The upper curve
gives the total transmitted power, and the lower curve the power collected when the receiving
antenna is optimized in the absence of the filter. The points are experimental with the
receiving antenna optimized in the absence of the filter. The grid reflectivity is 85% at 70
GHz. (From Arnaud *et al.*,[39] by permission of the IEEE.)

We shall only give the main results. Fabry–Perot étalons incorporate two parallel plane mirrors having equal field reflectivity r and spacing d. The total power transmitted through the filter is

$$P = \frac{2\,\mathrm{Re}(Z) - 1 + R}{1 + R} \qquad (2.284)$$

Fig. 2-31 Photograph of a three-grid Fabry–Perot diplexer. The meshes are in copper. The useful diameter is 0.5 m. This large area helps reduce the walk-off effect and increase the power dissipation. (After Saleh and Ruscio.[41])

where

$$Z = T \sum_{s=0}^{\infty} r^{2s} \exp(2isk_0 d \cos i')(1 + \tfrac{1}{2} isTD)^{-1}$$

$$\times \exp\left[-\tfrac{1}{4} s^2 T^2 G^2 (1 + \tfrac{1}{2} isTD)^{-1} \right] \qquad (2.285)$$

$$D \equiv \frac{2d \cos(i')}{k_0 \xi_0^2 T}, \qquad G \equiv \frac{2d \sin(i')}{\xi_0 T}$$

$$T \equiv 1 - R, \qquad R \equiv rr^*$$

The variation of r with the incidence angle is neglected. The incident beam is assumed gaussian with waist radius ξ_0, and the angle of incidence is denoted i'. (The mirror power transmissivity T should not be confused with the transmission T_0 evaluated earlier.) Dissipation losses have been neglected.

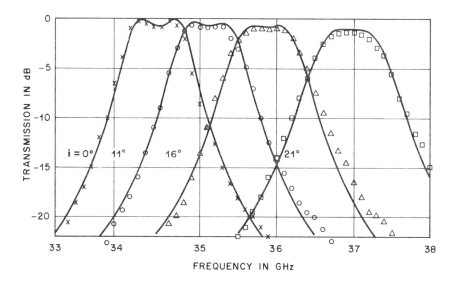

Fig. 2-32 Transmission of the three-grid filter in Fig. 2-31 as a function of frequency for different incidence angles. The plain curves are theoretical, and the points are experimental. The incident (near-gaussian) beam diameter is 107 mm, and the grid reflectivities are 65, 90, and 65%. (After Saleh and Ruscio.[41])

A result identical to (2.284) can be obtained using a plane wave expansion of the incident beam. When the incident beam waist radius $\xi_0 \to \infty$ (the plane incident wave), we have $D = G = 0$. Thus, $P = 1$ at the frequency of resonance: $k_0 d \cos(i') = l\pi$. (For simplicity, we have omitted the effect of the phase of r and set $r = \sqrt{R}$.)

If the transmitted beam is detected by optical heterodyning or by an antenna in the microwave range, we are interested not in the total transmitted power but in the field that has the same structure as the field that would be transmitted in the absence of the filter. The corresponding detected power is found to be

$$\bar{P} = ZZ^* \qquad (2.286)$$

where Z is given by (2.285). For small values of the parameters D and G in

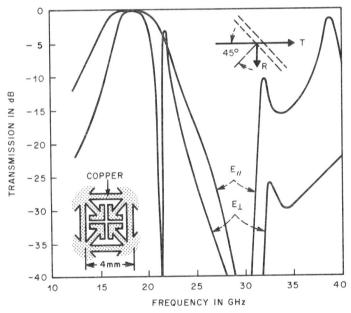

Fig. 2-33 Response of a Fabry–Perot diplexer incorporating two self-resonant grids. Each grid exhibits a transmission band and a rejection band. The transmission loss is less than 0.2 dB from 17.5 to 19.5 GHz, and the peak rejection at 30 GHz exceeds 68 dB. Depolarization is less than -30 dB at 19 GHz. The sharp absorption dip for one polarization is also observed with individual grids. Its cause is unknown. (From Arnaud and Pelow.[41] Reprinted with permission from the *Bell System Technical Journal*, Copyright 1975, the American Telegraph and Telephone Co.)

(2.285), the total transmitted power at resonance is

$$P \approx 1 - \tfrac{1}{2}(G^2 + D^2), \qquad D, G \ll 1 \qquad (2.287)$$

This simple expression provides an estimate of the degradation that results from exciting a Fabry–Perot étalon by a gaussian beam of finite radius instead of exciting it by an infinite plane wave.

The theoretical result (2.286) is in excellent agreement with experiment as shown in Fig. 2-30. The theory can be generalized to maximally flat or Tchebitcheff filters incorporating three or more mirrors.[41] A photograph of a filter operating at 30 GHz and incorporating three copper meshes is shown in Fig. 2-31. The comparison between the theory (not given here) and experiment is again excellent, as shown in Fig. 2-32. The response of a Fabry–Perot resonator, incorporating two self-resonant grids of the "self-supported" type with gaussian-beam excitation, is shown in Fig. 2-33.

2.20 Three-Dimensional Resonators; Anisotropic Resonators

The three-dimensional resonators that were considered earlier possessed two mutually perpendicular meridional planes of symmetry. Because for such systems the wave equation is separable, generalization of the two-dimensional results was straightforward. The field of a mode m, n has the form

$$\psi_{m,n}(x, y) = \psi_m(x)\psi_n(y) \qquad (2.288)$$

where $\psi_m(x)$ and $\psi_n(y)$ are obtained independently in the two mutually orthogonal meridional planes of symmetry. The expression for the resonance frequency is

$$k_0 L - (m + \tfrac{1}{2})\theta_x - (n + \tfrac{1}{2})\theta_y = 2l\pi, \qquad m, n, l = 0, 1, 2, \ldots \qquad (2.289)$$

where θ_x and θ_y are defined as before.

The problem is more difficult when the system does not possess meridional planes of symmetry. Such resonators are called nonorthogonal resonators. (See, for example, Fig. 2-22b and d). Early investigations of nonorthogonal resonators were made by Popov[42] and Kahn and Nemit.[43] The complete solution was first given by Arnaud.[33] It turns out that the results obtained earlier for two-dimensional resonators remain formally applicable, provided we change the scalar quantities U, V, W, to 2×2

matrices **U**, **V**, **W**, where **U** and **W** are symmetrical. The main general results of the theory of nonorthogonal optical resonators are as follows:

1. The modes $\psi_{m,n}(x, y)$ are expressible as products of a function of Gauss and a Hermite polynomial in *two complex variables*.

2. The fundamental equation (2.289) for the resonance frequencies remains formally unchanged. θ_x, θ_y are replaced by the phases of two eigenvalues of the 4×4 round-trip ray matrix, which we shall denote θ_1 and θ_2 (the other two phases are $-\theta_1$ and $-\theta_2$).

3. If the resonator is lossless, the angles θ_1 and θ_2 are either real or imaginary. If both are real, the resonator is stable. Otherwise, that is, if *either* θ_1 or θ_2, or both, are imaginary, the resonator suffers from geometrical optics losses and is unstable.

4. If the resonator is lossless and has a plane of symmetry (this is always the case for a resonator with folded optical axis if it incorporates only reciprocal media and is free of internal reflection) and is not degenerate, the wavefronts are the same for modes of all order m, n. They coincide with the plane of symmetry. The mode fields at that plane are expressible in the separable form (2.288) in an oblique coordinate system.

In general, even if the resonator is lossless, the wavefronts are different for different modes and do not have simple shapes.

If an optical resonator is filled with an anisotropic medium, for instance, a uniaxial crystal, a magnetized plasma, or a moving glass rod, we need to define four point-eikonals, corresponding, respectively, to clockwise and counterclockwise waves and to the two states of polarization (e.g., the ordinary and extraordinary waves). For acoustical waves we need to define as many as six point-eikonals because, in addition to the two transverse (shear) waves, longitudinal waves can also propagate. There is no particular difficulty in evaluating these point-eikonals as long as the medium parameters vary sufficiently slowly in space, the local eigenstates of polarization of the field being uncoupled as discussed in the introduction. The previous results are applicable separately to any one of the four (or six) point-eikonals. We have noted before that if the medium is reciprocal, there is identity between the losses and resonant frequencies of clockwise and counterclockwise modes. In laser gyroscopes (to be discussed next), one would like clockwise and counterclockwise modes to have distinct resonant frequencies or losses in order to avoid the locking of the two waves (the coupling being due, for instance, to backscattering). This, however, cannot be achieved with the help of apertures only, unless nonlinear effects take place. Nonreciprocal elements (Faraday rotators, continuous spinning of the laser) therefore need to be introduced.

A laser gyroscope can be viewed as a nonreciprocal ring-type resonator. A simple model of laser gyroscope is that in Fig. 1-25. The guiding structure is a perfectly conducting cylinder in free space. Clockwise (CW) and counterclockwise (CCW) waves cling to the concave side of the boundary (whispering-gallery modes) with a velocity almost equal to the speed of light in free space c. (As we shall see in Chapter 5, the phase velocity slightly exceeds c at the boundary. This is of no consequence, however, for the present discussion.) Let us now assume that the fields of the CW and CCW waves are mixed in a rotating nonlinear device (detector) having a linear velocity v. Because of the Doppler effect, the relative beat frequency between CW and CCW waves is

$$\frac{\Delta f}{f} = \frac{2v}{c} \tag{2.290}$$

The rotation of the (perfectly conducting) cylinder itself being immaterial, (2.290) is applicable to the case where the cylinder and the detector are held fixed with respect to each other and rotate at the same angular velocity. Such an arrangement is called a laser gyroscope. If the gyroscope is immersed in a medium with refractive index n moving with it, the drag of the CW and CCW waves resulting from the rotation of the medium has to be taken into account. It is evaluated by using the formula of addition of velocities of special relativity: $u = (v + v')/(1 + vv'/c^2)$, with $v' = \pm \, c/n$. The relative beat frequency is found to be, after some rearranging,

$$\frac{\Delta f}{f} = \frac{2v}{cn} \tag{2.291}$$

to first order in v/c. Note, incidentally, that the gyroscopic effect vanishes in the limit of large n. We now observe that the rotation can be ignored, provided the isotropic medium is replaced by a fictitious nonreciprocal medium with refractive indices $n_{CW} = n + v/c$ and $n_{CCW} = n - v/c$ applicable to waves propagating in the CW and CCW directions, respectively. The lack of isotropy resulting from spinning is usually very small ($v \ll c$) and does not significantly affect the focusing properties of the medium. If the laser gyroscope is located in a gravitational potential $V < 0$, the frequency measured by a distant observer where $V = 0$ is red-shifted. The relative change of frequency $\Delta f/f = V/c^2$. (Note that the gravitational *field* may be the same at both locations and perhaps be equal to zero. This is the case, for example, if the first location is at the center of a planet, and the second location is far from masses.)

2.21 The WKB Approximation of Gaussian Beams and Other Representations

The purpose of this section is to recall a few elementary results of geometrical optics and to indicate how (real) ray manifolds can be associated with gaussian beams. Because exact solutions are available for the propagation of gaussian beams in square-law media, as we have seen, the WKB approximation is not essential. However, the discussion of square-law media helps clarify the significance of that approximation, which is useful for more complicated problems.

One often makes use in geometrical optics of manifolds of rays called normal congruences. Ray congruences are defined by the condition that no more than one ray passes through any point in space. These congruences are called normal if a surface $S(\mathbf{x}) = $ const exists that is perpendicular to all of the rays of the manifold. The theorem of Malus and Dupin [e.g., ref. 3 (Chapter 1), p.131] grants that a normal congruence remains normal after an arbitrary number of refractions. Geometrical optics fields can be associated with such manifolds, the phase of the field being obtained by integrating $k\,ds$ along a ray from some reference surface, where ds denotes the elementary length. The field amplitude is obtained by specifying that the power flowing in a narrow tube of rays is invariant.

It often happens that the rays of the manifold are tangent to a curve called a caustic. (For simplicity, we now restrict ourselves to the two-dimensional case, exemplified in Fig. 2-34a.) Two neighboring rays intersect each other near the caustic. If ϵ denotes the distance between two such rays, the field amplitude must be proportional to $\epsilon^{-1/2}$ to satisfy the law of conservation of power. Thus, the field amplitude is infinite at the caustic where ϵ vanishes. We further observe that ϵ is an algebraic quantity, which changes sign at the caustic. Because of the exponent $\frac{1}{2}$ in the expression of the field amplitude, the change in sign of ϵ results in a phase shift equal to $\pi/2$. (This phase shift turns out to be a phase retardation.) The difficulty that we now encounter is that two rays instead of one are passing through any given point. To overcome this difficulty, Keller[44] suggested that we view the plane of the figure as double sided and assume that, at the caustic, the rays actually change side, i.e., pass behind the plane. Rays are represented in Fig. 2-34b by dashed lines when they are located behind the plane and by solid lines when they are in front of it. The manifold of rays is now a normal congruence with respect to the double-sided surface.

Let us consider as an example a square-law medium with wavenumber

$$k(x) = k_0\left(1 - \tfrac{1}{2}\,\Omega^2 x^2\right) \tag{2.292}$$

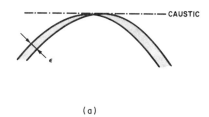

(a)

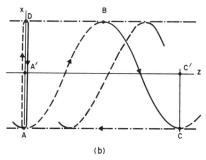

(b)

Fig. 2-34 (a) Illustration of the fact that two neighboring rays meet at the caustic when such a caustic is present. (b) Ray congruences in square-law media. (From Arnaud,[16] by permission of North-Holland Publ. Co.)

where Ω is a constant. Paraxial rays $\bar{q}(z)$ obey the equation [as in (2.11)]

$$\ddot{\bar{q}} + \Omega^2 \bar{q} = 0 \qquad (2.293)$$

where overdots denote differentiation with respect to z. The general solution of (2.293) is

$$\bar{q}(z) = \bar{\xi} \cos(\Omega z + \theta) \qquad (2.294)$$

where $\bar{\xi}$ and θ are constants of integration. Consider now the manifold of rays obtained by letting θ assume all values from 0 to 2π. According to Keller's representation, we view the strip $-\bar{\xi} < x < \bar{\xi}$, bounded by the two caustic lines $x = \bar{\xi}$, $x = -\bar{\xi}$, as double sided. The variation of the phase between two points such as A and C (see Fig. 2-34b) can be evaluated along two independent paths. The Fermat principle readily shows that the optical length of the ray ABC is equal to the optical length of the ray $A'C'$. The geometric distance from A' to C' is $2\pi/\Omega$. Consider next the path AC, which coincides with the caustic line. Because each elementary section of the caustic coincides with a ray, the variation of the eikonal along the caustic line is $(2\pi/\Omega)k(\bar{\xi}) = (2\pi/\Omega)k_0(1 - \frac{1}{2}\Omega^2\bar{\xi}^2)$. Thus the difference $\overline{ABC} - \overline{AC}$, where overbars denote optical lengths, is equal to $\pi\Omega\bar{\xi}^2$. To evaluate the total phase shift along the closed path $\overline{ABCA}$, we must take into account the fact that two caustic lines are crossed and introduce a phase retardation equal to π. Thus the phase is single valued if $k_0\pi\Omega\bar{\xi}^2 - \pi = 2m\pi$, where m denotes an integer, i.e., if

$$\tfrac{1}{2} k_0\Omega\bar{\xi}^2 = m + \tfrac{1}{2} \tag{2.295}$$

Only discrete values $\bar{\xi}_m$ or $\bar{\xi}$ are therefore permissible, those corresponding to $m = 0, 1, 2, \ldots$ in (2.295). For these values of $\bar{\xi}$, a geometrical optics field can be associated to the ray manifold by adding the contributions of the two congruences shown by solid and dashed lines, respectively, in Fig. 2-34b. The field is obtained by keeping track of the phase along the rays of the manifold and using the law of conservation of energy. We obtain

$$\psi_m(x,z) = (1 - \chi^2)^{-1/4} \cos\left\{\left(m + \frac{1}{2}\right)\left[\cos^{-1}(\chi) - \chi(1 - \chi^2)^{1/2}\right] - \frac{\pi}{4}\right\}$$

$$\times \exp\{i[k_0 - (m + \tfrac{1}{2})\Omega]z\} \tag{2.296}$$

where $\chi \equiv x/\bar{\xi}_m$. The function $\psi(\chi)$ is represented by a dashed line in Fig. 2-35 for $m = 6$. $\psi(\chi)$ resembles the exact mode field shown by solid lines,

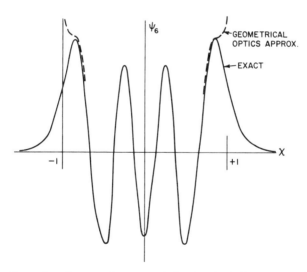

Fig. 2-35 Comparison between the exact solution for the field ψ of a mode with mode number $m = 6$ (solid line) and the geometrical optics approximation (dashed line). The latter departs from the former only in the close neighborhood of the caustics $\chi = \pm 1$. (From Arnaud,[16] by permission of the North-Holland Publ. Co.)

except for the fact that, unlike the exact field in (2.192), $\psi(\chi)$ tends to infinity as the caustic line is approached.

For $m = 0$ the geometrical optics field (2.296) departs sharply from the exact modal solution; however, the caustic $x = \bar{\xi}_0$ coincides with the beam profile defined in (2.49). Furthermore, the expression for the propagation constant of a mode of order m given in (2.296) turns out to be exact for any m.

Note that, instead of considering the closed path $\overline{ABCA}$, we could have considered as well the path $\overline{ADA}$, shown in Fig. 2-34b, taken at a fixed value of z. Setting $k_x \equiv \partial S / \partial x$, the variation of S along this path is

$$\int_{\overline{ADA}} k_x \, dx = 2\pi(m + \tfrac{1}{2}) \tag{2.297}$$

Equation (2.297) is known in quantum mechanics as the quantum condition of Bohr–Sommerfeld.

To evaluate the left-hand side of (2.297), it is convenient to represent the position of a ray at some plane z in the phase space with coordinates $k_x/(\Omega k_0)$, x. In the paraxial approximation, and for k close to k_0, k_x coincides with $k_0\bar{\dot{q}}$, where $\bar{\dot{q}}$ denotes the slope of the ray q, whose maximum value is $k_0\Omega\bar{\xi}$. The rays of the ray manifold representing a mode m are located, in phase space, on a circle. The left-hand side of (2.297) is equal to the area $\pi k_0\Omega\bar{\xi}_m^2$ enclosed by that circle. Thus (2.297) leads again to condition (2.295). The rays of the ray manifold that describes a beam mode, on the other hand, are located on an ellipse rather than on a circle.[16,45] Because this ellipse rotates at a uniform rate Ω as z varies, the beamwidth varies as illustrated in Fig. 1-10.

Slow Variations

The phase space representation just described is particularly useful when Ω is a slowly varying function of z. It can be shown that the area covered by any continuous set of rays remains the same as z varies. This result follows from the fact that the determinant of ray matrices (which map phase spaces at different planes) has determinant unity. Thus, as Ω varies, the ray manifold expands or contracts, but $\Omega\bar{\xi}_m^2$ remains the same. This is the law of adiabatic invariance for rays. According to (2.295), the constancy of $\Omega\bar{\xi}_m^2$ for slow variations of Ω implies that if a beam is at some plane z in a single mode m, it will remain in that same mode as Ω varies. If Ω varies rapidly with z, mode conversion does occur. However, it turns out that the general form of the field remains the same (e.g., gaussian). What is happening is that the beam wavefront does not remain plane, and its half-width $\bar{\xi}_m$ ceases to follow the law of adiabatic invariance.

Resonators

Let us now consider closed systems called resonators. Rayleigh noted that in a large room, sound waves tend to follow closed curves, either clinging to the boundaries (whispering-gallery modes) or bouncing back and forth between opposite walls (bouncing-ball modes). Keller and Rubinow[23] considered the case of perfectly reflecting elliptical boundaries, for which exact solutions are known. Using the geometrical theory of diffraction (or asymptotic forms of the exact solution), they obtained an approximate expression for the resonant frequencies associated with beams bouncing back and forth along the small axis of the ellipse (see Fig. 2-36) in the form

$$k_0 L = 2l\pi + 8(m + \tfrac{1}{2})\left[\tan^{-1}(\exp R_0) - \frac{\pi}{4}\right] \qquad (2.298)$$

where $L/2$ denotes the length of the ellipse minor axis and $\tanh(R_0)$ denotes the minor to major axis length ratio. k_0 is the free-space propagation constant associated with a mode with axial number l and transverse number m. If we introduce the radius of curvature R of the ellipse at the intersection with the small axis, (2.298) becomes

$$k_0 L = 2l\pi + 2(m + \tfrac{1}{2}) \cos^{-1}\left(1 - \frac{L}{2R}\right) \qquad (2.299)$$

an expression that coincides with a result obtained independently by Boyd and Kogelnik[24] for the resonance of an open cavity incorporating two

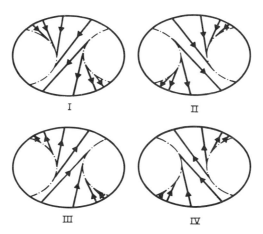

Fig. 2-36 Ray manifolds associated with bouncing-ball modes in resonators with elliptical boundaries. The normal congruences shown in I, II, III, and IV are to be viewed as superimposed and "sewed" together at the boundaries. (After Keller and Rubinow,[46] by permission of the Physical Society.)

circular mirrors of radius R facing each other. This agreement proves that within the paraxial approximation no distinction should be made between circular and elliptic mirrors having the same curvature.

The geometrical theory of modes shows that any ray launched in an optical cavity generates, after a large number of round trips, a double congruence of rays, bounded by two caustics. These caustics outline the beam profile. It can be shown further that the bisectrix of two rays of the ray congruence intersects the axis at the wavefront center. If a ray happens to retrace its own path after a number N of round trips, this ray can be chosen as the closed path defining the quantum condition (2.297). The resonator is then said to be $1/N$ degenerate.

According to the laws of conventional geometrical optics, the optical field vanishes beyond the caustics. By giving consideration to imaginary rays, Keller[44] was able to evaluate the penetration of the field beyond these classical limits, thereby taking into consideration diffraction effects. Keller's theory suffers from the defect that the field obtained is infinite at the caustic. A number of refinements have been proposed to "flesh a geometrical optics skeleton with diffraction substance," as Kravtsov[1] vividly pictures it, while avoiding the singularities at the caustics. One approach consists of assuming that the refractive index varies linearly in the neighborhood of the caustic and in matching the solutions obtained in various regions; this is an extension to three dimensions of the WKB method alluded to before. Kravtsov[47] and Ludwig[48] succeeded in making the asymptotic expansion of the field uniform in the neighborhood of caustics. These modifications of the geometrical theory of diffraction will not be discussed further.

Gaussian Beams as Ray Pencils with Complex Centers

Gaussian beams can be viewed as ray pencils whose centers have complex locations.[2,3,49,50] This alternative representation is useful for homogeneous media because it goes beyond the paraxial approximation. A complex orientation, as we have seen earlier, generates a generalized kind of astigmatism, for circularly symmetric media.[36]

Gaussian Pulses

The field of gaussian pulses varies as a function of time according to a law of the form $\exp[-\frac{1}{2}(t/t_0)^2]$, where t_0 denotes a constant. The complex-ray technique used earlier to describe the propagation of time-harmonic gaussian beams in anisotropic media is convenient also to analyze the evolution in time of gaussian pulses (see Ref. 38, Chapter 1).

We now consider time (t) as a fourth coordinate, to be appended to the three spatial coordinates x, y, z. Specifically, we define events in space–time by ($\mathbf{x}, it$), and we define 4-wave vectors by ($\mathbf{k}, i\omega$). For simplicity, c is set equal to unity. The (hyper) surface of 4-wave vectors in free space is shown in Fig. 1.31a, with the y coordinate omitted. This surface is a hypercone. Only a small region of that hypersurface is of interest, namely, the neighborhood of the optical carrier wave vector, defined by its angular frequency and its direction in space. The normal to the surface of wave vector gives the direction of motion of the pulse center in space–time. The rate of expansion (or contraction) of the pulse, on the other hand, depends on the local *curvature* of the surface of wave vector in the ω, $k \equiv (k_x^2 + k_y^2 + k_z^2)^{1/2}$ space. There is an important physical difference between the transformation of gaussian beams in free space and the evolution of gaussian pulses in time. Because the surface of wave vector in the k_x, k_y, k_z space is spherical and has curvature, gaussian beams eventually spread out in space, as we have seen in detail in previous sections. The curve of wave vector in the ω, k space, on the contrary, being a straight line free of curvature as Fig. 1.31a shows, gaussian wave packets do not spread in time.

For some media, however, the curve of wave vector in the ω, k space is slightly curved (media with material dispersion). In that case, gaussian pulses do spread in time or get focused, depending on the sign of the curvature of the curve of wave vector, and on the initial conditions. (A quadratic variation of the frequency of the optical carrier during the pulse is the time-equivalent of a curvature of the wavefront of time-harmonic optical beams.) Because the mathematical formalism applicable to gaussian pulses is identical to that given earlier for time-harmonic gaussian beams propagating in anisotropic media, the details will be omitted.

Advanced Theories of Gaussian-Beam Propagation

The theory given in previous sections was essentially limited to linear waves in the approximation of Gauss. Contributions that go beyond those approximations are briefly cited below. The behavior of near-gaussian beams in aberrated or truncated near square-law media, or that of widely diverging gaussian beams in homogeneous media, is discussed, for instance, in Marcuse and Miller,[51] Gordon,[52] Matsuhara,[53] Marcatili,[54] Suematsu and Nagashima,[55] Kawakami and Nishizawa,[56] Hull and Julius,[57] Milder,[58] Smith,[59] and Carter.[60] The effect of truncation by one or two apertures on gaussian beam propagation is considered in Takeshita,[61] Campbell and Deshazer,[62] Hadley,[63] Tanaka and Fukumitsu,[64] and Tanaka.[65] The propagation of gaussian beams in random media or through lenses with random defects is evaluated, for example, in Papanicolaou *et al.*,[66] Yoneyama and Nishida,[67] Schmeltzer,[68] deWolf,[69] and Bremmer.[70] Advanced or detailed results concerning ring-type resonators can be found in Volkov and Kiselev.[71] Nonlinear effects are considered in Lugovoi[72] and transverse-mode locking in Auston.[73]

2.22 Gaussian Beams in Quasi-Optics

Quasi-optics is the application of optical techniques to the microwave or millimeter wave range of frequencies.[74] This range can be defined, somewhat arbitrarily, from 1 to 1000 GHz. Quasi-optics deals with beams of electromagnetic radiation rather than with the more conventional co-axial cables and single-mode waveguides. Diffraction effects are, of course, of even greater importance in quasi-optics than in conventional optics, because most devices, though much larger than their optical counterparts, are usually smaller in comparison to the wavelength. For that reason, the laws of geometrical optics should be applied with even more caution than in conventional optics. We are in this section primarily concerned with the transition between conventional waveguides and quasi-optical beams and with the media available, particularly artificial dielectrics. The difference between quasi-optics and optics is of a technical rather than fundamental nature. In fact, the boundary between these two fields becomes fuzzy as new techniques of fabrication appear. For instance, complicated periodic structures that were first conceived for application to the microwave range are now found of interest in the infrared region, as the result of the development of accurate etching techniques using scanning electron microscopes or X rays. Sophisticated three-dimensional gratings can be fabricated, for instance, by freezing the interference pattern of ultraviolet light beams in polymers.[75]

Materials are available in optics that, unfortunately, have no counterpart in the microwave region. In particular, no dielectric material has yet been found in the microwave region, even at low temperatures, that has loss (expressed in decibels per kilometer or as a loss factor) as low as that of quartz in the 0.6- to 1.2-μm wavelength range. A material with a 12-dB/km loss at 12 GHz, however, has been recently reported.[76] Crystals, such as lithium niobate, exhibit electrooptic effects in the microwave region, but the phase shift that they provide is small, not because of a drastically different behavior, but because the number of wavelengths for any reasonable size crystal is small.

On the other hand, the range of relative permittivities of solid dielectrics available to the microwave engineer (from about 100 for rutile down to

1.05 for plastic foams) is much larger than that available to the optical engineer ($\epsilon/\epsilon_0 \sim 3$ to 1.5 in the visible region, and ~ 16 to 1.6 in the infrared). Artificial dielectrics provide greater flexibility in the design of components in microwave than in optics. A plastic foam consisting of hollow dielectric spheres distributed randomly in space is an example of a semiartificial dielectric. Unfortunately, randomness causes Rayleigh scattering. A diffusion takes place whose intensity is proportional to the fourth power of the ratio of the particle dimension to the wavelength.

Waveguide to Beam Transducers

Most microwave sources are terminated by single-mode rectangular waveguides. The problem then is to convert the waveguide field into a collimated beam of radiation, ideally a gaussian beam. This can be accomplished in essentially two ways. The first method consists in slowly increasing the waveguide cross section until the desired beam size is reached as shown in Fig. 2-37a. The second method consists in letting the opened end of the waveguide or a horn radiate into free space and focusing the beam with a lens or a curved reflector [77] as shown in Fig. 2-38. Both approaches have advantages and drawbacks. The flaring of a horn must be extremely slow if we do not want to introduce higher-order modes. Thus the horn is long and bulky. The losses, however, are not

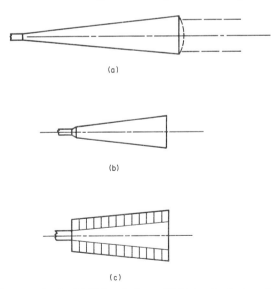

(a)

(b)

(c)

Fig. 2-37 Microwave feeds: (a) high-gain horn, (b) high-gain dual-mode feed, (c) corrugated feed.

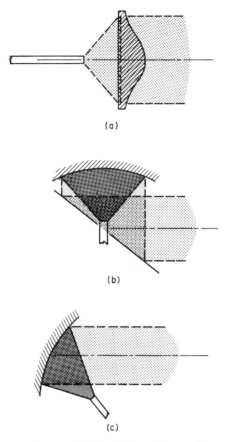

Fig. 2-38   Microwave antennas: (a) dielectric lens focuser, (b) periscopic launcher, (c) offset launcher.

necessarily high because the field intensity at the conductor becomes small, particularly if the horn is a reactive capacitive surface. Such a capacitive surface is obtained by introducing corrugations with a depth slightly exceeding a quarter of a wavelength[78] (see Fig. 2-37c). Because the field tends to detach itself from the metallic parts, the metallic boundary tends to lose control of the wave, and the dissipation losses can be low. It is also possible to excite two modes with the help of a fast transition shown as a steep cone angle in Fig. 2-37b. The relative phase and amplitude of these two modes are chosen such that the current vanishes at the aperture.[79] These so-called dual-mode feeds exhibit almost the same pattern in the E and H planes.[80] A specific design of a dual-mode feed is shown in Fig. 2-39, together with the measured field patterns.

A lens may be corrected for spherical aberration (the only aberration that matters for a properly located feed) by proper shaping as shown in Fig. 2-38a. It is difficult, however, to prevent the Fresnel reflection from taking place at the dielectric–air boundary. The periscopic arrangement[81] shown in Fig. 2-38b suffers from a small amount of feed blockage and spill-over. The offset configuration in Fig. 2-38c is free of blockage, but the beam pattern lacks symmetry and cross-polarized components are intense. An interesting compromise is that shown in Fig. 2-40. This is basically an offset launcher as in Fig. 2-38c. However, the ratio of focal length to reflector diameter is large, and therefore the lack of symmetry is acceptable, and the cross-polarized components have low intensity.[82] To maintain a small spill-over, the feed must have a moderately large gain. The

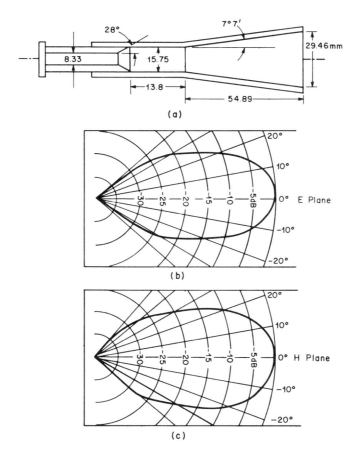

Fig. 2-39 (a) Moderate-gain dual-mode feed. (b) *E* plane pattern. (c) *H* plane pattern. Frequency = 34.5 GHz.

Fig. 2-40   Photograph of an offset launcher fed by a moderate-gain dual-mode feed. (From Gans and Semplak.[80] Reprinted with permission from the *Bell System Technical Journal*, copyright 1975, The American Telegraph and Telephone Co.)

measured far-field radiation pattern of this beam launcher is shown in Fig. 2-41 for different scanning angles. It is compared to gaussian field distributions. The cross-polarized components are below -44 dB.

Weak Focusers

The focusing properties of a pair of cylindrical mirrors are discussed in detail in Chapter 4. Curved metallic mirrors have been used to focus microwave gaussian beams of radiation (see Fig. 2-42). The transmission loss is shown in Fig. 2-43 as a function of the mirror curvature for a path incorporating two focusers (total path length $=80$ m). The experimental points are in good agreement with the theoretical curve obtained from the theory of diffraction of gaussian beams discussed earlier. Further tests were made at 105 GHz over a 800-m-long path incorporating 20 mirrors. In clear weather, the total loss does not exceed 2 dB/km (see Fig. 2-44).[83,84]

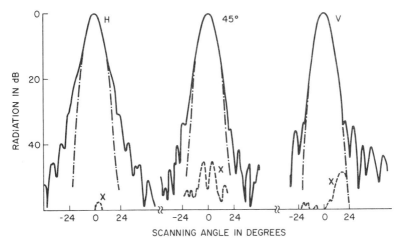

Fig. 2-41   Far-field pattern by the launcher in Fig. 2-40: (a) horizontal scan, (b) 45° scan, (c) vertical scan. The cross-polarized components (X) are shown. Also shown by a dotted–dashed line is the best-fitted gaussian beam. (From Gans and Semplak.[80] Reprinted by permission from *Bell System Technical Journal*, copyright 1975, The American Telegraph and Telephone Co.)

Fig. 2-42   Photograph of a two-cylindrical mirror weak focuser for 100-GHz beams. The mirrors are 1.2 m × 1.2 m. Nominal spacing between adjacent focusers is 80 m. (For a detailed discussion of the focusing properties, see Section 4.21.) (From Arnaud and Ruscio,[83] by permission of the American Optical Society.)

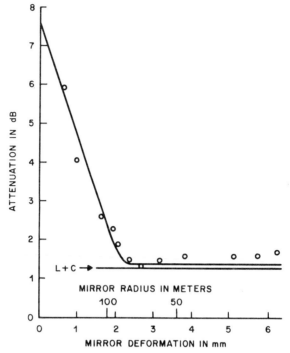

Fig. 2-43   Attenuation of a 80-m-long triangular beam guide path at 50 GHz, incorporating two focusers, as a function of the focuser focal length. Each focuser incorporates two cylindrical mirrors whose radii of curvature are shown on the horizontal axis. The circles are experimental. The curve is theoretical from the theory of coupling of gaussian beams. $L + C$ represents the launcher plus collector loss. (From Arnaud *et al.*,[84] by permission of the IEEE.)

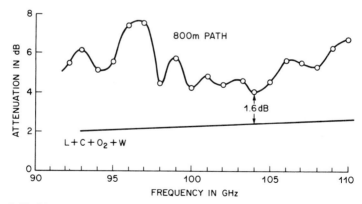

Fig. 2-44   Measured attenuation of a 800-m-long beam guide path as a function of frequency. $L + C + O_2 + W$ represents the launcher, collector, oxygen line absorption, and window loss. (From Arnaud *et al.*,[83] by permission of the IEEE.)

Periodic Metallic Structures

In the microwave range of frequency, low-loss structures can be made of metal, typically silver, copper, or gold-plated stainless steel. (These metals are cited in decreasing order of electrical conductivity and increasing order of mechanical strength and resistance to corrosion.) Here we shall discuss the use of periodic metallic structures[85] for the purpose of guiding or filtering beam waves at microwave frequencies.[39]

A perfectly periodic structure extends to infinity. It is by definition invariant under translations x_1, x_2, x_3, where x_1, x_2, x_3 denote three noncoplanar vectors, which need not be perpendicular to one another, as shown in Fig. 2-45. The propagation of waves in periodic media is characterized by a surface of wave vectors, as is the case for homogeneous

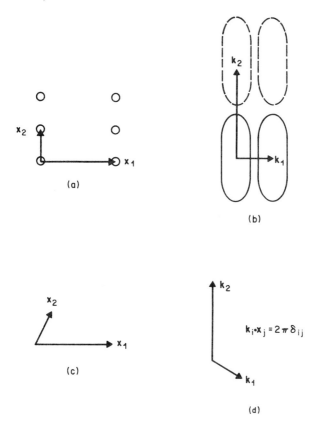

Fig. 2-45 General properties of periodic structures defined by (a) orthogonal, (c) non-orthogonal periods. The corresponding wave vector planes are shown in (b) and (d), respectively. The propagation of gaussian beams in such structures can be defined.

dielectrics. However, because the phase shift per period is defined only to within an integral multiple of 2π, the wave vector is defined only to within the addition of a factor $2\pi l / x_1$, where l denotes an integer and x_1 the period. In three dimensions, the surface of wave vectors can be displaced by integral numbers times the vectors $\mathbf{k}_1$, $\mathbf{k}_2$, $\mathbf{k}_3$, which define the so-called reciprocal lattice. These vectors are defined in Fig. 2-45d. The displaced surfaces of wave vectors are called the "space harmonics" of the structure. They describe the fine structure of the field.

If the wavelength is very large compared with the period, the periodic medium behaves very much like a homogeneous medium. It often becomes possible to define a permittivity and a permeability. This is the case, in particular, for a periodic array of conductive spheres, shown in Fig. 2-46a. If the three periods are mutually perpendicular and equal in length, the surface of wave vector is a sphere as is the case for isotropic dielectrics. If only two periods are equal, the surface of wave vector is a prolate or oblate ellipsoid, as is the case for uniaxial crystals. When the wavelength is not much larger than the period, effects of spatial dispersion show up. The tensor permittivities and permeabilities become functions of the wave vector. Related effects are found in optics in quartz. The slow rotation of the successive lattice planes in a quartz crystal breaks the polarization degeneracy along the optical axis. In that case, the two eigenstates of polarization are circular.

The propagation of gaussian beams is unaffected when the surface of wave vector is translated without deformation in k-space by some fixed vector. Thus it is unnecessary, in order to discuss beam propagation in periodic media, to specify which surface of wave vector is considered. The theory of beam propagation remains applicable if the artificial dielectric is made inhomogeneous by slowly varying the parameters, for instance, the lattice periods. In case of abrupt changes, reflections take place, which may differ somewhat in nature from the conventional Fresnel reflection if spatial dispersion is significant. The phenomenon of radiation into free space or into surface waves can be observed with periodic structures. For instance, slowly sinusoidally modulated corrugated plates have been successfully used as traveling wave antennas.[86]

When the period is comparable to the wavelength, the concepts of permittivity and permeability are not applicable, and a more direct approach must be used. This is the case, for example, for the meander line shown in Fig. 2-46b. For that structure, the mechanism of propagation is very simple: A TEM wave follows the meander wire at the speed of light. The wavenumber is obtained by multiplying the wavenumber along the wire by the ratio of the wire length to the meander period. Clearly the concept of effective permittivity has no relevance for such structures. It is interesting that artificial dielectrics sometimes have refractive indices that are less than unity, a circumstance that is found in optics only for plasmas, for instance, in metals, and for X rays in crystals. An example of an artificial dielectric with refractive index less than unity is the array of waveguides shown in Fig. 2-46c. The wavenumber of each individual waveguide, as is well known, is less than the free-space wavenumber.

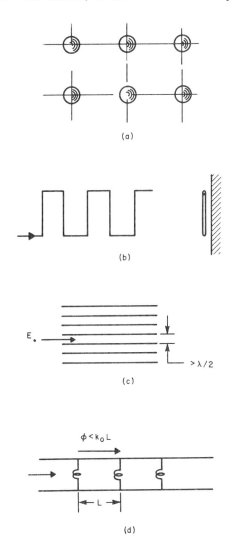

Fig. 2-46 Examples of periodic structures: (a) array of metallic spheres, (b) meander line, (c) array of waveguides, (d) inductance-loaded transmission line.

Another example is a bifilar line loaded periodically with inductances as shown in Fig. 2-46d. In the absence of the inductances, the refractive index is unity since a bifilar line carries TEM modes. The inductances have the effect of reducing the phase shift per period, thereby making the refractive index less than unity. Besides their usefulness in microwave engineering, artificial dielectrics are useful to model optical systems and to analyze problems in electrodynamics. In Chapter 1, for instance, the difference between canonical momenta and mass-carrying momenta was illustrated with the help of an oblique bifilar line model.

It is instructive to see how the multigrid Fabry–Perot diplexers discussed in Section 2.19 can be related to artificial dielectrics. In Section 2.19, the transmission of gaussian beams through a pair of plane-parallel grids was discussed. We have shown that the solution can be obtained by adding the fields of the successive passes of the incident beam or, alternatively, by

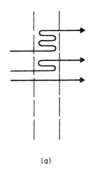

(a)

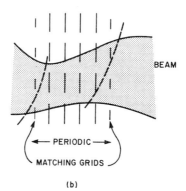

(b)

Fig. 2-47   (a) Transmission through a two-grid Fabry–Perot filter. (b) For a large number of grids, the filter can be considered a three-periodic artificial dielectric with matching layers (grids).

expanding the incident beam in a spectrum of plane waves and adding the contribution of each transmitted plane wave (see Fig. 2-47a). We have also indicated that a more uniform response in the passband can be obtained if three or more grids are used. Under normal incidence, standard filter design procedures can be used. For a large number of grids, these procedures lead to an almost periodic series of grids as shown in Fig. 2-47b. The first and last few grids are different from the others. They essentially have the function of matching the periodic section of grids to free space. Thus, when a Fabry–Perot diplexer incorporates a large number of grids, the problem can be considered from the point of view of a gaussian beam transmitted through a slab of artificial dielectric with matching layers at both ends. From that point of view, the frequency selectivity of the filter results from the very fast dependence of the angular frequency ω on the wavenumber k in the medium. As for any multisection filter, the matching at the edges of the band is always imperfect, which causes oscillations to appear in the frequency response. The three-dimensional geometry discussed here exhibits features that are unknown in conventional filter design. For instance, if the axis of the incident gaussian beam is not perpendicular to the grids, the beam may be steered significantly as the frequency varies. The consequence of this steering is that the transmitted beam is not collected in its entirety by the receiving dish, since the location of the receiving dish can be optimized only at one frequency in the band. This beam-steering effect is of the same nature as the beam pattern distortion that creates the "mismatch" loss discussed in Section 2.19. This beam-steering (or mismatch) loss is to be added to the power loss suffered on reflection at the filter boundaries and to the dissipation losses. In conclusion, a Fabry–Perot-type quasi-optical diplexer incorporating a large number of grids can be viewed as a highly dispersive anisotropic slab of artificial dielectric. This example further motivates our interest in periodic structures.

The progagation of electromagnetic waves along the surface of biperiodic structures has been thoroughly investigated for application to microwave traveling-wave tubes by Doehler et al.[87] These biperiodic structures are particularly attractive for crossed-field tubes because much larger powers can be attained than with more conventional structures. Experimental results concerning the multiple meander line shown in Fig. 2-48 are reported in Doehler et al.[87] The theory of propagation consists in assuming that TEM waves propagate along the wires. The transverse electrical connections are taken into account either by writing the boundary conditions for each cell or by considering first the Green function of an infinite array of parallel wires and periodic sources.[88] These two approaches lead to identical results.

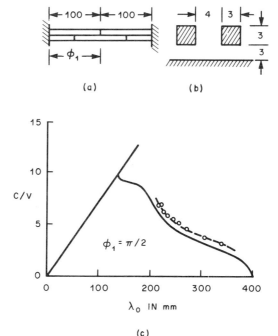

(c)

Fig. 2-48 (a) Multiple meander line that can be deposited by photolithographic techniques on high thermal conductivity ceramics.[89] (b) Cross section of the line. (c) Variation of the ratio c/v of the velocity of light in free space and the phase velocity, as a function of the wavelength λ_0 in free space, for a particular transverse mode $\phi_1 = \pi/2$. (From Arnaud,[88] by permission of the Société des Radioélectriciens.)

The use of thin metallic structures deposited by photographic techniques on slabs of high thermal conductivity ceramics, alumina (Al_2O_3), or beryllia (BeO) in high-power crossed-field tubes, was first reported by Arnaud and Epsztein.[89] Sputtering from the negatively biased electrode facing the biperiodic structure is tolerable, provided a good vacuum is maintained in the electron tube (pressure $\leqslant 10^{-8}$ Torrs). High power, gain, and efficiency at 3 GHz were demonstrated.[89] In this particular application, one seeks to achieve synchronism between a (truncated) plane wave, guided by the biperiodic structure, and the electron beam, guided between the biperiodic structure and the negatively biased electrode facing it. The biperiodic structure may also guide gaussian beams in its plane that obey the laws of diffraction discussed earlier.

Various coupling experiments made in optics with dielectric slabs or thin films can be performed also with biperiodic structures, as Ulrich[90] has shown recently.

Polarization Effects

A single metallic grid with parallel conducting wires constitutes a very efficient polarization-separator at millimeter wavelengths. Waves are transmitted if the electric field is perpendicular to the grid wires and reflected if the electric field is parallel to the grid wires. Saleh[91] discovered that a sequence of grid-type polarizers has pass-band characteristics and that the bandwidth can be varied by changing the angular orientation of some of the grids, an operation that is mechanically simple.

In some other applications, one needs partial reflectors (e.g., beam splitters) that have the same reflection for all polarization for waves at nonnormal incidence. This result can be achieved either with stacked dielectric plates[92] or with rectangular grids having the proper periods.[93] Usually, the condition of polarization insensitivity depends critically on the incident angle, and therefore on the angular spread of the incident (e.g., gaussian) beam. Beams of large cross section are usually required.

References

1. Y. A. Kravtsov, *Radiophys. and Quantum Electr.* **10**, 719 (1967).
2. J. A. Arnaud, *Appl. Opt.* **8**, 189, 1909 (1969).
3. G. A. Deschamps, *Electron. Lett.* **7**, No. 23, 684 (1971); *Proc. IEEE* **60**, 1022 (1972).
4. V. A. Fock, *in* "Electromagnetic Diffraction and Propagation Problems," Chapter 14. Pergamon, Oxford, 1965.
5. J. A. Arnaud, *Amer. J. Phys.* **41**, 549 (1973).
6. G. A. Deschamps and P. E. Mast, *in Symp. Quasi-Opt.* Polytech. Press, Brooklyn (1964), p. 379.
7. H. Kogelnik, *Appl. Opt.* **4**, 1562 (1965).
8. S. A. Collins, Jr., *Appl. Opt.* **3**, 1263 (1964).
9. T. S. Chu, *Bell Syst. Tech. J.* **45**, 287 (1966).
10. P. K. Tien, J. P. Gordon, and J. R. Whinnery, *Proc. IEEE* **53**, 129 (1965).
11. H. G. Unger, *Arch. Elek. Uber.* **19**, 189 (1965).
12. E. A. J. Marcatili, *Bell Syst. Tech. J.* **45**, 105 (1966).
13. H. Kogelnik, *Bell Syst. Tech. J.* **44**, 455 (1965). A. Gerrard and J. M. Burch, "Introduction to Matrix Methods in Optics." Wiley, New York, 1975.
14. N. G. Vakhimov, *Radio Eng. Electron Phys. (USSR)* **10**, 1439 (1965).
15. H. Kogelnik, *in Proc. Symp. Quasi-Opt.* Polytech. Press, Brooklyn (1964), p. 333.
16. J. A. Arnaud, *in* "Progress in Optics," p. 247 (E. Wolf, ed.), Vol. XI. North-Holland Publ., Amsterdam, 1973.
17. J. A. Arnaud, *J. Opt. Soc. Amer.* **61**, 751 (1971).
18. H. G. Booker and W. Walkinshaw, "Meteorological Factors in Radio Wave Propagation," p. 80. Physical Soc., London, 1946.
19. D. W. Berreman, *Bell Syst. Tech. J.* **44**, 2117 (1965).
20. J. A. Arnaud, Graded index fibers versus cladded fibers; a mechanical analogy, unpublished, Feb. 1972. D. Gloge, *Appl. Opt.* **11**, 2506 (1972).
21. K. Husimi, *Progr. Theor. Phys.* **9**, 381 (1953).

22. G. Goubau, U.S. Patent No. 3,101,472, filed 1958. J. R. Christian and G. Goubau, *IEEE Trans. Antennas Propagat.* **AP9**, 256 (1961).
23. J. B. Keller and S. I. Rubinow, *Ann. Phys.* **9**, 24 (1960).
24. G. D. Boyd and J. P. Gordon, *Bell Syst. Tech. J.* **40**, 489 (1961). G. D. Boyd and H. Kogelnik, *Bell Syst. Tech. J.* **41**, 1347 (1962). H. Kogelnik and T. Li, *Proc. IEEE* **54**, 1312 (1966).
25. A. E. Siegman, *Appl. Opt.* **13**, 353 (1974).
26. L. A. Vainshtein, *Sov. Phys. Tech. Phys.* **9**, 157 (1964).
27. A. G. Fox and T. Li, *Proc. IRE* **48**, 1904 (1960).
28. T. Li, *Bell Syst. Tech. J.* **44**, 917 (1965).
29. G. Toraldo di Francia, *Opt. Acta* **13**, 323 (1966).
30. A. G. Fox and T. Li, *IEEE J. Quantum Electron.* **QE4**, 460 (1968).
31. J. A. Arnaud, *Proc. IEEE* **62**, 1561 (1974).
32. W. W. Rigrod, *Appl. Phys. Lett.* **2**, 51 (1963).
33. J. A. Arnaud, *Bell Syst. Tech. J.* **49**, 2311 (1970).
34. R. D. Standley and V. Ramaswamy, *Appl. Phys. Lett.* **25**, 711 (1974).
35. J. H. Van Vleck, *Proc. Nat. Acad. Sci. U.S.* **14**, 178 (1928).
36. J. A. Arnaud and H. Kogelnik, *Appl. Opt.* **8**, 1687 (1969).
37. P. Marié, *Ann. Telecommun.* **24**, 177 (1969); **25**, 320 (1970).
38. J. A. Arnaud, *Appl. Opt.* **11**, 2514 (1972). I. M. Mason and E. Ash, *J. Appl. Phys.* **42**, 5343 (1971). J. A. Arnaud, *J. Opt. Soc. Amer.* **62**, 290 (1972).
39. J. A. Arnaud, A. A. M. Saleh, and J. T. Ruscio, *IEEE Trans. Microwave Theory Tech.* **MTT22**, 486 (1974).
40. U. Ganiel and Y. Silberberg, *Appl. Opt.* **14**, 306 (1975).
41. A. A. M. Saleh and J. T. Ruscio (private communication). J. A. Arnaud and F. Pelow, *Bell Syst. Tech. J.* **54**, 263 (1975).
42. M. M. Popov, *Opt. Spectrosc.* **25**, 170, 213 (1968).
43. W. K. Kahn and J. T. Nemit, *in Proc. Symp. Mod. Opt.* (J. Fox, ed.), Polytech. Press, Brooklyn, 1967, p. 501.
44. J. B. Keller, *in Proc. Symp. Appl. Math* **8**, 27 (1956).
45. S. Kawakami and J. Nishizawa, *Res. Inst. Elec. Comm. Tech. Rep.* **TR-25**, October, 1967.
46. J. B. Keller and S. I. Rubinow, *Ann. Phys.* **9**, 24 (1960).
47. Y. A. Kravtsov, *Sov. Radiophys.* **8**, 467 (1965).
48. D. Ludwig, *Comm. Pure Appl. Math.* **19**, 215 (1966).
49. *Special Iss. Proc. IEEE* **62** (Nov. 1974).
50. S. Y. Shin and L. B. Felsen, *Appl. Phys.* **5**, 239 (1974).
51. D. Marcuse and S. E. Miller, *Bell Syst. Tech. J.* **43**, 1759 (1964).
52. J. P. Gordon, *Bell Syst. Tech. J.* **45**, 321 (1966).
53. M. Matsuhara, *J. Opt. Soc. Amer.* **63**, 135, 1514 (1973).
54. E. A. J. Marcatili, *Bell Syst. Tech. J.* **45**, 105 (1966); **46**, 1733 (1967).
55. V. Suematsu and H. Nagashima, *Elec. Commun. Japan* **52B**, 76 (1969).
56. S. Kawakami and J. Nishizawa, *J. Appl. Phys.* **38**, 4807 (1967).
57. T. E. Hull and R. S. Julius, *Can J. Phys.* **34**, 914 (1956)
58. D. M. Milder, *J. Acoust. Soc. Amer.* **46**, Part 2, 1259 (1969).
59. R. Smith, *Phil. Trans. Roy. Soc.* **268A**, 289 (1970).
60. W. H. Carter, *Opt. Commun.* **11**, 410 (1974). A. L. Cullen, P. Nagenthiram, and A. D. Williams, *Proc. Roy. Soc.* **329**, 153 (1972).
61. S. Takeshita, *IEEE Trans. Antennas Propagat.* **AP17**, 90 (1969).
62. J. P. Campbell and L. G. Deshazer, *J. Opt. Soc. Amer.* **59**, 1427 (1969).
63. G. R. Hadley, *IEEE J. Quantum Electron.* **QE10**, 603 (1974).

64. K. Tanaka and O. Fukumitsu, *IEEE Trans. Microwave Theory Tech.* **MTT22**, 81 (1974).
65. K. Tanaka, *Opt. Commun.* **12**, 168 (1974).
66. G. C. Papanicolaou, D. McLaughlin, and R. Burridge, *J. Math. Phys.* **14**, 84 (1973).
67. T. Yoneyama and S. Nishida, *Rep. Res. Inst. Elec. Commun. Tohoku Univ.* **25**, 67 (1973).
68. R. A. Schmeltzer, *Quart. Appl. Math.* **24**, 339 (1967).
69. D. A. deWolf, *Radio Sci.* **10**, 53 (1975).
70. H. Bremmer, *Radio Sci.* **8**, 511 (1973).
71. A. M. Volkov and V. A. Kiselev, *Sov. Phys.-JETP* **31**, 996 (1970). V. E. Privalov and S. A. Fridrikhov, *Sov. Phys.-Usp.* **97**, 153 (1969). V. F. Boitsov, T. A. Murina, and E. E. Fradkin, *Opt. Spectrosc.* **36**, 311 (1974).
72. V. N. Lugovoi, *Sov. Phys.-JETP* **38**, 439 (1974).
73. D. H. Auston, *IEEE J. Quantum Electron.* **QE4**, 420 (1968).
74. A. F. Harvey, "Coherent Light." Wiley (Interscience), London, 1970.
75. G. Ward, "Integrated Optics and Optical Communication." MSS Inform. Corp., New York, 1974.
76. S. Ayers, G. J. Davies, J. Haigh, D. Marr, and A. E. Parker, *Proc. IEE(London)* **121**, 1447 (1974).
77. R. J. Chaffin and J. B. Beyer, *IEEE Trans. Microwave Theory Tech.* **MTT-12**, 555 (1964).
78. A. Harris, A. D. Olver, and P. J. B. Clarricoats, *Electron. Lett.* **10**, 304 (1974).
79. P. D. Potter, *Microwave J.* **6**, 71 (1963).
80. M. J. Gans and R. A. Semplak, *Bell Syst. Tech. J.* **54**, 1319 (1975).
81. A. B. Crawford and R. H. Turrin, *Bell Syst. Tech. J.* **48**, 1605 (1969).
82. T. S. Chu and R. H. Turrin, *IEEE Trans. Antennas Prop.* **AP-20**, 339 (1973).
83. J. A. Arnaud and J. T. Ruscio, *IEEE Trans. Microwave Theory Tech.* **MTT-23**, 377 (1975). I. Anderson and J. T. Ruscio, *Electron Lett.* **11** (15 May 1975).
84. J. A. Arnaud, D. C. Hogg, and J. T. Ruscio, *IEEE Trans. Microwave Theory Tech.* **MTT-20**, 344 (1972).
85. J. A. Arnaud, *in* "Crossed Field Microwave Devices" (E. Okress, ed.). Academic Press, New York, 1961.
86. A. Hessel, *in* "Antenna Theory" (R. E. Collin and F. J. Zucker, eds.), Part II, p. 151. McGraw-Hill, New York, 1969.
87. O. Doehler, B. Epsztein, and J. A. Arnaud, *Onde Elec.* **36**, 937 (1956).
88. J. A. Arnaud, *Ann. Radio Elec.* **21**, 232 (1966). J. A. Weiss, *IEEE Trans. Microwave Theory Tech.* **MTT.22**, 1194 (1974).
89. J. A. Arnaud and B. Epsztein, *in Conf. OTAN sur les Tech. Micro-Ondes.* EL1/712.01/769 (1961). Program for Development of X-Band Crossed-Field Amplifiers (September 1966). Prepared by Warnecke Electron Tubes, Inc., Project Serial Number SF0100201 (unpublished).
90. R. Ulrich, *in Symp. Opt. Acoust. Microelectron., Polytech. Inst. New York April 16, 1974.*
91. A. A. M. Saleh, *IEEE. Trans. Microwave Theory Tech.* **MTT-22**, 728 (1974).
92. A. A. M. Saleh, *Bell Syst. Tech. J.,* **54**, 1027 (1975).
93. A. A. M. Saleh and R. Semplak (private communication).

Wave Equations

In the first section of this chapter, the notation is set up and a mathematical discussion of various types of equations is given. The wave equation most relevant to beam optics is the Fock equation, a Schrödinger-like equation (parabolic with imaginary first derivative in the axial coordinate). It is nevertheless useful to acquire some knowledge of the properties of elliptic and hyperbolic equations as well.

3.1 Hyperbolic, Parabolic, and Elliptic Equations

Most physical phenomena are described by wave functions ψ satisfying partial differential equations.[1] The differentiations are with respect to the three spatial coordinates x, y, z (or x_1, x_2, x_3) and time t (or $x_4 \equiv ict$). We shall mostly consider time-harmonic fields of the form

$$\psi(x, y, z, t) = \text{Re}[\psi(x, y, z) \exp(-i\omega t)] \tag{3.1}$$

For the moment, however, we keep t an independent variable.

The wave function ψ can be defined, with respect to a given coordinate system, by one or more numbers. If ψ is characterized by just one number, this number is usually invariant under a coordinate transformation. In that case, it is called a scalar. Sometimes, three numbers are necessary to characterize a wave fully. If these numbers transform as the coordinates of a point in space in a rotation of the x, y, z coordinate system, the wave is said to be a vector field. This is the case, for example, for the deformation (strain) of an elastic medium. The displacement of a point of the medium with respect to the unstrained state is a vector. For more complicated systems, waves are described by second-rank tensors (electromagnetic waves) or bispinors (electron waves). Let us discuss as an example small vibrations in homogeneous nonviscous compressible fluids. These vibrations are described by a velocity potential ψ satisfying the scalar Helmholtz equation

$$\mathbf{\nabla \cdot \nabla}\psi \equiv \left(\frac{\partial^2}{\partial x^2} + \frac{\partial^2}{\partial y^2} + \frac{\partial^2}{\partial z^2} \right)\psi = u^{-2}\frac{\partial^2\psi}{\partial t^2} \tag{3.2}$$

where u denotes the velocity of plane waves in the medium. ψ itself does not have a direct physical significance, but its gradient is the velocity of a point of the fluid at x, y, z and time t, that is,

$$\mathbf{v}(x, y, z, t) = \mathbf{\nabla}\psi(x, y, z, t) \tag{3.3}$$

We have introduced in the above equation the gradient operator $\mathbf{\nabla}$, which operates on the scalar ψ to give a vector with components

$$\mathbf{\nabla}\psi \equiv \begin{bmatrix} \partial\psi/\partial x \\ \partial\psi/\partial y \\ \partial\psi/\partial z \end{bmatrix} \tag{3.4}$$

The divergence operator $\mathbf{\nabla}\cdot$, on the other hand, operates on a vector such as $\mathbf{a}$, with components (a_x, a_y, a_z), to give a scalar

$$\mathbf{\nabla}\cdot\mathbf{a} \equiv \frac{\partial a_x}{\partial x} + \frac{\partial a_y}{\partial y} + \frac{\partial a_z}{\partial z} \tag{3.5}$$

Thus the operator $\mathbf{\nabla}\cdot\mathbf{\nabla}$ introduced in (3.2), operating on the scalar ψ, gives the scalar $\partial^2\psi/\partial x^2 + \partial^2\psi/\partial y^2 + \partial^2\psi/\partial z^2$. The operator $\mathbf{\nabla}\cdot\mathbf{\nabla}$ is sometimes denoted Δ and called the Laplacian. We shall also make use of the rotational operator $\mathbf{\nabla}\times$, which operates on a vector such as $\mathbf{a}$, to give a three-component object

$$\mathbf{\nabla}\times\mathbf{a} \equiv \begin{bmatrix} \partial a_z/\partial y - \partial a_y/\partial z \\ \partial a_x/\partial z - \partial a_z/\partial x \\ \partial a_y/\partial x - \partial a_x/\partial y \end{bmatrix} \equiv \begin{bmatrix} 0 & -\partial/\partial z & \partial/\partial y \\ \partial/\partial z & 0 & -\partial/\partial x \\ -\partial/\partial y & \partial/\partial x & 0 \end{bmatrix}\begin{bmatrix} a_x \\ a_y \\ a_z \end{bmatrix} \tag{3.6}$$

The second expression in (3.6) is written in the form of a matrix product. The first matrix is antisymmetrical ($M_{ij} = -M_{ji}$, $i, j = 1, 2, 3$), and the second is a column vector representing the vector **a**. Let us recall that the product **L** of a matrix **M** and a matrix **N** has elements $L_{ij} = M_{ik}N_{kj}$, where summation over k is implied. In this expression, it is assumed that **M** and **N** are conformable, that is, the number of columns of **M** equals the number of rows of **N**.

A little bit more should be said about the rotational operator. $\nabla \times$ **a** is not a vector if **a** is a vector. Indeed, the components of a vector change sign if we change the x, y, z coordinate system to the coordinate system $-x, -y, -z$. The components of $\nabla \times$ **a**, however, according to their definition, do not change sign. $\nabla \times$ **a** is sometimes called an axial vector. It should more properly be considered an antisymmetrical tensor with components

$$(\nabla \times \mathbf{a})_{ij} = \frac{\partial a_j}{\partial x_i} - \frac{\partial a_i}{\partial x_j}, \qquad i, j = 1, 2, 3 \tag{3.7}$$

Because of its antisymmetry, the tensor $\nabla \times$ **a** has only three distinct components in three-dimensional space. The definition (3.7) is applicable, however, to spaces of arbitrary dimension.

Similarly, the vector product **a** $\times$ **b** of two vectors **a** and **b** is an axial vector. It is formed from the components of the antisymmetrical tensor

$$(\mathbf{a} \times \mathbf{b})_{ij} = a_i b_j - a_j b_i, \qquad i, j = 1, 2, 3 \tag{3.8}$$

The flow of incompressible fluids is free of divergence, $\nabla \cdot \mathbf{v} = 0$. The flow may be either irrotational, $\nabla \times \mathbf{v} = \mathbf{0}$ (Fig. 3-1a), or carry vortices where $\nabla \times \mathbf{v} \neq \mathbf{0}$ (Fig. 3-1b). In the former case we can express **v** as the gradient of a scalar: $\mathbf{v} = \nabla \psi$; the relation $\nabla \times \mathbf{v} = \mathbf{0}$ is then automatically satisfied [using the matrix form in (3.6) and observing that, for instance, $\partial^2/\partial x\,\partial y = \partial^2/\partial y\,\partial x$]. This example is given because it explains the physical origin of the word "rotational." A similar formalism is used in optics. The wave vector $\mathbf{k}(x, y, z)$ derives from a potential S called the eikonal, with $\mathbf{k} = \nabla S$ and, therefore, **k** is irrotational: $\nabla \times \mathbf{k} = \mathbf{0}$.

A partial differential equation such as (3.2) has a unique solution if the proper boundary conditions are specified. The way boundary conditions need to be specified depends on the type of equation considered. The scalar Helmholtz equation (3.2) is of the hyperbolic type[§] in space–time. In that case, one must impose the so-called Cauchy conditions on ψ, for instance, specify ψ and $\partial\psi/\partial t$ at every x, y, z and $t = 0$. Then ψ is defined

---

[§] If an equation has, in two dimensions (x, y), the form

$$A \frac{\partial^2 \psi}{\partial x^2} + 2B \frac{\partial^2 \psi}{\partial x\,\partial y} + C \frac{\partial^2 \psi}{\partial y^2} = F\left(x, y, \psi, \frac{\partial\psi}{\partial x}, \frac{\partial\psi}{\partial y}\right)$$

it is called hyperbolic, parabolic, or elliptic depending on whether $AC - B^2$ is negative, zero, or positive, respectively. The function F is not relevant to this classification. For the equation in (3.10), we have $C = B = 0$. Thus $AC - B^2 = 0$, and the equation is parabolic irrespective of the values of $a \neq 0$ and b.

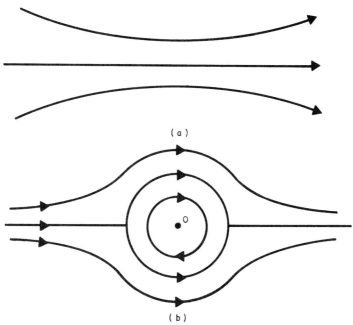

Fig. 3-1 (a) Irrotational fluid flow, $\nabla \times \mathbf{v} = \mathbf{0}$, where $\mathbf{v}$ is the fluid velocity. (b) Flow with $\nabla \times \mathbf{v} \neq \mathbf{0}$ within the circle. In geometrical optics we have $\nabla \times \mathbf{k} = \mathbf{0}$, $\mathbf{k}$ being the wave vector.

at all times. This boundary condition is similar to the one used for ordinary second-order differential equations when one specifies ψ and $d\psi/dx$ at $x = 0$.

Let us now assume that ψ has the time-harmonic form in (3.1). Then the scalar Helmholtz equation (3.2) becomes

$$\left(\frac{\partial^2}{\partial x^2} + \frac{\partial^2}{\partial y^2} + \frac{\partial^2}{\partial z^2} \right)\psi + \left(\frac{\omega}{u} \right)^2 \psi = 0 \tag{3.9}$$

where ω denotes the angular frequency, which equals 2π times the frequency. Equation (3.9) is of the elliptic type (see previous footnote) and different initial conditions must be specified for it. One can, for instance, specify the value of ψ on some closed surface (Dirichlet condition). Then ψ has a unique solution in the interior of this surface. Instead, one can specify the behavior of ψ at infinity.

In Chapter 2 we used an approximate form of the scalar Helmholtz equation applicable to waves propagating in a direction close to the z-axis. This equation, first proposed by Fock, is of the parabolic type (see previous footnote)

$$\frac{\partial^2 \psi}{\partial x^2} = a \frac{\partial \psi}{\partial z} + b\psi \tag{3.10}$$

The solution of parabolic equations is essentially defined from the specification of $\psi(x, 0)$.

It is useful to consider the finite-difference equation that goes to (3.10) in the limit of small differences. x and z are assumed to vary only by fixed increments h, and ψ is defined only at these points [$x = mh$, $z = nh$, $\psi(x, z) \to \psi(m, n)$, m, n integers]. Assuming for simplicity $b = 0$ in (3.10) and setting $d \equiv (ha)^{-1}$, the difference equation

$$\psi(m, n + 1) = d[\psi(m + 1, n) + \psi(m - 1, n) - 2\psi(m, n)] + \psi(m, n)$$

$$(3.11)$$

is easily seen to reduce to (3.10) in the limit where $h \to 0$. Equation (3.11) shows that if we know $\psi(m, 0)$, we can calculate in succession $\psi(m, 1)$, $\psi(m, 2), \ldots$. Whether the solution obtained is stable, that is, insensitive to small changes in the initial $\psi(m, 0)$, depends on the value of d. If d is real positive and not too large, irregularities in $\psi(m, n)$ tend to smooth out as n increases. Indeed, let us assume that for some $m = m_0$, $\psi(m_0, n)$ exceeds both $\psi(m_0 + 1, n)$ and $\psi(m_0 - 1, n)$, that is, the function $\psi(m, n)$ has a "bump" at $m = m_0$. Then, according to (3.11), $\psi(m_0, n + 1)$ is smaller than $\psi(m_0, n)$, and the bump is reduced in amplitude as n increases. If, however, we go to negative n, irregularities amplify. Thus it becomes more and more difficult to predict $\psi(m, n)$ from the knowledge of $\psi(m, 0)$ if n is a large negative number. This discussion is applicable, for instance, to the heat equation. In the Fock equation, which is of primary interest to us, the constant d is imaginary, and a different behavior is to be expected with respect to stability. Knowledge of $\psi(m, 0)$ is, in fact, sufficient to evaluate $\psi(m, n)$ everywhere. For that case, (3.11) provides us with a method for numerical integration of (3.10).

After this brief review, we consider in succession the Maxwell equations, the Helmholtz equation, and the Fock equation. The Fock equation is most easily constructed in a heuristic manner from the equation of geometrical optics. It is nevertheless useful to know how it can be derived from the Helmholtz equation and how the Helmholtz equation can in turn be derived from the Maxwell equations. These alternative derivations provide a better understanding of the approximations made.

3.2 The Maxwell Equations

Most optical phenomena are described by a pair of vectors **E**, **B** satisfying, in a region free of sources, the equations

$$\nabla \times \mathbf{E} = i\omega\mathbf{B}, \qquad \nabla \times \mathbf{H} = -i\omega\mathbf{D} \qquad (3.12)$$

where $\nabla \times$ denotes the rotational operator and $(\mathbf{D}, \mathbf{H})$ is related to $(\mathbf{E}, \mathbf{B})$ by a linear operator that characterizes the medium. If the relation is local, that is, if $(\mathbf{D}, \mathbf{H})$ depends only on the value of $(\mathbf{E}, \mathbf{B})$ at the same point and same instant, this linear operator can be represented by a 6×6 matrix independent of ω. We may have, for instance,

$$\mathbf{D}(\mathbf{x}) = \epsilon(\mathbf{x})\mathbf{E}(\mathbf{x}), \qquad \mathbf{B}(\mathbf{x}) = \mu(\mathbf{x})\mathbf{H}(\mathbf{x}) \tag{3.13}$$

ϵ and μ are 3×3 matrices (or second-rank tensors) called the permittivity and permeability of the medium, respectively. They are functions of space $(\mathbf{x})$ if the medium lacks homogeneity. Such a medium, with ϵ, μ independent of ω, is called nondispersive. If, however, $(\mathbf{D}, \mathbf{H})$ depends on $(\mathbf{E}, \mathbf{B})$ through time convolutions, ϵ and μ depend on ω. The medium is then called dispersive. Introducing (3.13) in (3.12), we have

$$\nabla \times \mathbf{E} = i\omega\mu\mathbf{H}, \tag{3.14a}$$

$$\nabla \times \mathbf{H} = -i\omega\epsilon\mathbf{E} \tag{3.14b}$$

If ϵ (respectively, μ) has a finite discontinuity, the tangential components of $\mathbf{E}$ (respectively, $\mathbf{H}$) remain continuous. Indeed, let us consider the closed path shown in Fig. 3-2a. The integral of $\mathbf{E}$ (respectively, $\mathbf{H}$) over the path is proportional to the flux of $i\omega\mathbf{B}$ (respectively, $-i\omega\mathbf{D}$). This flux vanishes together with the area enclosed by the path.

Furthermore, we observe that, because for any vector $\mathbf{A}$, $\nabla \cdot (\nabla \times \mathbf{A}) = 0$,

$$\nabla \cdot \mathbf{D} = 0 \tag{3.15a}$$

$$\nabla \cdot \mathbf{B} = 0 \tag{3.15b}$$

From these equations it follows that the normal component of $\mathbf{D}$ (respectively, $\mathbf{B}$) is continuous. This conclusion is reached by integrating (3.15) over the volume of a small cylinder whose axis is perpendicular to the surface of discontinuity (see Fig. 3-2b), and letting the height of this cylinder go to zero.

In this section, we derive a basic property of the solutions of Maxwell's equations, namely, the fact that they satisfy a reciprocity theorem. Let $\mathbf{E}$, $\mathbf{H}$ be a solution of (3.14) and let $\mathbf{E}^\dagger$, $\mathbf{H}^\dagger$ be a solution of the Maxwell equations for a transposed medium whose permittivity and permeability $\epsilon^\dagger$, $\mu^\dagger$ are the transposes $\tilde{\epsilon}$, $\tilde{\mu}$ of ϵ, μ, respectively (that is, $\epsilon_{ij}^\dagger = \epsilon_{ji}$, $\mu_{ij}^\dagger = \mu_{ji}$, where $i, j = 1, 2, 3$). We have in that new medium

$$\nabla \times \mathbf{E}^\dagger = i\omega\tilde{\mu}\mathbf{H}^\dagger, \qquad \nabla \times \mathbf{H}^\dagger = -i\omega\tilde{\epsilon}\mathbf{E}^\dagger \tag{3.16}$$

The relation

$$\nabla \cdot \mathbf{J}_M = 0 \tag{3.17}$$

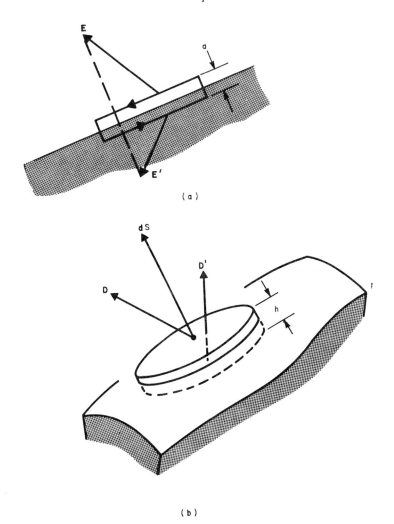

(a)

(b)

Fig. 3-2   (a) Because the line integral of **E** along the rectangular contour (flux of $\nabla \times \mathbf{E}$) vanishes in the limit $a \to 0$, the tangential component of **E** is continuous. (b) Because the flux of **D** through the cylinder (volume integral of $\nabla \cdot \mathbf{D}$) vanishes in the limit $h \to 0$, the normal component of **D** is continuous.

where

$$\mathbf{J}_M \equiv \mathbf{E}^\dagger \times \mathbf{H} + \mathbf{H}^\dagger \times \mathbf{E} \qquad (3.18)$$

follows from (3.14) and (3.16) if we make use of the mathematical identity

$$\nabla \cdot (\mathbf{a} \times \mathbf{b}) = \mathbf{b} \cdot \nabla \times \mathbf{a} - \mathbf{a} \cdot \nabla \times \mathbf{b} \qquad (3.19)$$

By application of the Gauss theorem, we find that the flux of $\mathbf{J}_M$ over any closed surface S vanishes:

$$\int_S \mathbf{J}_M \cdot \mathbf{dS} = 0 \tag{3.20}$$

where $\mathbf{dS}$ has magnitude dS and is normal to the surface, pointing outward. This result is called the Lorentz reciprocity theorem. As we shall see, similar results are applicable to the Helmholtz and Fock wave equations.

In the next section, we consider the propagation of waves that are locally plane in gradually inhomogeneous media.

3.3 The Geometrical Optics Limit

When the medium is homogeneous (ϵ, μ being independent of $\mathbf{x}$), plane waves of the form

$$\mathbf{E} = \mathbf{E}_0 \exp(i\mathbf{k} \cdot \mathbf{x}), \qquad \mathbf{H} = \mathbf{H}_0 \exp(i\mathbf{k} \cdot \mathbf{x}) \tag{3.21}$$

can propagate. $\mathbf{k}$ is called the wave vector. Its magnitude $k \equiv |\mathbf{k}|$ is equal to $2\pi/\lambda$, if λ denotes the wavelength in the medium. Substituting (3.21) in (3.14), we find that the Maxwell equations are satisfied, provided

$$\mathbf{k} \times \mathbf{E}_0 = \omega\mu\mathbf{H}_0, \qquad \mathbf{k} \times \mathbf{H}_0 = -\omega\epsilon\mathbf{E}_0 \tag{3.22}$$

Equations (3.22) show that $\epsilon\mathbf{E}_0$ and $\mu\mathbf{H}_0$ are perpendicular to the wave vector $\mathbf{k}$. In general, the vector $\mathbf{E}_0 \times \mathbf{H}_0$ is not parallel to $\mathbf{k}$.

Equations (3.22) have nontrivial solutions only if the determinant of the coefficients of the unknown quantities vanishes, that is, if

$$\det \begin{bmatrix} \mathbf{k}\times & -\omega\mu \\ \omega\epsilon & \mathbf{k}\times \end{bmatrix} = 0 \tag{3.23}$$

where

$$\mathbf{k}\times \equiv \begin{bmatrix} 0 & -k_3 & k_2 \\ k_3 & 0 & -k_1 \\ -k_2 & k_1 & 0 \end{bmatrix} \tag{3.24}$$

and k_1, k_2, k_3 denote the x-, y-, z-components of $\mathbf{k}$. Equation (3.23) is called the dispersion equation. The tip of the vector $\mathbf{k}$ is constrained by (3.23) to remain on a surface called the surface of wave vectors.

In the special case where $\epsilon = \epsilon\mathbf{1}$, $\mu = \mu\mathbf{1}$, where ϵ and μ are scalars and $\mathbf{1}$ denotes the 3×3 unit matrix (with diagonal elements equal to unity, the other elements being equal to zero), (3.23) reduces to

$$|\mathbf{k}|^2 \equiv k^2 = \omega^2 \epsilon \mu \qquad (3.25)$$

It is customary to introduce a refractive index $n \equiv (\epsilon\mu/\epsilon_0\mu_0)^{1/2}$. Then we have $k = (\omega/c)n$, where $c^2\epsilon_0\mu_0 = 1$. However, in most calculations, it is preferable to deal directly with the wavenumber k.

Equation (3.25) has the form of the dispersion equation used in the major part of this book. The wave vector $\mathbf{k}$ has the same magnitude k in all directions, and the surface of wave vectors is a sphere. Such a medium is called isotropic. Going back to (3.22) with ϵ, μ scalars, we find that the power density $\mathbf{E}_0 \times \mathbf{H}_0$ has the direction of $\mathbf{k}$. The vectors $\mathbf{E}_0$ and $\mathbf{H}_0$ are perpendicular to $\mathbf{k}$ and are mutually perpendicular. (See Fig. 3-3a). We are still free to choose the direction of $\mathbf{E}_0$ in the plane perpendicular to $\mathbf{k}$. For a given $\mathbf{k}$, any wave can be represented by a linear superposition of two waves, one with $\mathbf{E}_0$ circularly polarized clockwise and one with $\mathbf{E}_0$ circularly polarized counterclockwise. Other representations, with linearly polarized waves, are also possible.

Let us now assume that the medium is inhomogeneous but that the wavelength λ is small compared with the scale of the inhomogeneities in the medium. We consider solutions of the Maxwell equations of the form

$$\mathbf{E} = \mathbf{E}_0(\mathbf{x}) \exp[iS(\mathbf{x})], \qquad \mathbf{H} = \mathbf{H}_0(\mathbf{x}) \exp[iS(\mathbf{x})] \qquad (3.26)$$

Because the variations of $\mathbf{E}_0$ and $\mathbf{H}_0$ are slow compared with those of S, we have approximately

$$\nabla S \times \mathbf{E}_0 = \omega\mu\mathbf{H}_0, \qquad \nabla S \times \mathbf{H}_0 = -\omega\epsilon\mathbf{E}_0 \qquad (3.27)$$

By comparing these expressions to (3.22), we see that the wave described by (3.26) is locally plane, with wave vector

$$\mathbf{k} = \nabla S \qquad (3.28)$$

As indicated before in this chapter, (3.28) implies that $\nabla \times \mathbf{k}(\mathbf{x}) = \mathbf{0}$.

By using (3.28), the dispersion equation (3.23) becomes

$$\det \begin{bmatrix} \nabla S \times & -\omega\mu \\ \omega\epsilon & \nabla S \times \end{bmatrix} = 0 \qquad (3.29)$$

and, for the case of isotropic media,

$$|\nabla S(\mathbf{x})| = k(\mathbf{x}) \qquad (3.30)$$

Equations (3.29) and (3.30) are partial differential equations for S, called eikonal equations. We shall now make use of (3.30) to obtain the Helmholtz and Fock equations.

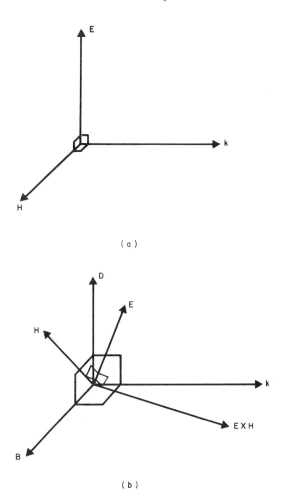

Fig. 3-3 (a) Isotropic medium; **E**, **H**, **k** are mutually perpendicular. (b) Anisotropic medium; **k** is not always parallel to **E** × **H**. Besides, **E**, **H**, **D**, **B** are not always linearly polarized.

3.4 The Helmholtz Equation

The Maxwell equations have a rather complicated form. Fortunately, for most problems of beam propagation, they can be replaced by simpler scalar wave equations, namely, the Helmholtz and Fock equations. Instead of deriving these equations directly from the Maxwell equations, we obtain them in a heuristic way, from the requirement that they have the same limit as the Maxwell equations for short wavelengths. This limit is the realm of geometrical optics, whose basic law was given in Section 3.3. This procedure, in general, is not free of ambiguity, because different wave equations may have the same geometrical optics limit. For instance, the Klein–Gordon and Dirac equations have the same classical limit: the equations of relativistic mechanics. Yet they differ in many significant ways. Nevertheless, this procedure is satisfactory for most of our problems. It makes it relatively easy to obtain a scalar wave equation applicable to anisotropic media when the geometrical optics behavior is known from experiments. A direct derivation from the Maxwell equations is, in that case, very complicated. The direct derivation will be given, for the sake of comparison, for isotropic media.

For isotropic media with wavenumber k, (3.30) is explicitly

$$\left(\frac{\partial S}{\partial x} \right)^2 + \left(\frac{\partial S}{\partial y} \right)^2 + \left(\frac{\partial S}{\partial z} \right)^2 = k^2(\mathbf{x}) \tag{3.31}$$

If the substitution $i\, \partial S/\partial x_i \rightarrow \partial/\partial x_i$ is made in (3.31), the scalar Helmholtz equation is obtained:

$$\nabla \cdot \nabla E + k^2(\mathbf{x})E = 0 \tag{3.32}$$

Let us show that each Cartesian component of the electric or magnetic field indeed obeys (3.32) approximately. Substituting $\mathbf{H}$ from (3.14a) in (3.14b), we obtain

$$\nabla \times (\mu^{-1}\nabla \times \mathbf{E}) - \omega^2\epsilon\mathbf{E} = 0 \tag{3.33}$$

Assuming that μ is $\mu_0\mathbf{1}$, where $\mathbf{1}$ denotes the unit matrix and μ_0 a constant, and recalling that the vector Laplacian $\Delta \equiv \nabla \nabla \cdot - \nabla \times \nabla \times$, (3.33) becomes

$$\Delta\mathbf{E} - \nabla\nabla \cdot \mathbf{E} + \omega^2\mu_0\epsilon(\mathbf{x})\mathbf{E} = 0 \tag{3.34a}$$

It can be shown that the rectangular components of $\Delta\mathbf{E}$ are $\nabla \cdot \nabla E_x$, $\nabla \cdot \nabla E_y$, and $\nabla \cdot \nabla E_z$. We now assume that ϵ is $\epsilon\mathbf{1}$ (isotropic medium). If the spatial variation of ϵ in (3.15a) can be neglected, the relation $\nabla \cdot (\epsilon\mathbf{E}) = 0$ implies $\nabla \cdot \mathbf{E} \approx 0$, and the second term in (3.34) vanishes. Thus each Cartesian component E_x, E_y, or E_z of $\mathbf{E}$ approximately obeys the Helmholtz equation (3.32). A similar derivation can be given for the components of $\mathbf{H}$.

The exact equations for $\mathbf{E}$ and $\mathbf{H}$ in an inhomogeneous dielectric are, respectively,

$$\Delta\mathbf{E} + k^2\mathbf{E} + \nabla(\epsilon^{-1}\nabla\epsilon \cdot \mathbf{E}) = \mathbf{0} \tag{3.34b}$$

$$\Delta\mathbf{H} + k^2\mathbf{H} + (\epsilon^{-1}\nabla\epsilon) \times (\nabla \times \mathbf{H}) = \mathbf{0} \tag{3.34c}$$

where $k^2(\mathbf{x}) \equiv \omega^2\mu_0\epsilon(\mathbf{x})$ and Δ is the vector Laplacian operator.

Whatever the physical significance of E in (3.32) might be, E and its first derivatives must remain continuous if the discontinuities of k^2 are finite. The normal component of ∇E is clearly continuous because its flux through the surface of a small cylinder such as the one shown in Fig. 3-2 vanishes in the limit of vanishing cylinder height, E being bounded. The continuity of E and of the tangential component of ∇E follows.

Let us now consider any two solutions E and $E^\dagger$ of the Helmholtz equation (3.32). We have

$$\nabla \cdot \mathbf{J}_H = 0 \tag{3.35}$$

where

$$\mathbf{J}_H = E^\dagger(\nabla E) - (\nabla E^\dagger)E \tag{3.36}$$

Indeed, using the identity $\nabla \cdot (\alpha \mathbf{a}) \equiv \nabla \alpha \cdot \mathbf{a} + \alpha \nabla \cdot \mathbf{a}$, we obtain

$$\nabla \cdot [E^\dagger(\nabla E) - (\nabla E^\dagger)E] = E^\dagger(\nabla \cdot \nabla E) - E(\nabla \cdot \nabla E^\dagger) \tag{3.37}$$

The right-hand side of (3.37) vanishes if (3.32) is used. Integrating (3.35) over a volume V bounded by a surface S gives the scalar form of the Lorentz reciprocity theorem,

$$\int_S [E^\dagger(\nabla E) - (\nabla E^\dagger)E] \cdot \mathbf{d}S = 0 \tag{3.38}$$

3.5 The Fock Equation

We now make one more approximation, applicable to waves that propagate approximately along the z-axis. This approximation reduces the (elliptic) Helmholtz equation to a (parabolic) wave equation,[3] which has the same form as the Schrödinger equation of quantum mechanics if the following changes are made: $t \to z$, $m/\hbar \to k(0)$ and $-(e/\hbar)U(x) \to k(x)$, where t denotes time, z the axial coordinate, m and e the particle mass and charge, respectively, U the electric potential, and $\hbar$ the Planck constant divided by 2π (see Chapter 1).

This equation can be obtained by solving (3.30) for $\partial S/\partial z$ and neglecting the terms $(\partial S/\partial x)^4$, $(\partial S/\partial y)^4$, ... We obtain

$$\frac{\partial S}{\partial z} = \left[k^2 - \left(\frac{\partial S}{\partial x} \right)^2 - \left(\frac{\partial S}{\partial y} \right)^2 \right]^{1/2}$$

$$\approx k - \frac{1}{2} k^{-1} \left[\left(\frac{\partial S}{\partial x} \right)^2 + \left(\frac{\partial S}{\partial y} \right)^2 \right] \tag{3.39}$$

We further assume that the transverse variation of k is small so that k can be replaced by its value $k_0(z)$ on axis ($x = y = 0$) in the second term on the rhs of (3.39). Introducing in (3.39) the formal substitution $i\, \partial S/\partial x_i \to \partial/\partial x_i$, the parabolic wave equation is obtained:

$$- i \frac{\partial \psi}{\partial z} = k(x, y, z)\psi + \frac{1}{2} k_0(z)^{-1} \left(\frac{\partial^2 \psi}{\partial x^2} + \frac{\partial^2 \psi}{\partial y^2} \right) \qquad (3.40)$$

The term $k_0(z)$ is essentially eliminated if we replace z by a new variable $\zeta(z)$ defined by

$$\frac{dz}{d\zeta} = k_0(z) \qquad (3.41)$$

With this change in axial variable, (3.40) becomes

$$- i \frac{\partial \psi}{\partial \zeta} = k(x, y, \zeta)k_0(\zeta)\psi + \frac{1}{2} \left(\frac{\partial^2 \psi}{\partial x^2} + \frac{\partial^2 \psi}{\partial y^2} \right) \qquad (3.42)$$

Only the product $k(x, y, \zeta)k_0(\zeta)$ now enters.

The above derivation does not provide us with the physical significance of ψ in relation to the electromagnetic field. Before discussing the relation of ψ to $(\mathbf{E}, \mathbf{H})$, let us establish the adjoint equation to (3.42) and the reciprocity theorem. The approximation (3.39) is applicable to waves whose $\mathbf{k}$ vector has a direction close to the $+z$-axis. A similar approximation can be used for waves whose $\mathbf{k}$ vector is directed in a direction close to the $-z$-axis (see Fig. 3-4). We obtain

$$\frac{\partial S}{\partial z} \approx -k + \frac{1}{2} k^{-1} \left[\left(\frac{\partial S}{\partial x} \right)^2 + \left(\frac{\partial S}{\partial y} \right)^2 \right] \qquad (3.43)$$

where we have changed $\partial S/\partial z$ to $-\partial S/\partial z$. Using the same method as before, we obtain a new (adjoint) parabolic wave equation:

$$i \frac{\partial \psi^\dagger}{\partial \zeta} = k(x, y, \zeta)k_0(\zeta)\psi^\dagger + \frac{1}{2} \left(\frac{\partial^2 \psi^\dagger}{\partial x^2} + \frac{\partial^2 \psi^\dagger}{\partial y^2} \right) \qquad (3.44)$$

We can now define a current $\mathbf{J}_F$ with transverse ($\mathbf{j}$) and axial (ρ) components,

$$\mathbf{J}_F \begin{cases} \mathbf{j} = -\tfrac{1}{2} i(\psi^\dagger \, \boldsymbol{\partial} \psi - \psi \, \boldsymbol{\partial} \psi^\dagger) \\ \rho = \psi^\dagger \psi \end{cases} \qquad (3.45)$$

where ψ is a solution of (3.42) and $\psi^\dagger$ a solution of (3.44). $\boldsymbol{\partial}$ denotes the

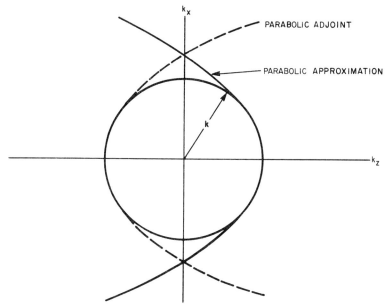

Fig. 3-4   For isotropic media, the surface of wave vectors is a sphere (the k_y-component is omitted). The Fock wave equation is obtained by approximating this sphere with a paraboloid. The adjoint wave equation corresponds to the dashed line.

gradient operator in the transverse plane, with components $\partial/\partial x$ and $\partial/\partial y$. It is not difficult to show that

$$\mathbf{\nabla} \cdot \mathbf{J}_F \equiv \mathbf{\partial} \cdot \mathbf{j} + \frac{\partial \rho}{\partial \zeta} = 0 \qquad (3.46)$$

by substituting for $\mathbf{j}$ and ρ their expressions in (3.45) and using for $\partial\psi/\partial\zeta$ and $\partial\psi^\dagger/\partial\zeta$ the expressions in (3.42) and (3.44).

Let us now integrate (3.46) over the volume bounded by plane ζ and plane $\zeta + d\zeta$. We obtain, using the Gauss theorem, and neglecting the contribution at infinity

$$\frac{d}{d\zeta} \int\int_{-\infty}^{+\infty} \psi^\dagger\psi \, dx \, dy = 0 \qquad (3.47)$$

This is the new form of the Lorentz reciprocity theorem, applicable to the Fock equation. For comparison, let us specialize our previous result (3.36) to paraxial waves. We assume that E and $E^\dagger$ describe waves propagating essentially in the $+z$- and $-z$-directions, respectively. Their z-dependence is therefore

$$E \sim \exp\left[i \int_0^z k_0(z) \, dz \right], \qquad E^\dagger \sim \exp\left[-i \int_0^z k_0(z) \, dz \right] \qquad (3.48)$$

and (3.35) can be written

$$\partial \cdot (E^\dagger \partial E - E \partial E^\dagger) + \frac{\partial}{\partial z} \left[2ik_0(z) E^\dagger E \right] = 0 \qquad (3.49)$$

Comparison of (3.49) and (3.46) shows that these expressions coincide if we define

$$\psi \equiv k_0^{1/2} E, \qquad \psi^\dagger \equiv k_0^{1/2} E^\dagger \qquad (3.50)$$

Relations (3.50) provide us with a link between the wave function ψ and the transverse components of the electric field.

For the case of isotropic media, relations (3.50) can be obtained more directly from the Helmholtz equation without invoking reciprocity. The Helmholtz equation (3.32) is

$$\nabla \cdot \nabla E + k^2(\mathbf{x})E = 0 \qquad (3.51)$$

Because for waves propagating in a direction close to the z-axis, the z-dependence of E is close to $\exp[i \int^z k_0(z) \, dz]$, we expect the z-variations of

$$E'(x, y, z) \equiv E(x, y, z) \exp\left[-i \int_0^z k_0(z) \, dz \right] \qquad (3.52)$$

to be slow. Substituting (3.52) in (3.51) and neglecting $\partial^2 E' / \partial z^2$, we obtain

$$\partial \cdot \partial E' + 2ik_0 \frac{\partial E'}{\partial z} + iE' \frac{dk_0}{dz} + (k^2 - k_0^2)E' = 0 \qquad (3.53)$$

To clarify the significance of this approximation, let us consider a plane wave propagating in a homogeneous medium at an angle α to the $+z$-axis. For such a wave

$$E(x, z) = E_0 \exp[ik(z \cos \alpha - x \sin \alpha)] \qquad (3.54a)$$

$$E'(x, z) = E_0 \exp\{ik[z(\cos \alpha - 1) - x \sin \alpha]\} \qquad (3.54b)$$

Consequently, for small α

$$\frac{\partial E'}{\partial z} \approx k\alpha^2 E_0, \qquad \frac{\partial^2 E'}{\partial z^2} \approx k^2\alpha^4 E_0, \qquad \frac{\partial^2 E'}{\partial x^2} \approx k^2\alpha^2 E_0 \qquad (3.55)$$

From this example, it is clear that the conditions

$$\frac{\partial^2 E'}{\partial z^2} \ll k \frac{\partial E'}{\partial z} \approx \frac{\partial^2 E'}{\partial x^2} \tag{3.56}$$

indeed hold if $\alpha \ll 1$. For lens-type waveguides, α is restricted to values as small as 10^{-3} rad, and (3.53) can be considered exact for all practical purposes. However, in fiber optics, α is of the order of 0.1 rad, and the Fock equation is insufficiently accurate. The scalar Helmholtz equation remains an adequate approximation in most cases.

Equation (3.53) simplifies further if we introduce the changes of function and variable

$$E'' = k_0^{1/2} E', \qquad \frac{dz}{d\zeta} = k_0(\zeta) \tag{3.57}$$

We obtain

$$\partial \cdot \partial E'' + 2i \frac{\partial E''}{\partial \zeta} + (k^2 - k_0^2)E'' = 0 \tag{3.58}$$

Let us make one more change of function,

$$\psi = E'' \exp\left[i \int_0^\zeta k_0(\zeta)\, d\zeta \right] \tag{3.59}$$

to restore the original dependence on z. We obtain

$$\partial \cdot \partial \psi + 2i \frac{\partial \psi}{\partial \zeta} + (k^2 + k_0^2)\psi = 0 \tag{3.60}$$

If we further observe that if k does not depart very much from k_0 in any transverse plane $(k - k_0)^2 \approx 0$ and therefore

$$k^2 + k_0^2 \approx 2k_0 k \tag{3.61}$$

our original parabolic equation (3.42) is recovered. As we can see, even for isotropic media, a direct derivation of (3.42) from the Maxwell equations is rather lengthy, and approximations must be introduced in many successive steps. It was nevertheless instructive to carry out this procedure for verification.

3.6 Axial Coupling

The fundamental conservation relation $\nabla \cdot \mathbf{J} = 0$, where $\mathbf{J}$ is given in (3.18), (3.36), or (3.45), can be used to evaluate the coupling between a matched emitter of radiation and a matched collector of radiation,[4] as

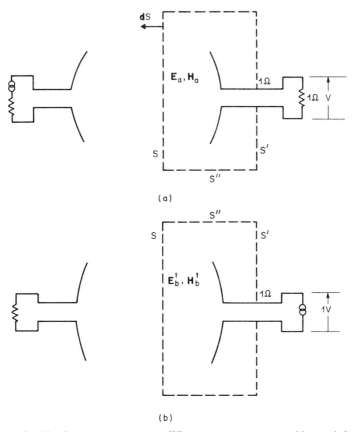

Fig. 3-5 Coupling between two antennas. When two antennas are weakly coupled and the receiving antenna is matched to its load, the voltage across the load can be expressed as an integral, over any surface such as S located between the two antennas, of the product of the field of the transmitting antenna and the field that the receiving antenna would transmit if it were used as a transmitter.

illustrated in Fig. 3-5. Let us consider the Maxwell field and assume that the medium is reciprocal ($\epsilon = \tilde{\epsilon}$, $\mu = \tilde{\mu}$). Let $(\mathbf{E}, \mathbf{H}) \equiv (\mathbf{E}_a, \mathbf{H}_a)$ denote the field from the radiator (Fig. 3-5a) and let $(\mathbf{E}^\dagger, \mathbf{H}^\dagger) \equiv (\mathbf{E}_b^\dagger, \mathbf{H}_b^\dagger)$ denote the field when the source of the radiator is turned off and a 1-V source is connected to the 1-Ω load of the collector (Fig. 3-5b). The feeder is assumed to have a 1-Ω characteristic impedance. The vector

$$\mathbf{J}_{ab} = \mathbf{E}_b^\dagger \times \mathbf{H}_a + \mathbf{H}_b^\dagger \times \mathbf{E}_a \qquad (3.62)$$

is divergenceless because both $(\mathbf{E}_a, \mathbf{H}_a)$ and $(\mathbf{E}_b^\dagger, \mathbf{H}_b^\dagger)$ satisfy the Maxwell equations (3.14).Let now $\nabla \cdot \mathbf{J}_{ab}$ be integrated over the volume V enclosed by the surface $S + S'$ shown in Fig. 3-5. This surface surrounds the receiving dish but intersects its feeder. From the divergence theorem, the surface integral

$$\tfrac{1}{2} \int (\mathbf{E}_b^\dagger \times \mathbf{H}_a + \mathbf{H}_b^\dagger \times \mathbf{E}_a) \cdot \mathbf{d}S \tag{3.63}$$

assumes the same value when the integration is performed at a plane located between the two antennas (plane S) and at a plane intersecting the feeder (plane S'), the contributions at infinity (S'') being neglected. Let us show that the integral over S' is the voltage across the 1-Ω load of the receiving feeder. We can assume, for concreteness, that the feeder with characteristic impedance 1 Ω is made up of a pair of conducting strips having unity spacing and width $w = (\mu_0/\epsilon_0)^{1/2} \approx 377$ unit lengths. In that case, the electric field is equal to the voltage between the strips, and the magnetic field is equal to the current flowing through the strips divided by w, the fringing fields being neglected.

The field $E_b^\dagger$ at S' is therefore equal to the source voltage, 1 V. The product $wH_b^\dagger$ is equal to the current in the feeder. This current is unity if we assume that the reflected wave can be neglected, that is, that the feeder is matched to the antenna. Going back to the configuration in Fig. 3-5a, the voltage V across the load that we wish to evaluate is equal to E_a, and we have (exactly) $wH_a = E_a$ because the 1-Ω load matches the feeder perfectly. In the above discussion, we have considered only the magnitudes of the fields $\mathbf{E}_a, \mathbf{H}_a, \mathbf{E}_b^\dagger, \mathbf{H}_b^\dagger$. The direction of these fields in the feeder is well known from the theory of TEM waves propagating along a pair of wires. Substituting $E_b^\dagger = 1$, $H_b^\dagger = 1/w$, $E_a = V$, and $H_a = V/w$, the integral in (3.63) evaluated over the cross section of the feeder is found equal to the voltage across the collector load. It should be noted that this voltage is obtained in phase as well as in amplitude. The phase reference is that of the 1-V source applied to the collector when the fields $\mathbf{E}_b^\dagger, \mathbf{H}_b^\dagger$ are defined. The motivation for the notation $\mathbf{E}_b^\dagger, \mathbf{H}_b^\dagger$ rather than $\mathbf{E}_b, \mathbf{H}_b$ is to call attention to the fact that these fields propagate in a direction opposite to that of $\mathbf{E}_a, \mathbf{H}_a$. For a generalization of the previous results to media with arbitrary ϵ, μ tensors, we would need to consider the field $\mathbf{E}_b^\dagger, \mathbf{H}_b^\dagger$ radiated in an "adjoint" medium having permittivity and permeability tensors that are the transposes $\tilde{\epsilon}$ and $\tilde{\mu}$ of ϵ and μ, respectively. Note that ϵ and μ need not be Hermitian; that is, the medium considered can be lossy or have gain. A different kind of adjoint medium will be considered later when the concept of losslessness is discussed.

Because the integral over S is equal to the integral over S', we have the coupling formula

$$V = \tfrac{1}{2} \int_S (\mathbf{E}_b^\dagger \times \mathbf{H}_a + \mathbf{H}_b^\dagger \times \mathbf{E}_a) \cdot d\mathbf{S} \qquad (3.64)$$

Thus the voltage across the load of the receiving antenna can be evaluated if we know the field $\mathbf{E}_a$, $\mathbf{H}_a$ radiated by the transmitting antenna at some surface, such as S, located between the two antennas and the fields $\mathbf{E}_b^\dagger$, $\mathbf{H}_b^\dagger$ that the receiving antenna would radiate at that plane, if it were excited by a 1-V source, on that same surface. Up to that point, only one assumption was made, namely, that the feeder of the receiving antenna, when used in transmission, is matched to the antenna.

The fields $\mathbf{E}_a$, $\mathbf{H}_a$, $\mathbf{E}_b^\dagger$, $\mathbf{H}_b^\dagger$ are radiated *in the presence* of the other antenna and are generally unknown. However, if the coupling between the two antennas is sufficiently small, either because of a large separation, or because an absorbing sheet of material effectively decouples them, the fields are approximately the same as the fields in the absence of the other antenna. Instead of two antennas, we may consider two glass fibers coupled end to end, as shown in Fig. 3-6. The coupling is given by the integral in (3.64), provided Fresnel reflection can be neglected.

For scalar fields ψ_a, $\psi_b^\dagger$, the coupling is defined in a similar way. We have

$$c_{ab} = \int_S \psi_b^\dagger \psi_a \, dS \qquad (3.65)$$

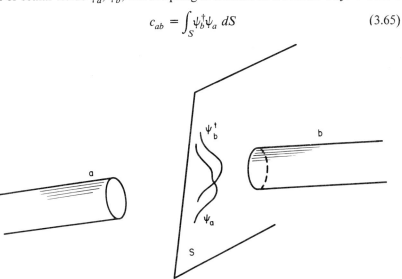

Fig. 3-6 Coupling between two weakly guiding glass fibers. The coupling is approximately equal to the integral of $\psi_b^\dagger \psi_a$ over any plane, such as S, located between the two fibers, where ψ_a, $\psi_b^\dagger$ denotes the fields radiated by the two fibers in the absence of the other. Fresnel reflection is neglected.

As we have indicated before, this coupling is independent of the surface S, provided the fields vanish sufficiently rapidly at infinity.

Extensive use has been made in Chapter 2 of expression (3.65), in particular for the case where ψ_a, $\psi_b^\dagger$ represent gaussian beams.

In some cases, we are interested in the power transferred rather than in the complex amplitudes. A result similar to (3.17) is obtained if we consider fields at the opposite angular frequency $-\omega$, that is, the complex-conjugate fields [for any real field such as $E(t)$ we have $E(-\omega) = E^*(\omega)$]. Let $\mathbf{E}_a$, $\mathbf{H}_a$ denote as before a solution of the Maxwell equations

$$\nabla \times \mathbf{E}_a = i\omega\mu\mathbf{H}_a, \qquad \nabla \times \mathbf{H}_a = -i\omega\epsilon\mathbf{E}_a \tag{3.66}$$

and assume that ϵ and μ are real (lossless medium). Let $\mathbf{E}_b$, $\mathbf{H}_b$ be another solution of the Maxwell equations (3.66). It is easily shown that

$$\nabla \cdot (\mathbf{J}_{ab}) = 0 \tag{3.67}$$

where

$$\mathbf{J}_{ab} \equiv \mathbf{E}_b^* \times \mathbf{H}_a - \mathbf{H}_b^* \times \mathbf{E}_a \tag{3.68}$$

The generalization of (3.67) to media with arbitrary ϵ, μ

$$\nabla \times \mathbf{E}_a = i\omega\mu\mathbf{H}_a, \qquad \nabla \times \mathbf{H}_a = -i\omega\epsilon\mathbf{E}_a \tag{3.69}$$

is obtained by considering fields $\mathbf{E}_b$, $\mathbf{H}_b$ satisfying the Maxwell equations for a medium whose permittivity and permeability are the Hermitian conjugate (complex-conjugate transpose) $\tilde{\epsilon}^*$ and $\tilde{\mu}^*$ of ϵ, μ,

$$\nabla \times \mathbf{E}_b = i\omega\tilde{\mu}^*\mathbf{H}_b, \qquad \nabla \times \mathbf{H}_b = -i\omega\tilde{\epsilon}^*\mathbf{E}_b \tag{3.70}$$

The conservation relation (3.67) then holds for arbitrary ϵ, μ. The two media considered above are identical if $\tilde{\epsilon}^* = \epsilon$ and $\tilde{\mu}^* = \mu$, that is, if the medium is lossless (but not necessarily reciprocal). Note that the form of $\mathbf{J}_{ab}$ in (3.68) can be written in the same form as (3.62) if we define the adjoint field $\mathbf{E}_b^\dagger \equiv \mathbf{E}_b^* = \mathbf{E}_b(-\omega)$ and $\mathbf{H}_b^\dagger \equiv -\mathbf{H}_b^* = -\mathbf{H}_b(-\omega)$.

The scalar form of the relation (3.67) is

$$\frac{d}{dz} \int_S \psi_b^* \psi_a = 0 \tag{3.71}$$

where S denotes a plane perpendicular to the system axis, with axial coordinate z. This relation has an important consequence for heterodyne optical receivers. Let us consider two paraxial optical beams with the same polarization and only slightly different frequencies, incident on the same detector, as shown in Fig. 3-7. Let the field of one of the two beams (say, the local oscillator) be denoted ψ_b and the field of the other beam (the

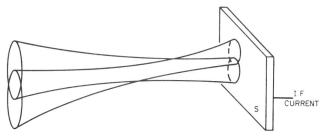

Fig. 3-7 The principle of heterodyne optical detection: the intermediate frequency current generated in a uniform square-law detector is proportional to the coupling between the two beams. This coupling is independent of the surface S of the detector, provided the detector covers both beams.

signal) be denoted ψ_a. The instantaneous intermediate frequency (i.f.) current density is proportional to the amplitude squared of the total field

$$I(x, t) \propto |\psi_a|^2 + |\psi_b|^2 - 2|\psi_a| \, |\psi_b| \cos(\text{phase } \psi_a - \text{phase } \psi_b) \quad (3.72)$$

The oscillating part of $I(x)$ is therefore, in complex notation,

$$I(x) \propto \psi_b^* \psi_a \quad (3.73)$$

and the total i.f. current is, to within a constant factor,

$$I = \int_{-\infty}^{+\infty} \psi_b^* \psi_a \, dx \quad (3.74)$$

assuming that the beams are located well within the area of the detector.

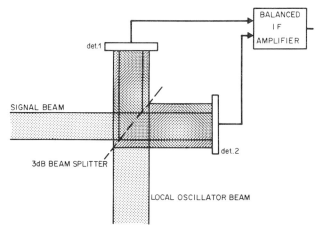

Fig. 3-8 Efficient mixing of two optical beams requires two detectors and a balanced i.f. amplifier.

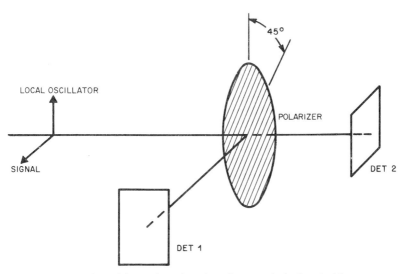

Fig. 3-9 Balanced heterodyne detection of cross-polarized optical beams.

The invariance relation (3.71) then tells us that the i.f. current generated by two optical beams in a heterodyne detector is independent of the location of the detector. This conclusion is at first surprising if we consider two optical beams crossing each other as in Fig. 3-7. One is tempted to think that the detection efficiency should be higher if the detector is located at the crossing point rather than far from it. This is not so, however, because at the crossing point the two beams have a strong *phase* mismatch. To be sure, far from the crossing point, the amplitude of one beam is small when the amplitude of the other beam is large and vice versa, but this mismatch is amplitude is compensated by an almost perfect match between the wavefronts.

This result rests on the assumption that the detector area covers both beams. Two uncoupled beams generate current if the detector covers only part of the beams, or, equivalently, if it has a nonuniform sensitivity. This is what is, in effect, accomplished in the balanced heterodyne mixer shown in Fig. 3-8. Because the i.f. currents generated on the two detectors have opposite phases, they would cancel out if they were simply added, in agreement with the general result discussed above. Because these currents are physically separated, they may be combined in a balanced i.f. amplifier. The balanced mixer is therefore an efficient method for mixing uncoupled optical beams.

The above result (3.74) for the scalar field can be generalized to the Maxwell field for arbitrary polarizations. The i.f. current density is proportional to the flux of $\mathbf{J}_{ab}$ given in (3.68), and the invariance condition proved above for the scalar field still applies. In particular, if two optical beams have, at some plane, orthogonal polarizations (for instance, circular clockwise and counterclockwise polarizations), the i.f. current is equal to zero and remains equal to zero for any location of the detector. Figure 3-9 shows a balanced mixer arrangement to mix two optical beams with orthogonal polarizations.

3.7 Transverse Coupling

The general conservation law $\nabla \cdot \mathbf{J}_{ab} = 0$ is used in this section to obtain the coupling between parallel open waveguides ("transverse" coupling).[5] Transverse coupling may be considered harmful when it takes place between fibers in a multichannel cable since it causes cross-talk. This coupling is useful, on the contrary, in integrated optics for the fabrication of directional couplers. A significant amount of coupling between open waveguides is observed only when the phase velocities of the uncoupled waves are almost equal. If the phase velocities are precisely equal and the losses can be neglected, the power incident on one waveguide is completely transferred to the other waveguide after a certain length and is subsequently transferred back to the first waveguide. A more complete discussion of that effect will be given in Chapter 5. Here, only general coupling formulas are derived. Let us consider a medium having only transverse anisotropy and uniform along the z-axis. If

$$(\mathbf{E}, \mathbf{H}) \equiv (E_z, \mathbf{E}_t, H_z, \mathbf{H}_t) \exp(ik_z z) \tag{3.75}$$

denotes a solution of the Maxwell equations, then

$$(\mathbf{E}^\dagger, \mathbf{H}^\dagger) \equiv (-E_z, \mathbf{E}_t, H_z, -\mathbf{H}_t) \exp(-ik_z z) \tag{3.76}$$

is also a solution of the Maxwell equations. In the above expressions, the subscript t refers to transverse components, and the arguments x, y have been omitted. This result can be verified by writing the Maxwell equations in rectangular coordinates and separating the axial and transverse components. The field $(\mathbf{E}^\dagger, \mathbf{H}^\dagger)$ describes a wave propagating in the medium at the same frequency as the field $(\mathbf{E}, \mathbf{H})$, but in the opposite direction. $(\mathbf{E}^\dagger, \mathbf{H}^\dagger)$ is called the adjoint field.

Let us now consider two open waveguides a and b uniform along the z-axis, and let the surface S in (3.20) be the surface $S_a + S_a' + C_a \, dz$ shown in Fig. 3-10. The field $(\mathbf{E}, \mathbf{H})$ is taken as the field $(\mathbf{E}_a, \mathbf{H}_a)$ of a trapped mode on waveguide a in the absence of waveguide b. The dependence of $(\mathbf{E}_a, \mathbf{H}_a)$ on z is denoted $\exp(ik_{za} z)$. The field $(\mathbf{E}^\dagger, \mathbf{H}^\dagger)$ is

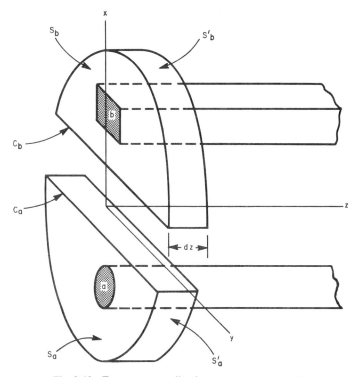

Fig. 3-10 Transverse coupling between open waveguides.

taken as the adjoint field of a trapped mode of the two coupled waveguides, with an $\exp(-ik_z z)$ dependence on z. Letting the spacing dz between S_a' and S_a tend to zero, (3.20) becomes

$$(ik_z - ik_{za})\int_{S_a}(\mathbf{E}^\dagger \times \mathbf{H}_a + \mathbf{H}^\dagger \times \mathbf{E}_a) \cdot d\mathbf{S}_a$$

$$= \int_{C_a}(\mathbf{E}^\dagger \times \mathbf{H}_a + \mathbf{H}^\dagger \times \mathbf{E}_a) \cdot d\mathbf{C}_a \qquad (3.77)$$

where $d\mathbf{C}_a$ is a vector in the transverse plane perpendicular to the contour C_a, pointing outward. Proceeding similarly for waveguide b, we obtain

$$(ik_z - ik_{zb})\int_{S_b}(\mathbf{E}^\dagger \times \mathbf{H}_b + \mathbf{H}^\dagger \times \mathbf{E}_b) \cdot d\mathbf{S}_b$$

$$= \int_{C_b}(\mathbf{E}^\dagger \times \mathbf{H}_b + \mathbf{H}^\dagger \times \mathbf{E}_b) \cdot d\mathbf{C}_b \qquad (3.78)$$

Because the coupling between the two waveguides is small, we can assume that the field $(\mathbf{E}, \mathbf{H})$ at plane $z = 0$ is the sum of the fields of the two waveguides, that is,

$$\mathbf{E} = \mathbf{E}_a + \mathbf{E}_b, \qquad \mathbf{H} = \mathbf{H}_a + \mathbf{H}_b \qquad (3.79)$$

the relation between $(\mathbf{E}, \mathbf{H})$ and $(\mathbf{E}^\dagger, \mathbf{H}^\dagger)$ being that given in (3.76). Similar relations hold for $\mathbf{E}_a$, $\mathbf{H}_a$ and $\mathbf{E}_b$, $\mathbf{H}_b$.

Let us substitute these expressions (3.79) in (3.77) and (3.78) and observe that the cross terms can be neglected on the lhs because $(\mathbf{E}_b, \mathbf{H}_b)$ is small when $(\mathbf{E}_a, \mathbf{H}_a)$ is large and vice versa. On the rhs, on the contrary, only the cross terms remain, as we can verify by applying the Lorentz reciprocity theorem to each waveguide independently. Multiplying together the lhs and rhs of (3.77) and (3.78), the desired equation for k_z is obtained in the form of a second-degree equation[5]

$$(ik_z - ik_{za})(ik_z - ik_{zb}) = \frac{c_a c_b}{P_a P_b} \qquad (3.80)$$

where

$$c_{a,b} = \tfrac{1}{4} \int_{C_{a,b}} (\mathbf{E}^\dagger_{b,a} \times \mathbf{H}_{a,b} + \mathbf{H}^\dagger_{b,a} \times \mathbf{E}_{a,b}) \cdot d C_{a,b} \qquad (3.81)$$

$$P_{a,b} = \tfrac{1}{4} \int_{S_{a,b}} (\mathbf{E}^\dagger_{a,b} \times \mathbf{H}_{a,b} + \mathbf{H}^\dagger_{a,b} \times \mathbf{E}_{a,b}) \cdot d S_{a,b} \qquad (3.82)$$

Because the coupling takes place only if $k_{za} \approx k_{zb}$, the coupling c_a (respectively, c_b) is independent of the choice of the contour C_a (respectively, C_b) as long as it surrounds only one waveguide. By choosing the two contours as coincident in the region where the fields of the two trapped modes have a significant intensity and using (3.20), we find $c_a = c_b$.

Thus the axial propagation constants of a system of two coupled open waveguides can be obtained from the sole knowledge of their normalized fields (in the absence of the other), along any line that separates them. It should be kept in mind that our result (3.80) is a perturbation formula applicable only when the coupling is small.

A similar derivation can be made for the parabolic wave equation with $\mathbf{J}$ given in (3.45) and $\psi^\dagger(x, z) = \psi(x, -z)$:

$$(k_z - k_{za})(k_z - k_{zb}) = \left\{ \tfrac{1}{2} \int_C [\psi_b(\partial \psi_a) - \psi_a(\partial \psi_b)] \cdot d C \right\}$$

$$\times \left\{ \int \psi_a^2 \, dS \int \psi_b^2 \, dS \right\}^{-1} \qquad (3.83a)$$

where ∂ denotes the transverse gradient. The contributions from terms in the integral over C in (3.83a) are approximately equal. This can be seen by taking for C a straight line located between the two waveguides, for instance, the x-axis, where k is a constant, and expressing $\psi_a(x)$ and $\psi_b(x)$ as Fourier integrals. Only the components of $\psi_a(x)$ and $\psi_b(x)$ of the spectrum that have the same k_x contribute to the coupling integral. Because these components have the same k_x and (almost) the same k_z, the quantities $\psi_a^{-1} \partial \psi_a / \partial y$ and $\psi_b^{-1} \partial \psi_b / \partial y$ are equal and opposite. If we define further normalized fields by dividing ψ by $(k \int_S \psi^2 \, dS)^{1/2}$, the coupling formula (3.83a) simplifies to

$$(k_z - k_{za})(k_z - k_{zb}) = \left[\int_C \psi_a \, \partial \psi_b \cdot \mathbf{d}C \right]^2 \equiv c^2 \qquad (3.83b)$$

The theory of coupling between two anisotropic surface waves can be found in Arnaud.[12] The approach is similar to that considered above.

This completes the derivation of the results that are most often needed in beam and fiber optics. In the next sections, other general properties of linear media are derived. For generality and conceptual clarity, we let the electric and magnetic inductions depend on both the electric and magnetic fields. These magnetoelectric media are not often encountered in practice. However, by giving consideration to more general media than is usually done, the significance of the results is, in fact, clarified.

3.8 Bianisotropic Media

The most general linear relation that may exist in electromagnetism between the inductions $\mathbf{D}$, $\mathbf{H}$ and the fields $\mathbf{E}$, $\mathbf{B}$ is a convolution in both space and time. For example,

$$\mathbf{D}(\mathbf{x}, t) = \int_{-\infty}^{t} dt' \int (d\mathbf{x}') \boldsymbol{\epsilon}(\mathbf{x}, t; \mathbf{x}', t') \mathbf{E}(\mathbf{x}', t') \qquad (3.84a)$$

$$\mathbf{H}(\mathbf{x}, t) = \int_{-\infty}^{t} dt' \int (d\mathbf{x}') \boldsymbol{\mu}^{-1}(\mathbf{x}, t; \mathbf{x}', t') \mathbf{B}(\mathbf{x}', t') \qquad (3.84b)$$

The restriction of the time integrals to $-\infty$, t is a consequence of the principle of causality. The volume integrals are assumed to extend over all space. However, in most cases, $\mathbf{D}(\mathbf{x}, t)$ (respectively, $\mathbf{H}$) depends only on the values of $\mathbf{E}(\mathbf{x}', t')$ (respectively, $\mathbf{B}$) for $\mathbf{x}'$ close to $\mathbf{x}$ and t' close to t. Equations (3.84) are still not the most general, because $\mathbf{D}$ may depend on $\mathbf{B}$ as well as on $\mathbf{E}$, and $\mathbf{H}$ may depend on $\mathbf{E}$ as well as on $\mathbf{B}$.[6] The coupling between electric and magnetic quantities may result, for instance, from the motion of the medium with respect to the observer through special relativity effects. It is also observed in some materials, called "magnetoelectrics," such as chromium oxide. Spatial dispersion (e. g., natural activity) is found in materials such as quartz. It is sometimes described by postulating a coupling between the electric and magnetic field quantities. This is because spatial dispersion implies that $\mathbf{D}$ depends on $\nabla \times \mathbf{E}$. This term, in turn, can be written $i\omega\mathbf{B}$ with the help of one of the Maxwell equations. This description is convenient, but it may obscure the difference between the coupling that results from the constitutive relations and the coupling that results from the field equations.

Without investigating further the physical origin of the electromagnetic coupling, let us write the Maxwell equations for a time-harmonic source [a factor $\exp(-i\omega t)$ is omitted]:

$$\nabla \times \mathbf{E} = i\omega\mathbf{B}, \qquad \nabla \times \mathbf{H} = -i\omega\mathbf{D} \tag{3.85}$$

and postulate the formal linear relation

$$\begin{bmatrix} i\omega\mathbf{D} \\ \mathbf{H} \end{bmatrix} = \mathbf{MF} \tag{3.86a}$$

$$\mathbf{F} \equiv \begin{bmatrix} \mathbf{E} \\ i\omega\mathbf{B} \end{bmatrix} \tag{3.86b}$$

where $\mathbf{M}$ denotes an arbitrary 6×6 matrix, which depends on the angular frequency ω. Spatial dispersion is neglected here.

A number of important properties of the solution of (3.85) can be derived in compact form using the $\mathbf{M}$ matrix. Let $\mathbf{F}_1(\omega_1)$ denote a solution of (3.85) in a medium characterized by $\mathbf{M}_1(\omega_1)$, and $\mathbf{F}_2(\omega_2)$ a solution of (3.85) in a medium characterized by $\mathbf{M}_2(\omega_2)$. By application of the vector identity in (3.19), we find for any volume V bounded by a surface S,

$$\int_S \mathbf{V}_2(\omega_2)\mathbf{I}_1(\omega_1) \, dS = \int_V \mathbf{F}_2(\omega_2)\mathbf{M}_1(\omega_1)\mathbf{F}_1(\omega_1) \, dV \tag{3.87}$$

For brevity, transposition signs on the first vectors are omitted. $\mathbf{V}$ denotes the tangential component of $\mathbf{E}$ on S, and $\mathbf{I} = \mathbf{H} \times \mathbf{n}$, where $\mathbf{n}$ denotes the unit vector normal to S pointing outward, is also tangent to S. Exchanging the indices 1 and 2 in (3.87) and transposing, we obtain

$$\int_S \mathbf{I}_2(\omega_2)\mathbf{V}_1(\omega_1) \, dS = \int_V \mathbf{F}_2(\omega_2)\tilde{\mathbf{M}}_2(\omega_2)\mathbf{F}_1(\omega_1) \, dV \tag{3.88}$$

Important theorems follow readily from (3.87) and (3.88).

3.9 Reciprocity and Orthogonality

Let us set $\omega_1 = \omega_2$ in (3.87) and (3.88), set $\mathbf{M}_1 = \tilde{\mathbf{M}}_2$, and subtract. We obtain a generalized form of the Lorentz reciprocity theorem

$$\int_S (\mathbf{V}_2 \mathbf{I}_1 - \mathbf{I}_2 \mathbf{V}_1) \, dS = 0 \tag{3.89}$$

A medium is reciprocal when $\mathbf{M}$ matrix is symmetrical ($\mathbf{M} = \tilde{\mathbf{M}}$).

Let us now set $\omega_2 = -\omega_1$ in (3.87) and (3.88) and add. Because in the time-dependent form of the Maxwell equations the fields are real quantities, changing ω to $-\omega$ in the field components is equivalent to complex conjugation. Equation (3.86) defining $\mathbf{M}$, on the other hand, shows that $\mathbf{M}(-\omega)$ is the complex conjugate of $\mathbf{M}(\omega)$. Therefore we have

$$\int_S (\mathbf{V}_2^* \mathbf{I}_1 + \mathbf{I}_2^* \mathbf{V}_1) \, dS = \int_V \mathbf{F}_2^* (\mathbf{M}_1 + \tilde{\mathbf{M}}_2^*) \mathbf{F}_1 \, dV \tag{3.90}$$

Assuming now that $\mathbf{M}_1 + \tilde{\mathbf{M}}_2^* = 0$, we obtain

$$\int_S (\mathbf{V}_2^* \mathbf{I}_1 + \mathbf{I}_2^* \mathbf{V}_1) \, dS = 0 \tag{3.91}$$

a generalization of (3.67). This relation can be used to demonstrate the invariance of the intermediate frequency current generated in a square-law detector by two beams having almost the same frequency, as we have seen before.

If we drop the subscripts 1 and 2 in (3.87) and (3.88), we obtain

$$\int_S (\mathbf{V}^* \mathbf{I} + \mathbf{I}^* \mathbf{V}) \, dS = \int_V \mathbf{F}^* (\mathbf{M} + \tilde{\mathbf{M}}^*) \mathbf{F} \, dV \tag{3.92}$$

Because the lhs of (3.92) is proportional to the time-average power flowing out of the volume V, we have the following characterizations of the medium:

$$\mathbf{M} + \tilde{\mathbf{M}}^* = 0: \quad \text{lossless medium}$$

$$\mathbf{M} + \tilde{\mathbf{M}}^* \text{ negative definite}: \quad \text{passive medium}$$

$$\mathbf{M} + \tilde{\mathbf{M}}^* \text{ positive definite}: \quad \text{active medium}$$

When $\mathbf{M} + \tilde{\mathbf{M}}^*$ is indefinite, the medium amplifies or attenuates depending on the source configuration.

3.10 Equality of the Group and Energy Velocity

Let us show now that the group velocity is equal to the energy velocity. We consider, for generality, a periodic medium.[6] The unit cell is defined by three vectors x_α, $\alpha = 1, 2, 3$, which need not be perpendicular to one another (see Fig. 3-11). The existence of waves satisfying a relation of the form

$$\mathbf{F}(\mathbf{x} + \mathbf{x}_\alpha) = \mathbf{F}(\mathbf{x}) \exp(i\phi_\alpha) \tag{3.93}$$

where the phase shifts ϕ_α, $\alpha = 1, 2, 3$, satisfy a dispersion equation

$$\omega = H(\phi_1, \phi_2, \phi_3) \tag{3.94}$$

is guaranteed by the Floquet theorem. The ϕ's are here assumed to be real numbers. Let us now apply (3.87) and (3.88) to the volume V of a cell, with $\omega_1 = \omega + d\omega$, $\omega_2 = -\omega$, add, and drop the subscripts. Using (3.93) and the fact that $\phi_\alpha(-\omega) = -\phi_\alpha(\omega)$, we obtain, to first order in $d\omega$,

$$\sum_{\alpha=1}^{3} d(i\phi_\alpha) \int_{S_\alpha} (\mathbf{V}^*\mathbf{I} + \mathbf{I}^*\mathbf{V}) \, dS_\alpha = \int_V \mathbf{F}^*(d\mathbf{M})\mathbf{F} \, dV \tag{3.95}$$

provided the medium is lossless: $\mathbf{M} + \tilde{\mathbf{M}}^* = \mathbf{0}$. In (3.95) S_3 denotes the parallelogram with sides x_1 and x_2. Similar definitions apply to S_1 and S_2. $d\mathbf{M}$ represents the variation of $\mathbf{M}$ resulting from the variation $d\omega$ of ω. If the medium has spatial dispersion, we need consider also the dependence of $\mathbf{M}$ on the ϕ's, as well as on ω.

Let us now define the power through the cell by

$$\mathbf{S} = \sum_{\alpha=1}^{3} \frac{1}{4} \mathbf{x}_\alpha \left[\int_{S_\alpha} (\mathbf{V}^*\mathbf{I} + \mathbf{I}^*\mathbf{V}) \, dS_\alpha - \int_V \mathbf{F}^* \frac{\partial \mathbf{M}}{\partial(i\phi_\alpha)} \mathbf{F} \, dV \right] \tag{3.96}$$

The volume integral in (3.96) is associated with spatial dispersion. It can be omitted for most media. The energy stored in the cell is defined by

$$W = \frac{1}{4} \int_V \mathbf{F}^* \frac{\partial \mathbf{M}}{\partial(i\omega)} \mathbf{F} \, dV \tag{3.97}$$

and the wave vector $\mathbf{k}$ by

$$\mathbf{k} \cdot \mathbf{x}_\alpha = \phi_\alpha, \qquad \alpha = 1, 2, 3 \tag{3.98}$$

Introducing these definitions, relation (3.95) becomes

$$\mathbf{S} \cdot d\mathbf{k} = W \, d\omega \tag{3.99}$$

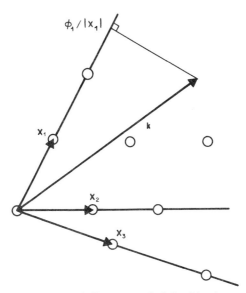

Fig. 3-11 The periodicity of a periodic structure is defined by three vectors x_1, x_2, and x_3 not necessarily perpendicular to one another. The orthogonal projections of the wave vector **k** on x_1, x_2, and x_3 are $\phi_1/|x_1|$, $\phi_2/|x_2|$, and $\phi_3/|x_3|$, where ϕ_1, ϕ_2, and ϕ_3 denote the phase shifts.

Because the direction of $d\mathbf{k}$ is arbitrary, (3.99) shows that the group velocity $\partial\omega/\partial\mathbf{k}$ is equal to the energy velocity $\mathbf{S}/W$. This is a general result for linear waves.

A homogeneous medium can, of course, be considered a special case of a periodic medium. By neglecting spatial dispersion, (3.99) becomes

$$\mathbf{u} \equiv \frac{\partial\omega}{\partial\mathbf{k}} = \frac{\mathbf{E}^* \times \mathbf{H} + \mathbf{E} \times \mathbf{H}^*}{\mathbf{F}^*[\partial\mathbf{M}/\partial(i\omega)]\mathbf{F}} \tag{3.100}$$

Another special case of interest is a periodic array of open waveguides, perhaps an array of glass fibers. In the limit where the period tends to infinity, we are left with just one waveguide. The above theorem thus says that the group velocity of trapped modes is equal to the ratio of the energy flow through the waveguide to the stored energy per unit length.

Similar results can be derived for the complex field. Let $\mathbf{E}$, $\mathbf{H}$, ϕ_α be a solution of the Maxwell equations for a material matrix $\mathbf{M}$, and $\mathbf{E}^\dagger$, $\mathbf{H}^\dagger$, $-\phi_\alpha$ be a solution of the Maxwell equation for $\mathbf{M}$ changed to $\tilde{\mathbf{M}}$. Proceeding as before, we obtain

$$\frac{\partial\omega}{\partial\mathbf{k}} = \frac{\mathbf{E}^\dagger \times \mathbf{H} + \mathbf{H}^\dagger \times \mathbf{E}}{\mathbf{F}^\dagger[\partial\mathbf{M}/\partial(i\omega)]\mathbf{F}} = \frac{d\mathbf{x}}{dt} \tag{3.101}$$

where **F** is defined as in (3.86b). In the present formulation, the material need not be assumed lossless. $\mathbf{x}(t)$ in (3.101) may therefore be complex valued.

3.11 Moving Media

In this section, we restrict ourselves to nondispersive media. When the dispersion of a dielectric medium is negligible, the polarization depends only on the value of the electric field at the same time and point. A dielectric medium can be pictured as a collection of dipoles. Each dipole consists of two charges of opposite signs attracting each other. In the absence of an external field, the two charges coincide, and the polarization is equal to zero. If the electric field is applied sufficiently slowly compared to the natural frequency of oscillation of the charges, the separation between the two charges, and therefore the polarization, is determined solely by the strength of the field at the same time. If the field variations are not slow, the complete equations of motion have to be solved. They involve all the previous values of the field. The stronger the spring constant, the shorter is the "memory" of the pair of charges. Thus if we neglect dispersion, it is not necessary to follow a particular piece of material along its trajectory. If the source is time harmonic and the medium is stationary, with $d\mathbf{x}/dt$, $d^2\mathbf{x}/dt^2$, etc., independent of t at any $\mathbf{x}$, the relevant field is time harmonic with the same angular frequency as that of the source. However, the strengths of the fields that are felt by the passing piece of material are not the fields **E**, **B**, defined in the rest frame. They are the transformed fields **E'**, **B'** in the medium frame obtained from **E**, **B** by a Lorentz transformation. Once the inductions **D'**, **H'** in the medium frame have been obtained, the inductions **D**, **H** in the rest frame are obtained from a second Lorentz transformation. Thus **D**, **H** can be obtained from **E**, **B**. It should be noted that what we call "medium frame" for short is, in fact, an inertial frame momentarily at rest with the medium. The two are different if the medium is accelerating. We restrict ourselves to small accelerations.

Let us assume that the medium is characterized in its own frame by a material matrix **M'** and show that if the medium is moving at a constant velocity **u**, the material matrix seen by a fixed observer has the form

$$\mathbf{M} = \tilde{\mathbf{T}}^{-1}(\mathbf{u})\mathbf{M}'\mathbf{T}(\mathbf{u}) \tag{3.102}$$

where **T** satisfies, for obvious reasons of consistency, the relation

$$\mathbf{T}(\mathbf{u})\mathbf{T}(-\mathbf{u}) = \mathbf{1} \tag{3.103a}$$

Furthermore, we have

$$T(u)T^*(u) = 1 \tag{3.103b}$$

The transformation of the E, B field follows from the theory of special relativity.[78] Setting, for brevity, the velocity of light equal to unity, $c = 1$, the field in the medium frame (primed quantities) is

$$E' = UE + \gamma u \times B, \qquad B' = UB - \gamma u \times E \tag{3.104}$$

where

$$U \equiv 1 + (1 - \gamma)\hat{u} \times \hat{u} \times \tag{3.105}$$

$$\gamma \equiv (1 - u^2)^{-1/2}, \qquad u^2 \equiv u \cdot u, \qquad \hat{u} \equiv \frac{u}{u} \tag{3.106}$$

The same relations are applicable to the pair D, H:

$$D' = UD + \gamma u \times H, \qquad H' = UH - \gamma u \times D \tag{3.107}$$

These relations can be written in matrix form

$$\begin{bmatrix} E' \\ i\omega B' \end{bmatrix} = T(u) \begin{bmatrix} E \\ i\omega B \end{bmatrix}$$

$$\begin{bmatrix} i\omega D' \\ H' \end{bmatrix} = \tilde{T}(u) \begin{bmatrix} i\omega D \\ H \end{bmatrix} \tag{3.108}$$

if we define

$$T(u) \equiv \begin{bmatrix} U & (\gamma/i\omega)u \times \\ -i\omega\gamma u \times & U \end{bmatrix} \tag{3.109}$$

Note that $(u \times)^\sim = -u \times$ and $\tilde{U} = U$. Introducing now the definitions of M and M' from (3.86)

$$\begin{bmatrix} i\omega D \\ H \end{bmatrix} = M \begin{bmatrix} E \\ i\omega B \end{bmatrix}, \qquad \begin{bmatrix} i\omega D' \\ H' \end{bmatrix} = M' \begin{bmatrix} E' \\ i\omega B' \end{bmatrix} \tag{3.110}$$

[§] Note the identity $\hat{u}\hat{u} \cdot \equiv 1 + \hat{u} \times \hat{u} \times$, which is a special case of the identity $A \times (B \times C) \equiv B(A \cdot C) - C(A \cdot B)$. The medium velocity u should not be confused with the group velocity of a pulse, denoted u in other sections.

the relation (3.102) between **M** and **M'** given above follows. The consistency relation (3.103a) can be proved directly from (3.109), using (3.106) and the identity

$$\hat{\mathbf{u}} \times \hat{\mathbf{u}} \times \hat{\mathbf{u}} \times \hat{\mathbf{u}} \times = -\hat{\mathbf{u}} \times \hat{\mathbf{u}} \times \qquad (3.111)$$

Relation (3.103b) follows by inspection because changing the sign of **u** in (3.109) amounts to a complex conjugation. This completes the proof of the general results in (3.102) and (3.103).

Relation (3.102) shows that if a medium is reciprocal in its own frame (**M'** = **M̃'**), the material matrix seen by a fixed observer is changed to its transpose when the velocity is reversed. Two dielectric rods moving at the same velocity in opposite directions, for instance, are characterized by matrices that are the transposes of each other. The Lorentz reciprocity theorem in (3.20) is applicable to the field (**E**, **H**) relative to a rod moving in one direction and the fields (**E**†, **H**†) for the direction of motion of the rod reversed. This gives a concrete example of what is meant by "adjoint" fields.

The law of causality implies that a nondispersive medium is also lossless. Let us verify that a moving nondispersive lossless medium appears lossless to a fixed observer. We assume that **M'** + **M̃'*** = **0**. Let us evaluate **M** + **M̃***. We have, as expected, using (3.102) and (3.103),

$$\mathbf{M} + \tilde{\mathbf{M}}^* = \tilde{\mathbf{T}}^{-1}(\mathbf{M'} + \tilde{\mathbf{M'}}^*)\mathbf{T} = \mathbf{0} \qquad (3.112)$$

Equation (3.112) says, for instance, that the total energy flow entering a dispersionless lossless torus spinning about its axis is equal to zero. However, if the torus material is slightly conductive, and therefore lossy in its own frame, it may exhibit gain above a certain speed of rotation; that is, there may be a net *outflow* of energy. This energy may originate from the torque that needs to be exerted on the torus to maintain its speed of rotation.

3.12 The Surface of Wave Vectors

We now assume that the medium is both time invariant and homogeneous; that is, the material matrix does not depend on **x** or *t*. If the fields have the form $\mathbf{E}(\mathbf{x}) = \mathbf{E}_0 \exp(i\mathbf{k} \cdot \mathbf{x})$, the Maxwell equations (3.85) become

$$i\mathbf{k} \times \mathbf{E}_0 = i\omega\mathbf{B}_0 \qquad (3.113a)$$

$$i\mathbf{k} \times \mathbf{H}_0 = -i\omega\mathbf{D}_0 \qquad (3.113b)$$

where

$$\begin{bmatrix} i\omega\mathbf{D}_0 \\ \mathbf{H}_0 \end{bmatrix} = \mathbf{M} \begin{bmatrix} \mathbf{E}_0 \\ i\omega\mathbf{B}_0 \end{bmatrix} \qquad (3.114)$$

Equations (3.113) together with (3.114) are a system of equations that admits nontrivial solutions only if

$$\det\left\{ \begin{bmatrix} 1 & i\mathbf{k} \times \end{bmatrix} \mathbf{M} \begin{bmatrix} 1 \\ i\mathbf{k} \times \end{bmatrix} \right\} = 0 \qquad (3.115)$$

where $\mathbf{k} \times$ is defined in (3.24) an $\mathbf{1}$ denotes the 3×3 unit matrix. $\mathbf{M}$ being a function of ω, this (scalar) algebraic equation relates ω and the components of the wave vector $\mathbf{k}$.

To obtain (3.115), note that (3.113) can be written in matrix form

$$\begin{bmatrix} 1 & i\mathbf{k} \times \end{bmatrix} \begin{bmatrix} i\omega\mathbf{D}_0 \\ \mathbf{H}_0 \end{bmatrix} = \begin{bmatrix} 1 & i\mathbf{k} \times \end{bmatrix} \mathbf{M} \begin{bmatrix} \mathbf{E}_0 \\ i\omega\mathbf{B}_0 \end{bmatrix}$$

$$= \begin{bmatrix} 1 & i\mathbf{k} \times \end{bmatrix} \mathbf{M} \begin{bmatrix} 1 \\ i\mathbf{k} \times \end{bmatrix} \mathbf{E}_0 = \mathbf{0} \qquad (3.116)$$

This system of three homogeneous linear equations in E_{0x}, E_{0y}, and E_{0z} has nontrivial solutions ($\mathbf{E}_0 \neq 0$) only if the determinant of the unknown is equal to zero. Thus (3.115) follows.

It is interesting that the dispersion relation (3.115) remains the same if $\mathbf{k}$ is changed to $-\mathbf{k}$ and $\mathbf{M}$ is changed to its transpose $\tilde{\mathbf{M}}$. This means that the wavenumbers are equal in magnitude for waves propagating in opposite directions in media characterized by $\mathbf{M}$ and $\tilde{\mathbf{M}}$, respectively. This conclusion would also follow from the Lorentz reciprocity theorem discussed in Section 3.9. Note that the symmetry of the surface of wave vector is a necessary but not a sufficient condition for reciprocity. For a magnetized plasma, for instance, the surface of wave vector is symmetrical, but the medium is nonreciprocal. This is perhaps best seen by expressing $\mathbf{M}$ in terms of four 3×3 submatrices $\mathbf{M}_{11}$, $\mathbf{M}_{12}$, $\mathbf{M}_{21}$, and $\mathbf{M}_{22}$. Equation (3.115) becomes

$$\det\{\mathbf{M}_{11} + \mathbf{M}_{12}(i\mathbf{k} \times) + (i\mathbf{k} \times)\mathbf{M}_{21} + (i\mathbf{k} \times)\mathbf{M}_{22}(i\mathbf{k} \times)\} = 0$$

$$(3.117)$$

If $M_{12} = M_{21} = 0$, the surface of wavenormal remains the same when k is changed to $-k$, whether M_{11} and M_{22} are symmetrical or not. Thus M need not always be symmetrical (reciprocal medium) for the surface of the wave vector to be symmetrical. Equation (3.117) being of degree six, we expect that, along any given direction, six solutions can be found for the magnitude of k. In fact, provided M is independent of k, that is, provided spatial dispersion can be neglected, only four such waves exist. This result will be proved most conveniently in Section 3.15 when we discuss the transverse form of the Maxwell equations. In free space, the four waves are purely transverse, e.g., circularly polarized clockwise and counterclockwise, forward and backward. The absence of longitudinal waves is a consequence of the zero rest mass of the photon.[§] In media with spatial dispersion, however, we may be able to find all six waves, as is the case in acoustics, where both longitudinal waves and shear waves propagate.

Let us specialize now (3.117) to biaxial crystals, with $D = \epsilon E$, and $H = \mu_0^{-1} B$, where ϵ is symmetrical: $\epsilon = \tilde{\epsilon}$. In that case, $M_{12} = M_{21} = 0$, $M_{11} = i\omega\epsilon$, $M_{22} = 1/i\omega\mu_0$. Equation (3.117) becomes

$$\det\{\omega^2 \epsilon \mu_0 + k \times k \times \} = 0 \qquad (3.118)$$

The surface described by (3.118) and shown in Fig. 3-12b is discussed in great detail in Born and Wolf.[2] We shall only observe here that the k-surface is composed of two shells, corresponding to the two states of polarization. The name "biaxial" comes from the fact that the two shells are in contact for two directions in space (and the opposite directions since the surface is symmetrical). Along these directions, the k-vectors have the same magnitude for the two polarizations.[2]

Let us specialize our results one step further and assume that two of the eigenvalues of the permittivity matrix are equal. The medium is then called uniaxial because the two previously defined axes coalesce. Assuming that ϵ is real, we can always, with a proper orientation of coordinate system, write ϵ in the diagonal form

$$\epsilon = \begin{bmatrix} \epsilon & 0 & 0 \\ 0 & \epsilon & 0 \\ 0 & 0 & \epsilon' \end{bmatrix} \qquad (3.119)$$

[§] Laboratory tests of the Coulomb law have shown that the rest mass of the photon (which would entail a departure from the $1/r^2$ force law) does not exceed 10^{-47} g. This mass m corresponds to a Compton wavelength $(\hbar/mc)$ of more than 100,000 km. Much lower upper bounds for the photon mass have been conjectured on the basis of cosmological arguments (see Goldhaber *et al., Rev. Mod. Phys.* (July 1971), 227). Thus, we can safely assume that the rest mass of the photon is equal to zero. We can, however, attribute to the photon a moving mass $\hbar\omega/c^2$.

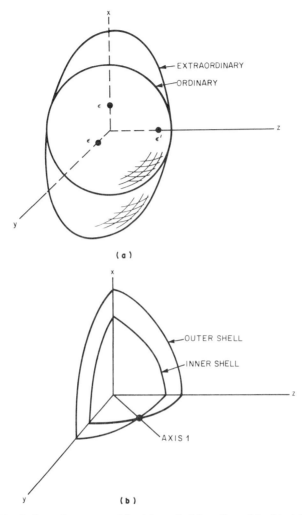

Fig. 3-12   Surface of wavenormal for (a) a uniaxial medium, (b) a biaxial medium.

where ϵ and ϵ' are real numbers. For such uniaxial media, one of the two shells is spherical and corresponds to the "ordinary" wave. For that particular state of polarization, the medium is equivalent to an isotropic medium with $k = \omega(\epsilon\mu_0)^{1/2}$. The physical reason for the existence of the ordinary wave is that for any direction of the wave vector **k**, it is possible to find a direction for **D** in the xy plane that is perpendicular to **k**. In that xy plane, the permittivity, according to (3.119), behaves as a scalar quantity, and we have simply $\mathbf{D} = \epsilon\mathbf{E}$. The other state of polarization corresponds to the so-called extraordinary wave. The surface of wave vector is an ellipsoid of revolution with axis z. It is oblate or prolate depending on whether $\epsilon' > \epsilon$ or $\epsilon' < \epsilon$, respectively (see Fig. 3-12a).

Let us now go back to the general case and consider the fields associated with a plane wave. Equations (3.113), from which all the previous results have been derived, show that $\mathbf{B}_0$ and $\mathbf{D}_0$ are perpendicular to the wave vector $\mathbf{k}$. Thus it is clear that the $\mathbf{k}$-vector has the direction of $\mathbf{D}_0 \times \mathbf{B}_0$. We shall show that we have, in fact,[8]

$$i\mathbf{k} = \frac{i\omega\mathbf{D}_0 \times i\omega\mathbf{B}_0}{\mathbf{E}_0 \cdot i\omega\mathbf{D}_0} \tag{3.120}$$

To prove this result, we first evaluate

$$\mathbf{E}_0 \cdot i\omega\mathbf{D}_0 = -\mathbf{E}_0 \cdot (i\mathbf{k} \times \mathbf{H}_0) \tag{3.121a}$$

and

$$\mathbf{H}_0 \cdot i\omega\mathbf{B}_0 = \mathbf{H}_0 \cdot (i\mathbf{k} \times \mathbf{E}_0) \tag{3.121b}$$

Using the property of the mixed product $\mathbf{a} \cdot (\mathbf{b} \times \mathbf{c}) = \mathbf{b} \cdot (\mathbf{c} \times \mathbf{a})$, it follows from (3.121) that

$$\mathbf{E}_0 \cdot i\omega\mathbf{D}_0 = \mathbf{H}_0 \cdot i\omega\mathbf{B}_0 \tag{3.122}$$

Next let us form the vector product of a vector $\mathbf{A}$ defined as

$$\mathbf{A} \equiv (\mathbf{E}_0 \cdot i\omega\mathbf{D}_0)i\mathbf{k} - (i\omega\mathbf{D}_0 \times i\omega\mathbf{B}_0)$$

and $\mathbf{E}_0$. We obtain

$$\mathbf{A} \times \mathbf{E}_0 = (\mathbf{E}_0 \cdot i\omega\mathbf{D}_0)(i\mathbf{k} \times \mathbf{E}_0) + \mathbf{E}_0 \times (i\omega\mathbf{D}_0 \times i\omega\mathbf{B}_0)$$

$$= (\mathbf{E}_0 \cdot i\omega\mathbf{D}_0)i\omega\mathbf{B}_0 + i\omega\mathbf{D}_0(\mathbf{E}_0 \cdot i\omega\mathbf{B}_0) - i\omega\mathbf{B}_0(\mathbf{E}_0 \cdot i\omega\mathbf{D}_0) = 0 \tag{3.123}$$

where we have used explicitly (3.113a) and the fact that $\mathbf{E}_0 \cdot \mathbf{B}_0 = 0$. We have also used the expression of the double-vector product $\mathbf{a} \times (\mathbf{b} \times \mathbf{c}) = \mathbf{b}(\mathbf{a} \cdot \mathbf{c}) - \mathbf{c}(\mathbf{a} \cdot \mathbf{b})$. Thus the vector $\mathbf{A}$ defined above has the direction of $\mathbf{E}_0$. A similar calculation can be made by multiplying $\mathbf{A}$ by $\mathbf{H}_0$ to show that $\mathbf{A}$ has also the direction of $\mathbf{H}_0$, this time using (3.113b). Since $\mathbf{E}_0$ and $\mathbf{H}_0$ have different directions, it follows that $\mathbf{A} = 0$. This proves that $\mathbf{k}$ has the form (3.120), which can be written alternatively

$$\frac{\mathbf{k}}{\omega} = \frac{\mathbf{D}_0 \times \mathbf{B}_0}{\mathbf{E}_0 \cdot \mathbf{D}_0} \tag{3.124}$$

3.13 Nondispersive Lossless Media

When a medium is nondispersive, the material matrix $\mathbf{M}$ is not arbitrary. For a lossless nondispersive dielectric crystal, in particular, the permittivity ϵ must be symmetrical. The generalization of this restriction to bianisotropic media is *not* that $\mathbf{M}$ be symmetrical, as one may think at first, but that the matrix that relates $\mathbf{D}$, $\mathbf{B}$ to $\mathbf{E}$, $\mathbf{H}$ be symmetrical. This matrix will be denoted $\mathbf{L}$. To prove this result, let us go back to the time-dependent Maxwell equations

$$\nabla \times \mathbf{E} = -\frac{\partial \mathbf{B}}{\partial t}, \qquad \nabla \times \mathbf{H} = \frac{\partial \mathbf{D}}{\partial t} \tag{3.125}$$

Using the vector identity

$$\nabla \cdot (\mathbf{a} \times \mathbf{b}) = \mathbf{b} \cdot \nabla \times \mathbf{a} - \mathbf{a} \cdot \nabla \times \mathbf{b} \tag{3.126}$$

we obtain

$$\nabla \cdot (\mathbf{E} \times \mathbf{H}) + \mathbf{H} \cdot \frac{\partial \mathbf{B}}{\partial t} + \mathbf{E} \cdot \frac{\partial \mathbf{D}}{\partial t} = 0 \tag{3.127}$$

In order that $\mathbf{E} \times \mathbf{H}$ be interpreted as the power density (the medium being assumed free of spatial dispersion), and because of conservation of power, the last two terms in (3.127) must be of the form $\partial W / \partial t$, where W denotes the energy density. Assuming a linear local relation of the form

$$\begin{bmatrix} \mathbf{D} \\ \mathbf{B} \end{bmatrix} = \mathbf{L} \begin{bmatrix} \mathbf{E} \\ \mathbf{H} \end{bmatrix} \equiv \begin{bmatrix} \epsilon & \xi \\ \zeta & \mu \end{bmatrix} \begin{bmatrix} \mathbf{E} \\ \mathbf{H} \end{bmatrix} \tag{3.128}$$

between $\mathbf{D}$, $\mathbf{B}$ and $\mathbf{E}$, $\mathbf{H}$, we have

$$\mathbf{H} \cdot \frac{\partial \mathbf{B}}{\partial t} + \mathbf{E} \cdot \frac{\partial \mathbf{D}}{\partial t} = \begin{bmatrix} \mathbf{E} \\ \mathbf{H} \end{bmatrix} \cdot \mathbf{L} \frac{\partial}{\partial t} \begin{bmatrix} \mathbf{E} \\ \mathbf{H} \end{bmatrix}$$

$$= \frac{\partial}{\partial t} \frac{1}{2} \begin{bmatrix} \mathbf{E} \\ \mathbf{H} \end{bmatrix} \cdot \mathbf{L} \begin{bmatrix} \mathbf{E} \\ \mathbf{H} \end{bmatrix} = \frac{\partial W}{\partial t} \tag{3.129}$$

provided $\mathbf{L}$ is symmetrical: $\mathbf{L} = \tilde{\mathbf{L}}$.[9] In particular, if $\mathbf{L}$ is block diagonal, the permittivity ϵ and the permeability μ must be symmetrical. Clearly the symmetry of **L** imposes some restrictions on the **M** matrix previously defined. **M**, however, must be symmetrical only if it is block diagonal.

By comparing the definition (3.128) of **L** to the definition (3.114) of **M**, we obtain with some matrix algebra

$$\mathbf{M} = \begin{bmatrix} i\omega(\epsilon - \xi\mu^{-1}\zeta) & \xi\mu^{-1} \\ -\mu^{-1}\zeta & (i\omega\mu)^{-1} \end{bmatrix} \tag{3.130}$$

Equation (3.130) shows that transposition of **M** amounts to the following transformation for **L**:

$$\mathbf{L} \equiv \begin{bmatrix} \epsilon & \xi \\ \zeta & \mu \end{bmatrix} \rightarrow \begin{bmatrix} \tilde{\epsilon} & -\tilde{\zeta} \\ -\tilde{\xi} & \tilde{\mu} \end{bmatrix} \tag{3.131}$$

Note further that the transformations

$$\mathbf{E} \rightleftharpoons i\omega\mathbf{D}, \qquad \mathbf{H} \rightleftharpoons i\omega\mathbf{B} \tag{3.132}$$

which change **M** to its inverse, change $i\omega\mathbf{L}$ to $(i\omega\mathbf{L})^{-1}$. These observations will be useful in the next section.

It can be shown also that the symmetry of the **L** matrix is preserved in a Lorentz transformation. This is to be expected since a moving nondispersive medium remains nondispersive for a fixed observer.

Another important property of nondispersive media is the following: If **E**, **H**, **k** denotes a plane wave solution of the Maxwell equations in a homogeneous medium with material matrix **M**, then **E**, $-\mathbf{H}$, $-\mathbf{k}$ is a plane wave solution of the Maxwell equations in the medium with material matrix $\tilde{\mathbf{M}}$. The proof will be left as a problem. Thus, for nondispersive media, we can write the group velocity defined in (3.101), setting $\mathbf{E}^\dagger = \mathbf{E}_0$ and $\mathbf{H}^\dagger = -\mathbf{H}_0$

$$\mathbf{u} \equiv \frac{\partial\omega}{\partial\mathbf{k}} = \frac{\mathbf{E}_0 \times \mathbf{H}_0}{\mathbf{E}_0 \cdot \mathbf{D}_0} \tag{3.133}$$

This very simple form is obtained in a different way in the next section, where ray vectors are discussed.

3.14 The Surface of Ray Vectors

We assume again in this section that the medium is nondispersive and define a ray vector **s** by

$$\frac{\mathbf{s}}{i} = \frac{\mathbf{E}_0 \times \mathbf{H}_0}{\mathbf{E}_0 \cdot i\omega\mathbf{D}_0} \tag{3.134}$$

or, equivalently,

$$s\omega = \frac{\mathbf{E}_0 \times \mathbf{H}_0}{\mathbf{E}_0 \cdot \mathbf{D}_0} \tag{3.135}$$

and derive a number of properties for s. Let us first show that

$$\frac{\mathbf{s}}{i} \times i\omega\mathbf{D}_0 = \mathbf{H}_0, \qquad \frac{\mathbf{s}}{i} \times i\omega\mathbf{B}_0 = -\mathbf{E}_0 \tag{3.136}$$

These relations are easily verified by substituting definition (3.134) of s in (3.136), using the vector identity (given in the previous section) for $\mathbf{a} \times (\mathbf{b} \times \mathbf{c})$, and the fact that, from the Maxwell equations, $\mathbf{D}_0$ is perpendicular to $\mathbf{H}_0$ and $\mathbf{E}_0$ to $\mathbf{B}_0$.

Comparison of (3.136) and (3.113) show that any relation obeyed by $i\mathbf{k}$ is obeyed as well by $\mathbf{s}/i$, provided the substitutions

$$\mathbf{E}_0 \rightleftharpoons i\omega\mathbf{D}_0, \qquad \mathbf{H}_0 \rightleftharpoons i\omega\mathbf{B}_0 \tag{3.137}$$

are made.

The definition (3.134) for $\mathbf{s}/i$, in particular, follows from (3.120) if the above substitutions are introduced. The surface of ray vector is obtained readily if we note that, with the substitution (3.137) introduced in (3.114), the material matrix $\mathbf{M}$ is changed to its inverse. Thus by changing $\mathbf{M}$ to $\mathbf{M}^{-1}$ and $i\mathbf{k}$ to $\mathbf{s}/i$ in (3.115), we obtain the equation defining the surface of ray vectors

$$\det\left\{ \left[\mathbf{1} \quad \frac{\mathbf{s}}{i} \times \right] \mathbf{M}^{-1} \left[\begin{array}{c} \mathbf{1} \\ (\mathbf{s}/i) \times \end{array} \right] \right\} = 0 \tag{3.138}$$

which, in the special case of a crystal of permittivity $\boldsymbol{\epsilon}$, reduces to

$$\det\left\{ (\omega^2\mu_0\boldsymbol{\epsilon})^{-1} + \mathbf{s} \times \mathbf{s} \times \right\} = 0 \tag{3.139}$$

This equation can be compared to the equation defining the surface of wave vector (3.118).

Let us prove now the important reciprocity relation

$$\mathbf{k} \cdot \mathbf{s} = 1 \tag{3.140}$$

This result is readily obtained by introducing (3.120) and (3.134) in (3.140) and making use of (3.122) and of the vector identity

$$(\mathbf{a} \times \mathbf{b}) \cdot (\mathbf{c} \times \mathbf{d}) = (\mathbf{a} \cdot \mathbf{c})(\mathbf{b} \cdot \mathbf{d}) - (\mathbf{a} \cdot \mathbf{d})(\mathbf{b} \cdot \mathbf{c}) \tag{3.141}$$

The assumption made before that the medium is nondispersive is now needed to prove that the ray vector s is perpendicular to the surface of wave vector. Let us show that, for any $d\mathbf{k}$ consistent with (3.115), we have

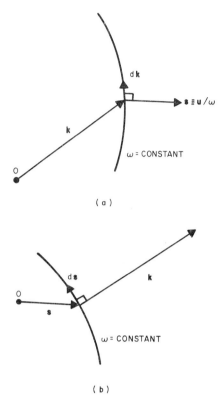

Fig. 3-13 (a) The tip of the wave vector **k** describes the surface of wave vector. The ray vector is perpendicular to that surface. (b) Conversely, the wave vector **k** is perpendicular to the surface of ray vector described by the tip of the ray vector **s**. The relation **k** · **s** = 1 holds.

$d\mathbf{k}$ · **s** = 0 (see Fig. 3-13). Instead of using directly the relation (3.115), we find it more convenient to go back to the original equations

$$i\mathbf{k} \times \mathbf{E}_0 = i\omega\mathbf{B}_0, \qquad i\mathbf{k} \times \mathbf{H}_0 = -i\omega\mathbf{D}_0 \qquad (3.142)$$

For generality, we allow ω to have a variation $d\omega$. By differentiation of (3.142), we obtain

$$d(i\mathbf{k}) \times \mathbf{E}_0 + i\mathbf{k} \times d\mathbf{E}_0 = d(i\omega)\mathbf{B}_0 + i\omega \, d\mathbf{B}_0$$

$$d(i\mathbf{k}) \times \mathbf{H}_0 + i\mathbf{k} \times d\mathbf{H}_0 = -d(i\omega)\mathbf{D}_0 - i\omega \, d\mathbf{D}_0 \qquad (3.143)$$

Let us multiply scalarly the first equation by $\mathbf{H}_0$, the second by $\mathbf{E}_0$, and subtract. Furthermore, let us use the cyclic property of the mixed products, and let us express $i\mathbf{k} \times \mathbf{H}_0$ and $i\mathbf{k} \times \mathbf{E}_0$ in terms of $\mathbf{B}_0$ and $\mathbf{D}_0$, respectively, according to (3.142). We obtain

$$2d(i\mathbf{k}) \cdot (\mathbf{E}_0 \times \mathbf{H}_0) + i\omega(d\mathbf{E}_0 \cdot \mathbf{D}_0 + d\mathbf{H}_0 \cdot \mathbf{B}_0)$$

$$= d(i\omega)(\mathbf{H}_0 \cdot \mathbf{B}_0 + \mathbf{E}_0 \cdot \mathbf{D}_0) + i\omega(d\mathbf{B}_0 \cdot \mathbf{H}_0 + d\mathbf{D}_0 \cdot \mathbf{E}_0) \quad (3.144)$$

Introducing the **L** matrix defined before in (3.128), we can write this equation as

$$2d(i\mathbf{k}) \cdot (\mathbf{E}_0 \times \mathbf{H}_0) = i\omega[d(\mathbf{L}\phi) \cdot \phi - d\phi \cdot \mathbf{L}\phi]$$

$$+ d(i\omega)(\mathbf{H}_0 \cdot \mathbf{B}_0 + \mathbf{E}_0 \cdot \mathbf{D}_0) \quad (3.145)$$

where we have defined

$$\phi \equiv \begin{bmatrix} \mathbf{E} \\ \mathbf{H} \end{bmatrix} \quad (3.146a)$$

The term in brackets on the rhs of (3.145) is equal to zero because **L** is symmetrical and independent of ω for nondispersive media, as we have seen in Section 3.13. Thus we have proved that, for nondispersive media, the vector **s**, which has the direction of $\mathbf{E}_0 \times \mathbf{H}_0$, is perpendicular to the surface of wave vector (taken at a fixed value of ω: $d\omega = 0$). Furthermore, the group velocity is, from (3.145),

$$\frac{d\mathbf{x}}{dt} \equiv \mathbf{u} = \frac{\partial \omega}{\partial \mathbf{k}} = \frac{\mathbf{E}_0 \times \mathbf{H}_0}{\frac{1}{2}(\mathbf{E}_0 \cdot \mathbf{D}_0 + \mathbf{H}_0 \cdot \mathbf{B}_0)} = \mathbf{s}\omega \quad (3.146b)$$

Thus (3.140) can be written $(d\mathbf{x}/dt) \cdot \mathbf{k} - \omega = 0$, or else $(d\mathbf{X}/d\sigma) \cdot \mathbf{K} = 0$, where $\mathbf{X} \equiv (\mathbf{x}, ict)$, $\mathbf{K} \equiv (\mathbf{k}, i\omega/c)$, and σ is an arbitrary parameter. It is a general property of nondispersive media that the 4-velocity is *perpendicular* to the 4-wave vector.

Reciprocally, the wave vector **k** is perpendicular to the surface of ray vector, described by the tip of the **s** vector. This follows from the relation $\mathbf{k} \cdot \mathbf{s} = 1$ in (3.140), which can be differentiated to give $d\mathbf{k} \cdot \mathbf{s} + \mathbf{k} \cdot d\mathbf{s} = 0$, and the result just obtained that $d\mathbf{k} \cdot \mathbf{s} = 0$.

In terms of the **L** matrix, the surface of wave vector can be written directly from (3.113) and (3.128)

$$\det\left\{ \begin{bmatrix} \mathbf{0} & i\mathbf{k} \times \\ -i\mathbf{k} \times & \mathbf{0} \end{bmatrix} + i\omega\mathbf{L} \right\} = 0 \quad (3.147)$$

and the surface of ray vector is, using the relation following (3.132),

$$\det\left\{\begin{bmatrix} 0 & (s/i) \times \\ -(s/i) \times & 0 \end{bmatrix} + (i\omega L)^{-1}\right\} = 0 \qquad (3.148)$$

These relations show that ω is a homogeneous function of degree 1 in $\mathbf{k}$ or of degree -1 in $\mathbf{s}$. We may assume that they have been solved for ω and write[§]

$$\omega = \omega(\mathbf{k}), \qquad \omega(\lambda\mathbf{k}) = \lambda\omega(\mathbf{k}) \qquad (3.149a)$$

$$\omega = \omega(\mathbf{s}), \qquad \omega(\lambda\mathbf{s}) = \lambda^{-1}\omega(\mathbf{s}) \qquad (3.149b)$$

where λ denotes an arbitrary number.

Using this notation, relations (3.133) and (3.135) become

$$\mathbf{s}(\mathbf{k}) = \frac{1}{\omega(\mathbf{k})} \frac{\partial\omega(\mathbf{k})}{\partial\mathbf{k}} \qquad (3.150)$$

The relation $\mathbf{k} \cdot \mathbf{s} = 1$, incidentally, follows from (3.150) and the fact that $\omega(\mathbf{k})$ is homogeneous of degree one in $\mathbf{k}$.[¶] Similarly, we have

$$\mathbf{k}(\mathbf{x}) = -\frac{1}{\omega(\mathbf{s})} \frac{\partial\omega(\mathbf{s})}{\partial\mathbf{s}} \qquad (3.151)$$

$\mathbf{k} \cdot \mathbf{s} = 1$ also follows from this relation. Relation (3.151) can be obtained in a way analogous to that used to obtain (3.150). It can also be obtained directly from (3.150) in the following manner. According to (3.149), we have the functional identity

$$\omega(\mathbf{s}) = \omega[\mathbf{k}(\mathbf{s})] \qquad (3.152)$$

$\mathbf{k}(\mathbf{s})$ being obtained by solving (3.150). Differentiating (3.152) with respect to s_α, we have

$$\frac{\partial\omega(\mathbf{s})}{\partial s_\alpha} = \frac{\partial\omega(\mathbf{k})}{\partial\mathbf{k}} \cdot \frac{\partial\mathbf{k}}{\partial s_\alpha} = \omega\mathbf{s} \cdot \frac{\partial\mathbf{k}}{\partial s_\alpha} = \omega\frac{\partial}{\partial s_\alpha}(\mathbf{k} \cdot \mathbf{s}) - \omega k_\alpha = -\omega(\mathbf{s})k_\alpha$$

[§] To avoid confusion between the function $\omega(\mathbf{k})$ and the (different) function $\omega(\mathbf{s})$, we always write the arguments $\mathbf{k}$ and $\mathbf{s}$. These two functions assume the same value ω, when $\mathbf{k}$ and $\mathbf{s}$ are solutions of (3.147) and (3.148), respectively.

[¶] Indeed, differentiating the relation $\omega(\lambda\mathbf{k}) = \lambda^\alpha\omega(\mathbf{k})$ with respect to λ, we obtain $\mathbf{k} \cdot \partial\omega(\lambda\mathbf{k})/\partial\mathbf{k} = \alpha\lambda^{\alpha-1}\omega(\mathbf{k})$. Setting $\lambda = 1$ and $\alpha = 1$, the result follows (Euler theorem).

because $\mathbf{k} \cdot \mathbf{s} = 1$. Thus the result (3.151) is established.

Relations (3.150) and (3.151) can be written, respectively,

$$\mathbf{s}(\mathbf{k}) = \tfrac{1}{2} [\omega(\mathbf{k})]^{-2} \frac{\partial [\omega(\mathbf{k})]^2}{\partial \mathbf{k}} \qquad (3.153a)$$

$$\mathbf{k}(\mathbf{s}) = \tfrac{1}{2} [\omega(\mathbf{s})]^2 \frac{\partial [\omega(\mathbf{s})]^{-2}}{\partial \mathbf{s}} \qquad (3.153b)$$

These alternative expressions are convenient when the surface of wave vectors is an ellipsoid, a situation encountered in uniaxial crystals. The surface of wave vectors has, in that special case, the form

$$\omega^2 = \mathbf{kWk} \qquad (3.154)$$

where $\mathbf{W}$ is a constant 3×3 symmetrical matrix. Substituting (3.154) in (3.153a), we obtain

$$\mathbf{s} = \omega^{-2}\mathbf{Wk} \qquad (3.155)$$

Substituting now (3.155) back in (3.154), the ray surface is described by another quadratic form, related to the previous one,

$$\omega^{-2} = \mathbf{sW}^{-1}\mathbf{s} \qquad (3.156)$$

Thus, if the surface of wave vector is an ellipsoid, the surface of ray vector is also an ellipsoid.

These relations will be useful in Chapter 4 dealing with geometrical optics.

3.15 Transverse Form of the Maxwell Equations

For studying beams propagating essentially along the z-axis in media free of spatial dispersion, it is convenient to rewrite the Maxwell equations in transverse form. In that form, only the transverse (x, y)-components of $\mathbf{E}$ and $\mathbf{H}$ need to be considered, and only first derivatives in z are involved.

The Maxwell equations

$$\nabla \times \mathbf{E} = i\omega\mathbf{B}, \qquad \nabla \times \mathbf{H} = -i\omega\mathbf{D} \qquad (3.157)$$

are explicitly, in a rectangular coordinate system, x_1, x_2, $x_3 \equiv z$,

$$
\begin{bmatrix}
\partial_z & 0 & 0 & 0 & \partial_1 & 0 \\
0 & \partial_z & 0 & 0 & \partial_2 & 0 \\
0 & 0 & \partial_z & 0 & 0 & \partial_2 \\
0 & 0 & 0 & \partial_z & 0 & -\partial_1 \\
\partial_2 & -\partial_1 & 0 & 0 & 0 & 0 \\
0 & 0 & \partial_1 & \partial_2 & 0 & 0
\end{bmatrix}
$$

$$
\times
\begin{bmatrix}
E_1 \\
E_2 \\
H_2 \\
-H_1 \\
-E_3 \\
-H_3
\end{bmatrix}
+ i\omega
\begin{bmatrix}
-B_2 \\
B_1 \\
-D_1 \\
-D_2 \\
B_3 \\
D_3
\end{bmatrix}
= 0
\qquad (3.158)
$$

where $\partial_z \equiv \partial/\partial z$, $\partial_1 \equiv \partial/\partial x_1$, $\partial_2 \equiv \partial/\partial x_2$. Introducing the elements ϵ_{ij}, ξ_{ij}, ζ_{ij}, μ_{ij} of the **L** matrix from (3.128), we have

$$
\begin{bmatrix}
-B_2 \\
B_1 \\
-D_1 \\
-D_2 \\
B_3 \\
D_3
\end{bmatrix}
$$

$$
\equiv
\begin{bmatrix}
-\zeta_{21} & -\zeta_{22} & -\mu_{22} & \mu_{21} & \zeta_{23} & \mu_{23} \\
\zeta_{11} & \zeta_{12} & \mu_{12} & -\mu_{11} & -\zeta_{13} & -\mu_{13} \\
-\epsilon_{11} & -\epsilon_{12} & -\xi_{12} & \xi_{11} & \epsilon_{13} & \xi_{13} \\
-\epsilon_{21} & -\epsilon_{22} & -\xi_{22} & \xi_{21} & \epsilon_{23} & \xi_{23} \\
\zeta_{31} & \zeta_{32} & \mu_{32} & -\mu_{31} & -\zeta_{33} & -\mu_{33} \\
\epsilon_{31} & \epsilon_{32} & \xi_{32} & -\xi_{31} & -\epsilon_{33} & -\xi_{33}
\end{bmatrix}
\begin{bmatrix}
E_1 \\
E_2 \\
H_2 \\
-H_1 \\
-E_3 \\
-H_3
\end{bmatrix}
$$

$$
(3.159)
$$

Let the 6×6 matrix in (3.159), denoted $\mathbf{K}$, be split into four submatrices, $\mathbf{K}_{11}$ (4×4), $\mathbf{K}_{12}$ (4×2), $\mathbf{K}_{21}$ (2×4), and $\mathbf{K}_{22}$ (2×2) as shown by dashed lines.

Setting

$$\mathbf{D}_1 \equiv \begin{bmatrix} \partial_1 & 0 \\ \partial_2 & 0 \\ 0 & \partial_2 \\ 0 & -\partial_1 \end{bmatrix} \tag{3.160a}$$

$$\mathbf{D}_2 \equiv \begin{bmatrix} \partial_2 & -\partial_1 & 0 & 0 \\ 0 & 0 & \partial_1 & \partial_2 \end{bmatrix} \tag{3.160b}$$

(3.158) becomes

$$\partial_z \psi - \mathbf{D}_1 \psi_z + i\omega(\mathbf{K}_{11}\psi - \mathbf{K}_{12}\psi_z) = 0 \tag{3.161a}$$

$$\mathbf{D}_2 \psi + i\omega(\mathbf{K}_{21}\psi - \mathbf{K}_{22}\psi_z) = 0 \tag{3.161b}$$

where we have defineu

$$\psi_z \equiv \begin{bmatrix} E_3 \\ H_3 \end{bmatrix} \tag{3.162a}$$

$$\psi \equiv \begin{bmatrix} E_1 \\ -E_2- \\ H_2 \\ -H_1 \end{bmatrix} \equiv \begin{bmatrix} \mathbf{V} \\ -- \\ \mathbf{I} \end{bmatrix} \tag{3.162b}$$

Assuming that $\mathbf{K}_{22}$ is independent of $\partial/\partial\mathbf{x}$ and nonsingular for all $\mathbf{x}$ as is almost always the case, it is easy to eliminate ψ_z from (3.161a) and (3.161b). We obtain the transverse form of the Maxwell equations

$$[\mathbf{H}(\partial_1, \partial_2, x_1, x_2, z) - \mathbf{1}\partial_z]\psi = \mathbf{0} \tag{3.163a}$$

where

$$\mathbf{H}(\partial_1, \partial_2, x_1, x_2, z) = -i\omega\mathbf{K}_{11} + (\mathbf{D}_1 + i\omega\mathbf{K}_{12})(i\omega\mathbf{K}_{22})^{-1}(\mathbf{D}_2 + i\omega\mathbf{K}_{21})$$

$$\tag{3.163b}$$

Note that the vector wave function ψ depends only on the transverse components of the fields **E**, **H**, and that the 4×4 matrix operator **H** depends only on the transverse gradient ($\partial \equiv \partial_1, \partial_2$) and space (**x**).

Let us now write the Maxwell equations for the adjoint[10] medium,

$$\nabla \times \mathbf{E}^\dagger = i\omega\mathbf{B}^\dagger \tag{3.164a}$$

$$\nabla \times \mathbf{H}^\dagger = -i\omega\mathbf{D}^\dagger \tag{3.164b}$$

in the form

$$
\begin{bmatrix}
\partial_z & 0 & 0 & 0 & \partial_2 & 0 \\
0 & \partial_z & 0 & 0 & -\partial_1 & 0 \\
0 & 0 & \partial_z & 0 & 0 & \partial_1 \\
0 & 0 & 0 & \partial_z & 0 & \partial_2 \\
\partial_1 & \partial_2 & 0 & 0 & 0 & 0 \\
0 & 0 & \partial_2 & -\partial_1 & 0 & 0
\end{bmatrix}
$$

$$
\times
\begin{bmatrix}
H_2^\dagger \\
-H_1^\dagger \\
-E_1^\dagger \\
-E_2^\dagger \\
-H_3^\dagger \\
E_3^\dagger
\end{bmatrix}
+ i\omega
\begin{bmatrix}
-D_1^\dagger \\
-D_2^\dagger \\
B_2^\dagger \\
-B_1^\dagger \\
D_3^\dagger \\
-B_3^\dagger
\end{bmatrix}
= 0 \tag{3.165}
$$

which differs from (3.158) only by the dagger superscripts and the order in which the equations are written. The sign is also changed for convenience. The four first components of the column vector on the lhs of (3.165) are

$$
\psi^\dagger \equiv
\begin{bmatrix}
H_2^\dagger \\
-H_1^\dagger \\
\hline
-E_1^\dagger \\
-E_2^\dagger
\end{bmatrix}
\equiv
\begin{bmatrix}
\mathbf{I}^\dagger \\
\hline
-\mathbf{V}^\dagger
\end{bmatrix} \tag{3.166}
$$

The adjoint medium can be characterized by the following 6×6 matrix

$K^\dagger$, which relates the two column vectors in (3.165),

$$
\begin{bmatrix}
-D_1^\dagger \\
-D_2^\dagger \\
B_2^\dagger \\
-B_1^\dagger \\
D_3^\dagger \\
-B_3^\dagger
\end{bmatrix}
=
\left[
\begin{array}{cccc:cc}
-\xi_{12}^\dagger & \xi_{11}^\dagger & \epsilon_{11}^\dagger & \epsilon_{12}^\dagger & \xi_{13}^\dagger & -\epsilon_{13}^\dagger \\
-\xi_{22}^\dagger & \xi_{21}^\dagger & \epsilon_{21}^\dagger & \epsilon_{22}^\dagger & \xi_{23}^\dagger & -\epsilon_{23}^\dagger \\
\mu_{22}^\dagger & -\mu_{21}^\dagger & -\zeta_{21}^\dagger & -\zeta_{22}^\dagger & -\mu_{23}^\dagger & \zeta_{23}^\dagger \\
-\mu_{12}^\dagger & \mu_{11}^\dagger & \zeta_{11}^\dagger & \zeta_{12}^\dagger & \mu_{13}^\dagger & -\zeta_{13}^\dagger \\ \hdashline
\xi_{32}^\dagger & -\xi_{31}^\dagger & -\epsilon_{31}^\dagger & -\epsilon_{32}^\dagger & -\xi_{33}^\dagger & \epsilon_{33}^\dagger \\
-\mu_{32}^\dagger & \mu_{31}^\dagger & \zeta_{31}^\dagger & \zeta_{32}^\dagger & \mu_{33}^\dagger & -\zeta_{33}^\dagger
\end{array}
\right]
\begin{bmatrix}
H_2^\dagger \\
-H_1^\dagger \\
-E_1^\dagger \\
-E_2^\dagger \\
-H_3^\dagger \\
E_3^\dagger
\end{bmatrix}
$$

$$(3.167)$$

The 6×6 matrix in (3.167) is denoted $\mathbf{K}^\dagger \equiv (\mathbf{K}_{11}^\dagger, \mathbf{K}_{12}^\dagger, \mathbf{K}_{21}^\dagger, \mathbf{K}_{22}^\dagger)$.

The Maxwell equations (3.164) in the transposed medium can therefore be written, eliminating the vector $(-H_3^\dagger, E_3^\dagger)$,

$$[\mathbf{H}^\dagger(\partial_1, \partial_2, x_1, x_2, z) + \mathbf{1}\partial_z]\psi^\dagger = 0 \qquad (3.168a)$$

where

$$\mathbf{H}^\dagger(\partial_1, \partial_2, x_1, x_2, z) = i\omega\mathbf{K}_{11}^\dagger - \left(\tilde{\mathbf{D}}_2 + i\omega\mathbf{K}_{12}^\dagger\right)\left(i\omega\mathbf{K}_{22}^\dagger\right)^{-1}\left(\tilde{\mathbf{D}}_1 + i\omega\mathbf{K}_{21}^\dagger\right)$$

$$(3.168b)$$

We have seen before that transposition of $\mathbf{M}$ has the effect of replacing ϵ by $\tilde{\epsilon}$, μ by $\tilde{\mu}$, ξ by $-\tilde{\zeta}$, and ζ by $-\tilde{\xi}$. Comparison of (3.159) and (3.167) shows that, under such a transformation, $\mathbf{K}$ becomes $-\tilde{\mathbf{K}}^\dagger$. That is, we have $\mathbf{K}_{11} = -\tilde{\mathbf{K}}_{11}^\dagger$, $\mathbf{K}_{12} = -\tilde{\mathbf{K}}_{21}^\dagger$, $\mathbf{K}_{21} = -\tilde{\mathbf{K}}_{12}^\dagger$, and $\tilde{\mathbf{K}}_{22}^\dagger = -\tilde{\mathbf{K}}_{22}^\dagger$. Comparison of (3.163b) and (3.168b) then shows that $\mathbf{H}^\dagger$ is obtained from $\mathbf{H}$ by matrix transposition and changing ∂ to $-\partial$ (that is, $\mathbf{D}_{1,2} \to -\mathbf{D}_{1,2}$). These two operations together amount to an operator transposition.[§] This observation proves that (3.168a) is the equation formally adjoint to (3.163a).

Therefore, assuming that the proper boundary (or radiation) conditions are satisfied, the scalar product of $\psi^\dagger$ and ψ is invariant in the sense that

$$\frac{d}{dz} \int \psi^\dagger \cdot \psi \, dS = 0 \qquad (3.169)$$

This is essentially the Lorentz reciprocity theorem given in (3.89). A general result has been obtained with the present formulation that had not been derived before: at most four waves (rather than six) can propagate in a given direction in a medium free of spatial dispersion. This follows from the fact that the **H** matrix operator is only 4 × 4.

For the special case of an isotropic medium with scalar ϵ, μ, (3.163) takes the form [using the notation in (3.162b)]

$$\frac{d\mathbf{V}}{dz} = \begin{bmatrix} -\partial_1(i\omega\epsilon)^{-1}\partial_1 + i\omega\mu & -\partial_1(i\omega\epsilon)^{-1}\partial_2 \\ -\partial_2(i\omega\epsilon)^{-1}\partial_1 & -\partial_2(i\omega\epsilon)^{-1}\partial_2 + i\omega\mu \end{bmatrix} \mathbf{I}$$

$$\frac{d\mathbf{I}}{dz} = \begin{bmatrix} -\partial_2(i\omega\mu)^{-1}\partial_2 + i\omega\epsilon & \partial_2(i\omega\mu)^{-1}\partial_1 \\ \partial_1(i\omega\mu)^{-1}\partial_2 & -\partial_1(i\omega\mu)^{-1}\partial_1 + i\omega\epsilon \end{bmatrix} \mathbf{V} \quad (3.170)$$

If the medium is stratified, that is, if **M** does not depend on x_1, x_2, the general solution of (3.170) is a superposition of plane waves with $\partial_1 = ik_1$, $\partial_2 = ik_2$. In Chapter 5 we shall discuss in detail the case where, in (3.170), $\partial_1 = 0$ and $\partial_2 = ik_2$, that is, where the field is independent of the x-coordinate. Then the upper (E_1, H_2) and lower (E_2, $-H_1$) components of **V** and **I** can be separated because the off-diagonal terms in (3.170) vanish. These components are associated with H-waves and E-waves, respectively.

Similarly, in a cylindrical coordinate system r, φ, z, we have

$$\frac{d\mathbf{V}}{dr} = \begin{bmatrix} -\partial_\varphi(i\omega\epsilon r)^{-1}\partial_\varphi + i\omega\mu r & -\partial_\varphi(i\omega\epsilon r)^{-1}\partial_z \\ -\partial_z(i\omega\epsilon r)^{-1}\partial_\varphi & -\partial_z(i\omega\epsilon r)^{-1}\partial_z + i\omega\mu r^{-1} \end{bmatrix} \mathbf{I}$$

$$\frac{d\mathbf{I}}{dr} = \begin{bmatrix} -\partial_z(i\omega\mu r)^{-1}\partial_z + i\omega\epsilon r^{-1} & \partial_z(i\omega\mu r)^{-1}\partial_\varphi \\ \partial_\varphi(i\omega\mu r)^{-1}\partial_z & -\partial_\varphi(i\omega\mu r)^{-1}\partial_\varphi + i\omega\epsilon r \end{bmatrix} \mathbf{V} \quad (3.171)$$

[§] In function space, the operator formally adjoint to $\partial/\partial x$ is $-\partial/\partial x$. Indeed,

$$\int_{-\infty}^{+\infty}\left(\frac{d}{dx}f\right)g \, dx + \int_{-\infty}^{+\infty}f\left(\frac{d}{dx}g\right)dx = \int_{-\infty}^{+\infty}\frac{d}{dx}(fg)\, dx = fg\Big|_{-\infty}^{+\infty} = 0$$

if fg vanishes at $\pm\infty$. Thus $\langle(d/dx)f, g\rangle = \langle f, -(d/dx)g\rangle$.

where

$$\mathbf{V} \equiv \begin{bmatrix} rE_\varphi \\ E_z \end{bmatrix}, \qquad \mathbf{I} \equiv \begin{bmatrix} H_z \\ -rH_\varphi \end{bmatrix} \qquad (3.172)$$

These relations go to (3.170) for large r. (Set $r = 1$, $\varphi \to 1$, $z \to 2$.) For dielectric rods with circular symmetry, $\mathbf{V}$ and $\mathbf{I}$ are continuous functions of r. The general solution of (3.171) is a superposition of waves with $\partial_\varphi = i\nu$, ν integer, and $\partial_z = ik_z$. If $\nu = 0$, the equations separate in H-waves and E-waves. This will be discussed further in Chapter 5.

3.16 The Coupled Mode Equations

We illustrated in Fig. 1-19 the coupling between optical fibers and pointed out that if the fibers are lossless and identical, all the power fed into one fiber is transferred to the other fiber after a certain characteristic length inversely proportional to the coupling strength. In Section 3.7 we showed that this coupling strength can be evaluated in term of the fields of the uncoupled fibers. Here we examine in some detail the general properties of the coupled mode equations.

For lossless media, the coupled mode equations can be written[11,12]

$$-i\frac{d\psi_a}{dz} = k_a\psi_a + c\psi_b, \qquad -i\frac{d\psi_b}{dz} = k_b\psi_b + c^*\psi_a \qquad (3.173)$$

where ψ_a, ψ_b denote the fields of the two guides normalized in such a way that $|\psi_a|^2$ and $|\psi_b|^2$ represent the power carried by waveguides a and b, respectively. k_a, k_b denote the (real) wavenumbers of the uncoupled guides. The subscripts z of k are omitted here for brevity. It can be verified, using (3.173), that the power conservation law

$$\frac{d}{dz}(\psi_a\psi_a^* + \psi_b\psi_b^*) = 0 \qquad (3.174)$$

holds. The modes are obtained by setting

$$\psi_a(z) = \Psi_a \exp(ikz), \qquad \psi_b(z) = \Psi_b \exp(ikz) \qquad (3.175)$$

where k is an as yet unknown constant. Substituting (3.175) in (3.173), we obtain

$$(k - k_a)\Psi_a - c\Psi_b = 0, \qquad -c^*\Psi_a + (k - k_b)\Psi_b = 0 \qquad (3.176)$$

This system of equations admits nontrivial solutions only if the determinant vanishes, that is, if

$$(k - k_a)(k - k_b) = cc^* \qquad (3.177)$$

Equation (3.177) is a second-degree equation for k that was obtained before [see, for example, (3.83)]. The solution of (3.177) is

$$k^\pm = \tfrac{1}{2}(k_a + k_b) \pm \Delta$$

where

$$\Delta \equiv \left[\tfrac{1}{4}(k_a - k_b)^2 + cc^* \right]^{1/2} \qquad (3.178)$$

Because k_a and k_b are real numbers, the two solutions in (3.178) are real. In the absence of coupling ($c = 0$), the solutions are $k^+ = k_a$ and $k^- = k_b$. If the uncoupled waves are synchronous ($k_a = k_b \equiv k_0$), the normal-mode wavenumbers are

$$k^\pm = k_0 \pm |c| \qquad (3.179)$$

and the beat wavelength is $2\pi/|c|$. Thus the length required for complete transfer of power from one waveguide to the other is

$$L = \frac{\pi}{2|c|} \qquad (3.180a)$$

Let us derive a more general expression for the coupling, defined as the power in waveguide b when a power unity is fed in waveguide a. The solutions of (3.173) are superpositions of normal modes

$$\psi_a(z) = \Psi_a^+ \exp(ik^+ z) + \Psi_a^- \exp(ik^- z)$$

$$\psi_b(z) = \Psi_b^+ \exp(ik^+ z) + \Psi_b^- \exp(ik^- z) \qquad (3.180b)$$

The initial conditions are $\psi_b(0) = 0$, $\psi_a(0) = 1$. From the latter condition and the first equation in (3.173), it follows that $d\psi_b/dz = ic^*$ at $z = 0$. Thus $\Psi_b^+ + \Psi_b^- = 0$, $\Psi_b^+(k^+ - k^-) = c$, and

$$\psi_b(z) = \frac{ic^*}{\Delta} \exp[\tfrac{1}{2}i(k_a + k_b)z] \sin(\Delta z) \qquad (3.181)$$

$$P_b(z) \equiv \psi_b(z)\psi_b^*(z) = \frac{cc^*}{\Delta^2} \sin^2(\Delta z) \qquad (3.182)$$

When $k_a = k_b$, P_b reaches unity at $z = \pi/2|c|$. The general expression for the transferred power, for arbitrary $k_a - k_b$, is

$$P_b = \left[1 + \left(\frac{k_a - k_b}{2|c|} \right)^2 \right]^{-1} \tag{3.183}$$

and for small z, (3.182) is simply [$\sin(\Delta z) \approx \Delta z$]

$$P_b(z) = cc^* z^2 \tag{3.184}$$

Thus, for small z, P_b is independent of $k_a - k_b$ (see Fig. 3–14). The approximation breaks down, however, when $z \gtrsim \Delta^{-1}$. As an example of an application of (3.184), let us investigate under what condition a 40-dB cross talk takes place between optical fibers in a 1-km-long bundle. Setting in (3.184) $P_b = 10^{-4}$ and $z = 10^3$ m, we obtain $|c| = 10^{-5}\mathrm{m}^{-1}$ provided $z \ll \Delta^{-1}$, that is, provided $|k_a - k_b|/k \ll \lambda/z \approx 10^{-9}$. It is quite unrealistic to expect that glass fibers will ever be made with such an accuracy.

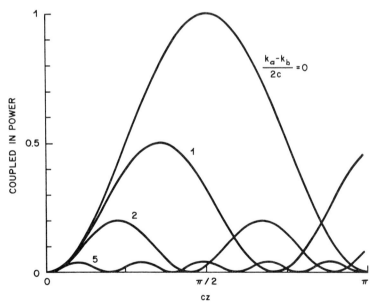

Fig. 3-14   Power in a coupled guide as a function of distance z for various differences $k_a - k_b$ of the wavenumbers of the uncoupled guides. For small z, the power is independent of $k_a - k_b$.

In any real fiber bundle, the cross talk is considerably smaller, by orders of magnitude, than the one evaluated above because the function $k_a - k_b$ fluctuates as a function of z by significant amounts. Coupling takes place only near the zero-point crossings, with the different contributions adding incoherently. This observation points to the need of considering the case where $k_a - k_b$ varies with z.

To do that, let us go back to our original equations (3.173) and assume that k_a and k_b vary linearly with z, that is, $k_a = k_0 + \alpha z$, $k_b = k_0 - \alpha z$. Synchronism takes place only near the origin, $z = 0$. Assuming that a power unity is fed into guide a at $z = -\infty$, we wish to find how much power is left in that guide after the interaction between the two guides has taken place. We now assume for simplicity that c is a real constant. Thus the equations are

$$- i \frac{d\psi_a}{dz} = (k_0 + \alpha z)\psi_a + c\psi_b, \qquad - i \frac{d\psi_b}{dz} = (k_0 - \alpha z)\psi_b + c\psi_a$$

(3.185)

In (3.185) let us set

$$\psi_a = \Psi_a \exp(ik_0 z), \qquad \psi_b = \Psi_b \exp(ik_0 z)$$ (3.186)

We obtain

$$- i \frac{d\Psi_a}{dz} = \alpha z \Psi_a + c\Psi_b$$ (3.187a)

$$- i \frac{d\Psi_b}{dz} = -\alpha z \Psi_b + c\Psi_a$$ (3.187b)

Let us differentiate (3.187a) and use (3.187b). An equation for Ψ_a alone is obtained:

$$\frac{d^2\Psi_a}{dz^2} + (\alpha^2 z^2 + c^2 - i\alpha)\Psi_a = 0$$ (3.188)

This is the equation for parabolic cylinder functions. The asymptotic form of the solution, valid for $-\pi \leqslant \arg(z) \leqslant \pi/2$, is[13]

$$\Psi_a(z) = \exp\left[i \frac{\alpha}{2} z^2 - i \frac{c^2}{2\alpha} \log(-z) \right]$$ (3.189)

indicating that the power is unity at $z = -\infty$. It is easily verified that when (3.189) is substituted in (3.188), all terms vanish with the exception of terms proportional to z^{-2}, which are very small for large $|z|$. For positive z, (3.189) becomes, because $\log(-z) = -i\pi + \log(z)$,

$$\Psi_a(z) = \exp\left[i\,\frac{\alpha}{2}\,z^2 - i\,\frac{c^2}{2\alpha}\,\log(z) - \frac{\pi c^2}{2\alpha}\right] \tag{3.190}$$

The phase term $(\alpha/2)z^2$ in (3.190) can be removed by defining the phase with respect to that of the uncoupled guide. The phase correction $-(c^2/2\alpha)\log(z)$ is also of little importance for our discussion. We are mostly interested in the last term. The power left in guide a after the interaction has taken place is

$$P_a = \Psi_a\Psi_a^* = \exp\left(-\frac{\pi c^2}{\alpha}\right) \tag{3.191}$$

We now discuss the physical consequences of this result.

Let us first assume that $\pi c^2/\alpha$ is very small compared with unity, that is, the k's crossing is very fast. In that case, guide a loses only a small amount of power equal to $\pi c^2/\alpha$, which is collected by guide b because of conservation of power. If the wavenumber of guide b oscillates about that of guide a, guide b collects at each crossing a power $\pi c^2/\alpha$ from guide a. An estimate of the total power collected by guide b is obtained by assuming that $k_a - k_b = \Delta k \sin(\Omega z)$. In that case $2\alpha = \Delta k\Omega$, and the number of crossings per unit length is Ω/π. Assuming incoherence, the total power collected by guide b over a length z is equal to $2c^2z/\Delta k$. This is the product of Ω/π, $\pi c^2/\alpha$, and the length z. For a maximum cross talk of 40 dB, selecting a realistic value $\Delta k/k = 0.01$, or $\Delta k = 2\pi \times 10^{-4}$ at $\lambda = 1$ μm, we find $c < 0.05$ m^{-1}, a result that differs by more than three orders of magnitude from the one obtained before under the assumption of rigorously synchronous fibers ($c < 10^{-5}$ m^{-1}).

Let us now go to the other extreme and assume that the k's crossing is very slow: $\pi c^2/\alpha \gg 1$. In other words, we assume that there are many beat wavelengths $2\pi/c$ over the length where $|k_a - k_b| < c$. Then, according to (3.191), P_a is very nearly zero, meaning that all the power has been transferred from guide a to guide b (see Fig. 3–15). This result rests on the principle of adiabatic invariance. The power launched in one normal mode remains in that same mode. For large negative z, the power of a given mode is mostly in guide a, while for large positive z, the power of that same mode is mainly in guide b. This is the principle of the directional coupler described by Cook.[14] It is applicable as well to optical waveguides, as recently discussed by Wilson and Teh.[15]

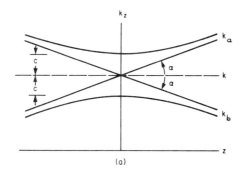

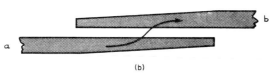

Fig. 3-15   (a) Wavenumbers of the normal modes when the wavenumbers of the uncoupled guides vary with z. (b) Complete transfer of power is effected in a coupler, provided the variations in wavenumbers are slow enough.[14,15] (Note $k \equiv k_0$ in the text.)

3.17   Perturbation Formulas and Variational Principles

Most often the wave equation cannot be solved exactly. Approximate solutions can be obtained, however, if we know a solution for a medium that does not differ too much from the medium considered. We give in this section a simple general result, applicable to scalar waves in unbounded media. This result is sufficient for most of the applications that we shall consider in subsequent chapters.

The required algebra is best understood by considering first a closely related result in matrix algebra. Let us consider the matrix eigenvalue equation

$$\mathbf{M}\mathbf{x} = \lambda \mathbf{x} \tag{3.192}$$

where $\mathbf{M}$ is a square matrix, $\mathbf{x}$ an eigenvector of $\mathbf{M}$, and λ the corresponding eigenvalue. Premultiplying both sides in (3.192) by the vector $\mathbf{x}$ transposed and omitting tildes on the first vector of a product for simplicity, we obtain

$$\lambda = \frac{\mathbf{x}\mathbf{M}\mathbf{x}}{x^2} \tag{3.193}$$

where $x^2 \equiv \mathbf{x}\mathbf{x}$. Let us increment $\mathbf{x}$:

$$\mathbf{x} \rightarrow \mathbf{x} + \delta\mathbf{x} \tag{3.194}$$

in (3.193) by a small vector $\delta\mathbf{x}$. The resulting change on λ is

$$\delta\lambda = \frac{(\mathbf{x} + \delta\mathbf{x})\mathbf{M}(\mathbf{x} + \delta\mathbf{x})}{(\mathbf{x} + \delta\mathbf{x})(\mathbf{x} + \delta\mathbf{x})} - \frac{\mathbf{x}\mathbf{M}\mathbf{x}}{x^2}$$

$$\propto \delta\mathbf{x}\left[(\mathbf{M} + \tilde{\mathbf{M}})\mathbf{x} - 2\lambda\mathbf{x}\right] + O(\delta x^2) \qquad (3.195)$$

Because of (3.192), $\delta\lambda$ vanishes to first order if $\mathbf{M}$ is symmetrical: $\tilde{\mathbf{M}} = \mathbf{M}$. Thus (3.193) is a variational expression for λ. An accurate value of λ is obtained if an approximate value of $\mathbf{x}$ is substituted in (3.193).

Similarly, the first-order variation of

$$\lambda' = \frac{\mathbf{x}^*\mathbf{M}\mathbf{x}}{x^2} \qquad (3.196)$$

where $x^2 \equiv \mathbf{x}^*\mathbf{x}$, vanishes if $\mathbf{M}$ is Hermitian: $\tilde{\mathbf{M}}^* = \mathbf{M}$.

Let us now assume that $\mathbf{M}$, and therefore $\mathbf{x}$ and λ, depends on a parameter ω. We have the Hellmann–Feynman theorem

$$\frac{d\lambda}{d\omega} = \frac{\mathbf{x}(d\mathbf{M}/d\omega)\mathbf{x}}{x^2} \qquad (3.197)$$

In the differentiation, the variation of $\mathbf{x}$ with ω can be ignored because λ in (3.193) is insensitive to small variations of $\mathbf{x}$.

Let us now go back to wave equations and consider the Helmholtz equation for a z-invariant medium

$$H(\partial_x, \partial_y, x, y)E \equiv \left[\partial_x^2 + \partial_y^2 + k^2(x, y)\right]E = k_z^2 E \qquad (3.198)$$

The operator H in (3.198) is symmetrical (self-adjoint) because, for any square integrable E_a, E_b, we have

$$\langle E_a H E_b \rangle = \langle E_b H E_a \rangle \qquad (3.199)$$

or, explicitly,

$$\int \left\{ E_a\left[\partial_x^2 + \partial_y^2 + k^2(x, y)\right]E_b \right.$$
$$\left. - E_b\left[\partial_x^2 + \partial_y^2 + k^2(x, y)\right]E_b \right\} dx\, dy = 0 \qquad (3.200)$$

This result follows from an integration by parts. Note that $k(x, y)$ is not required to be a real function of x, y.

The variational expression of k_z in (3.198) is therefore, by analogy with (3.193),

$$k_z^2 = \frac{\int E\left[\partial_x^2 + \partial_y^2 + k^2(x, y)\right]E\, dx\, dy}{\int E^2\, dx\, dy} \qquad (3.201)$$

where $E(x, y)$ is assumed close to the field of the eigenmode corresponding to the desired k_z. Let us now assume that k^2 is a function of the angular frequency ω. By application of the Hellmann–Feynman theorem (3.197), we have

$$\frac{dk_z^2}{d\omega^2} = \frac{\int E^2 (\partial k^2 / \partial \omega^2)\, dx\, dy}{\int E^2\, dx\, dy} \tag{3.202}$$

This result is extremely useful to evaluate the group velocity $d\omega / dk_z$ of an optical waveguide.

An expression for the group velocity of electromagnetic pulses in term of the fields $\mathbf{E}$, $\mathbf{H}$, and $\partial(\omega\epsilon)/\partial\omega$ was given in (3.100). This relation, which states the equality of the group and energy velocities, is only one component of a more general identity that says the canonical stress-energy tensor has zero 4-divergence. Using this more general relation, Brown (Ref. 36, Chapter 1) discovered an interesting relation applicable to dispersionless materials ($\partial\epsilon / \partial\omega = \partial\mu / \partial\omega = 0$). For any waveguide incorporating isotropic and dispersionless but possibly inhomogeneous media, the product of the group velocity u_z and phase velocity v_z is given by

$$(u_z v_z)^{-1} \equiv \frac{dk_z^2}{d\omega^2} = \frac{\int \epsilon\mu S_z\, dx\, dy}{\int S_z\, dx\, dy} \tag{3.203}$$

where S_z denotes the axial component of the energy flow density, or Poynting vector: $\mathbf{E} \times \mathbf{H}|_z$. The integrals in (3.203) are over the waveguide cross section. Clearly, (3.203) reduces to the scalar relation (3.202) for nondispersive media ($\partial k^2 / \partial\omega^2 = k^2 / \omega^2 = \epsilon\mu$) if we identify S_z and E^2. We shall refer to (3.203) as Brown's identity.

References

1. P. M. Morse and H. Feshback, "Methods of Theoretical Physics." McGraw-Hill, New York, 1963.
2. M. Born and E. Wolf, "Principles of Optics." Pergamon, Oxford, 1965. C. A. Greebe, *Philips Tech. Rev.* **33**, 311 (1973).
3. V. A. Fock, *in* "Electromagnetic Refraction and Propagation Problems," Chapter 4. Pergamon, Oxford, 1965.
4. J. Robieux, *Ann. Radioelec.* **15**, 28 (1960).
5. J. A. Arnaud, *Bell Syst. Tech. J.* **53**, 217 (1974).
6. J. A. Arnaud and A. A. M. Saleh, *Proc. IEEE* **60**, 639 (1972), **61**, 667 (1973). J. A. Kong, *Proc. IEEE* **60**, 1036 (1972). J. A. Arnaud, *J. Opt. Soc. Amer.* **63**, 238 (1973). K. Suchy and C. Altman, *J. Plasma Phy.* **13**, 299, 437 (1975).

7. D. S. Jones, "The Theory of Electromagnetism." Pergamon, New York, 1964.

8. M. Kline and I. W. Kay, "Electromagnetic Theory and Geometrical Optics." Wiley (Interscience), New York, 1965.

9. P. Penfield and H. A. Haus, "Electrodynamics of Moving Media." MIT Press, Cambridge, Massachusetts, 1967.

10. B. Friedman, "Principles and Techniques of Applied Mathematics," p. 207. Wiley, New York, 1960.

11. C. C. Johnson, "Field and Wave Electrodynamics." McGraw Hill, New York, 1965. S. E. Miller, *Bell Syst. Tech. J.* **33**, 661 (1954).

12. J. A. Arnaud, *in* "Crossed Field Microwave Devices" (E. Okress, ed.), Chapter 2. Academic Press, New York, 1961.

13. G. H. Wannier, *Physics* **1**, 251 (1965).

14. J. S. Cook, *Bell Syst. Tech. J.* **34**, 807 (1955).

15. M. G. F. Wilson and G. A. Teh, *IEEE J. Microwave Theory Tech.* **MTT23**, 85 (1975).

Geometrical Optics

For nondispersive media, the laws of geometrical optics can be obtained from an asymptotic expansion of the field equations in power series of the angular frequency ω. Alternatively, these laws have been defined as the laws of transport of discontinuities of the field.[1,2] Because we are interested in dispersive as well as in nondispersive media, we shall use a less rigorous approach. We assume that the laws of geometrical optics are applicable whenever the wavelength in the medium is small compared with the scale of the inhomogeneities and compared to the transverse dimension of the optical beams to be transported. In fact, significant departures from geometrical optics are observed even when these conditions are met. However, in most of this book, the laws of geometrical optics are used in a generalized form that accounts to some extent for diffraction, or consideration is given to highly multimoded fibers for which the above limitations are not important.

Elementary results in geometrical optics applicable to free space, conventional thin lenses, and lenslike media were given in Chapter 2. To go beyond the approximation of Gauss, it is convenient to make use of the powerful methods devised by Hamilton.[1-3] According to the conventional procedure, rays are traced throughout an optical system by repeated application of the Descartes–Snell law of refraction (if the medium is stepwise homogeneous). In Hamiltonian optics, a ray is defined by its starting point and its endpoint, rather than by its initial position and slope. The phase shift along a ray from the starting point $\mathbf{x}'$ to the endpoint $\mathbf{x}$ is the point-eikonal, denoted $S(\mathbf{x}; \mathbf{x}')$.[§] The concept of point-eikonal is quite fundamental.

Hamilton is perhaps most famous for the unification of mechanics and geometrical optics that he proposed in 1837. Although Hamilton himself was only concerned with classical mechanics and geometrical optics, wave effects are, in fact, implicit in his formulation.

The presentation of the Hamilton theory of light is complicated by the fact that some results have been often rediscovered, and different names have been given to quantities that are essentially identical. We have tried in this presentation to minimize the number of quantities introduced.

Readers who are unfamiliar with the methods of Hamiltonian optics should not find the first sections in this chapter too difficult because simple examples are discussed to illustrate the method. In Section 4.16, the problem of pulse broadening in optical fibers for incoherent sources is treated from the point of view of a mapping from the space–time phase space at the input of the fiber to the space–time phase space at the output of the fiber. This formalism may appear rather abstract until the reader has gained some familiarity with the problem. The reader may wish to read Section 4.17 first, where experimental results concerning the broadening of optical pulses are reported, and subsequently go back to the derivations in Section 4.16.

4.1 Time-Harmonic Plane Waves
in Homogeneous Stationary Media

In this section, we assume that the sources are time harmonic (e.g., the light is green) and the medium is stationary. In that case, the frequency of the waves and time can be ignored. We need to deal with only two fundamental quantities: the ray trajectories $\mathbf{x}(\sigma)$ and the wave vectors $\mathbf{k}(\sigma)$, where σ denotes an arbitrary parameter.

Let $\mathbf{x}$ be a vector with components x_1, x_2, x_3 (or x, y, z) in rectangular coordinates, which denotes a point in space. A ray trajectory is defined in parametric form $\mathbf{x} = \mathbf{x}(\sigma)$, implying three functions of the parameter σ:

$$x_1 = x_1(\sigma), \qquad x_2 = x_2(\sigma), \qquad x_3 = x_3(\sigma) \qquad (4.1a)$$

[§] In some media (e.g., moving media), the ray from $\mathbf{x}$ to $\mathbf{x}'$ may not coincide with the ray from $\mathbf{x}'$ to $\mathbf{x}$. The possible existence of such media had already been envisioned by Descartes (1637).[4] For simplicity, we ignore the possibility that more than one ray goes from $\mathbf{x}'$ to $\mathbf{x}$, and the case where $\mathbf{x}$ is the image of $\mathbf{x}'$.

Let us now assume that the medium is homogeneous. The distance
between adjacent crests of a plane wave is denoted λ (see Fig. 4-1a), and $\mathbf{k}$
is a vector perpendicular to the wavefronts with magnitude $2\pi/\lambda$. The field
is the real part of

$$\psi_0 \exp[i(\mathbf{k} \cdot \mathbf{x} - \omega t)] \tag{4.1b}$$

As usual, the $\exp(-i\omega t)$ term is omitted and the field ψ is a complex
function of $\mathbf{x}$. In general, the magnitude of $\mathbf{k}$ depends on its direction. The
propagation of plane waves in the medium is characterized by a *surface of
wave vectors*, shown in Fig. 4-1b, whose equation can be written

$$H(\mathbf{k}) = 0 \tag{4.2}$$

For an isotropic medium, only the magnitude k of $\mathbf{k}$ enters in (4.2). We
have in rectangular coordinates

$$H(\mathbf{k}) \equiv k_1^2 + k_2^2 + k_3^2 - k^2 = 0 \tag{4.3}$$

Equation (4.2) or (4.3) can be solved, in principle, for any particular
component of $\mathbf{k}$, k_3 for instance, and $H(\mathbf{k})$ can be taken to be of the form

$$H(\mathbf{k}) \equiv k_3 - k_3(k_1, k_2) = 0 \tag{4.4}$$

The ray trajectories are independent of the particular form for H that we
select.

The form (4.4) proves particularly convenient for systems that are
invariant under a translation along the x_3-axis because the component k_3
of $\mathbf{k}$ is, in that case, a constant of motion.

Let us consider again a plane wave and consider the distance $\Lambda > \lambda$
between adjacent crests in a direction different from that of the normal to
the wavefronts, as shown in Fig. 4-1a. The apparent wavenumber $2\pi/\Lambda$ is
obtained by projecting $\mathbf{k}$ on the chosen direction. Thus, if we consider a
packet of plane waves with slightly different $\mathbf{k}$ vectors, the spacings
between adjacent crests for these plane waves is the same in a direction
perpendicular to the surface of wave vectors, as shown in Fig. 4-1b. It is
therefore natural to define rays as lines perpendicular to the surface of
wave vectors. Mathematically, this condition is written

$$\frac{d\mathbf{x}}{d\sigma} = \frac{\partial H}{\partial \mathbf{k}} \tag{4.5}$$

Optical thickness can be measured with considerable accuracy by introducing slabs of the material investigated cut with various orientations

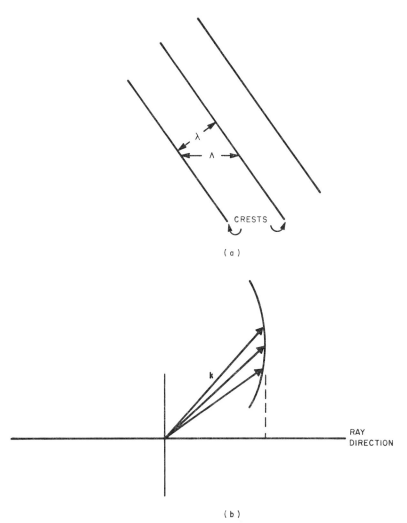

Fig. 4-1 (a) The distance between adjacent crests along the normal to the crest is the wavelength λ. The wavelength in some other direction is $\Lambda > \lambda$. (b) Waves in a wave packet have the same Λ in the direction of the ray, which is perpendicular to the surface of wave vector.

with respect to the crystal axes in interferometers such as the Mach-Zehnder interferometer. Thus **k** rather than n is the quantity that can be directly and accurately measured, from a distance measurement. In what follows, we assume that the surface of wave vector, described by $H(\mathbf{k}) = 0$, has been obtained, perhaps by interferometric measurements.

It is sometimes useful to define a ray vector **s**, which has the direction of the ray

$$\mathbf{s} = \frac{d\mathbf{x}/d\sigma}{\mathbf{k} \cdot (d\mathbf{x}/d\sigma)} = \frac{\partial H/\partial \mathbf{k}}{\mathbf{k} \cdot \partial H/\partial \mathbf{k}} \tag{4.6}$$

Clearly, from this definition, the scalar product of **s** and **k** is unity:

$$\mathbf{s} \cdot \mathbf{k} = 1 \tag{4.7}$$

The surface of ray vectors $\mathcal{L}(\mathbf{s}) = 0$, described by the tip of the s-vector, can be constructed from the surface of wave vector described by the tip of the k-vector as shown in Fig. 3-13. Because **s** is perpendicular to the surface $H(\mathbf{k}) = 0$, we have $\mathbf{s} \cdot \delta\mathbf{k} = 0$ for any small variation $\delta\mathbf{k}$ of **k**. Thus, from (4.7), we also have $\mathbf{k} \cdot \delta\mathbf{s} = 0$. This proves that **k** is perpendicular to the surface of ray vector. Thus

$$\mathbf{k} = \frac{\partial \mathcal{L}/\partial \mathbf{s}}{\mathbf{s} \cdot \partial \mathcal{L}/\partial \mathbf{s}} \tag{4.8}$$

The surface of ray vectors is the wavefront radiated by a time-harmonic point source. More precisely, $2\pi|\mathbf{s}|$ represents the distance between adjacent crests in the radial direction. The two surfaces $H(\mathbf{k}) = 0$ and $\mathcal{L}(\mathbf{s}) = 0$ can be obtained from the Maxwell equations if the medium permittivity and permeability are known. They are given in (3.147) and (3.148), respectively. With the notations in (3.149), we have $H(\mathbf{k}) \equiv \omega - \omega(\mathbf{k})$ and $\mathcal{L}(\mathbf{s}) = \omega - \omega(\mathbf{s})$. [Note that $\omega(\mathbf{s})$ is not the same function as $\omega(\mathbf{k})$.]

Let us consider as an example the surface of wave vector

$$H(\mathbf{k}) = \mathbf{V} \cdot \mathbf{k} + ak - \omega = 0 \tag{4.9}$$

where **V** is a constant vector and a, ω are constant scalars. This equation is applicable to the propagation of sound waves with angular frequency ω in the presence of a wind with velocity **V**. a denotes the magnitude of the sound wave velocity in the absence of wind. Dispersion is neglected. The surface of wave vectors is shown in Fig. 4-2a for the special case where the wind velocity **V** is directed along the z-axis and has a magnitude equal to a. In that case, the curve of wave vector is a parabola. Indeed, substituting $\mathbf{V} \cdot \mathbf{k} = ak_z$ in (4.9) and rearranging the terms, the surface of wave vector is found to be described by a new function[§]

$$H(k_x, k_z) = k_z - \frac{\omega}{2a} + \frac{a}{2\omega} k_x^2 = 0 \tag{4.10}$$

[§] Although the function $H(\mathbf{k})$ in (4.10) is not the same as the function $H(\mathbf{k})$ in (4.9), we use the same symbol H because they both describe the same surface of wave vector for the special case considered.

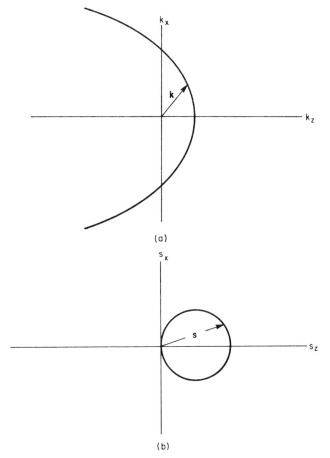

Fig. 4-2 (a) Surface of wave vector for sound waves with velocity a when the wind blows with a velocity equal to a in the direction of the z-axis. (b) Surface of ray vector.

To obtain the surface of ray vector, we evaluate

$$\frac{\partial H}{\partial k_x} = \frac{a}{\omega} k_x, \qquad \frac{\partial H}{\partial k_z} = 1 \tag{4.11}$$

Thus

$$s_x = \frac{\partial H/\partial k_x}{k_x \, \partial H/\partial k_x + k_z \, \partial H/\partial k_z} = \frac{(a/\omega)k_x}{(a/\omega)k_x^2 + k_z} \tag{4.12a}$$

$$s_z = \frac{\partial H/\partial k_z}{k_x \, \partial H/\partial k_x + k_z \, \partial H/\partial k_z} = \frac{1}{(a/\omega)k_x^2 + k_z} \tag{4.12b}$$

Solving (4.12) for k_x, k_z, and substituting back in (4.10), we obtain

$$\mathcal{L}(s_x, s_z) \equiv \frac{s_z - (\omega/2a)(s_x^2 + s_z^2)}{s_z^2} = 0 \tag{4.13}$$

This curve is most easily expressed in polar coordinates, with $s_x = s \sin(\alpha)$, $s_z = s \cos(\alpha)$. Substituting in (4.13), we obtain

$$s = \frac{2a}{\omega} \cos(\alpha) \tag{4.14}$$

Thus the curve of ray vectors is a circle, as shown in Fig. 4-2b.

4.2 Time-Harmonic Plane Waves in Inhomogeneous Stationary Media

We now assume that the properties of the medium vary from point to point, that is, the medium is inhomogeneous. The function $H(\mathbf{k}) = 0$, which describes the surface of wave vectors, becomes a function of $\mathbf{x}$, as well as a function of $\mathbf{k}$, which we denote

$$H(\mathbf{k}, \mathbf{x}) = 0 \tag{4.15}$$

What the Hamilton theory of light is going to tell us is how to patch together the elementary plane waves that can be defined in every small region of space. The local *relation* between direction and wavelength of the wave is known from (4.15), but the actual direction and wavelength are as yet unknown.

In inhomogeneous media, the field is denoted

$$\psi(\mathbf{x}) = \psi_0(\mathbf{x}) \exp[iS(\mathbf{x})] \tag{4.16}$$

where ψ_0 is assumed to vary slowly with $\mathbf{x}$ in comparison with the variations of $S(\mathbf{x})$. The wave vector $\mathbf{k}$ is the gradient of the function $S(\mathbf{x})$:

$$\mathbf{k} = \nabla S \tag{4.17}$$

because the slow variations of $\psi_0(\mathbf{x})$ are neglected.

Substituting (4.17) in (4.15), we have

$$H(\nabla S, \mathbf{x}) = 0 \tag{4.18}$$

Differentiating (4.18) totally with respect to x_α, we obtain, with the summation over repeated subscripts implied,

$$\frac{\partial H}{\partial k_\beta} \frac{\partial^2 S}{\partial x_\alpha \partial x_\beta} + \frac{\partial H}{\partial x_\alpha} = 0 \tag{4.19}$$

or, using (4.5) and interchanging the order of differentiations,

$$\frac{dx_\beta}{d\sigma} \frac{\partial^2 S}{\partial x_\beta \partial x_\alpha} + \frac{\partial H}{\partial x_\alpha} = 0 \tag{4.20}$$

Equation (4.20) is, in vector form,

$$\frac{d\mathbf{k}}{d\sigma} = - \frac{\partial H}{\partial \mathbf{x}} \tag{4.21}$$

Equation (4.21) tells us how the wave vector $\mathbf{k}$ varies with σ. Equations (4.5) and (4.21) completely define the ray trajectories for given initial conditions consistent with (4.15).

4.3 Time-Invariant, z-Invariant Media

We shall consider here in some detail the case of media that are independent of the z-coordinate because of our interest in optical fibers, which are usually uniform along the z-axis. We find it convenient in that case to write H in the form[§]

$$H(k_x, k_z, x) \equiv k_z - k_z(k_x, x) = 0 \tag{4.22}$$

For simplicity, the y-coordinate is omitted. Substituting this special form (4.22) of H in (4.5) and (4.21), we obtain a new form of the Hamilton equations:

$$\frac{dx}{d\sigma} = \frac{\partial H}{\partial k_x} = - \frac{\partial k_z(k_x, x)}{\partial k_x} \tag{4.23a}$$

$$\frac{dz}{d\sigma} = 1 \tag{4.23b}$$

$$\frac{dk_x}{d\sigma} = - \frac{\partial H}{\partial x} = \frac{\partial k_z(k_x, x)}{\partial x} \tag{4.23c}$$

$$\frac{dk_z}{d\sigma} = 0 \tag{4.23d}$$

[§] The reader is cautioned not to confuse the function $k_z(k_x, x)$ and the number k_z. da/dz denotes a total derivative, that is, in the present context, the variation of a along some given ray:

$$\frac{da(x, y)}{dz} \equiv \frac{\partial a}{\partial x} \frac{dx}{dz} + \frac{\partial a}{\partial y} \frac{dy}{dz}$$

for a ray $x(z), y(z)$.

Equation (4.23d) says that k_z is a constant of motion, that is, k_z remains the same along a ray trajectory. Equation (4.23b) allows us to identify σ with z, the axial coordinate.

Thus the Hamilton equations (4.23) are simply

$$\frac{dx}{dz} = -\frac{\partial k_z(k_x, x)}{\partial k_x} \tag{4.24a}$$

$$\frac{dk_x}{dz} = \frac{\partial k_z(k_x, x)}{\partial x} \tag{4.24b}$$

Note that these two equations imply the constancy of $k_z = k_z(k_x, x)$ along any ray trajectory (calculate dk_z / dz).

Let us assume now that the medium is isotropic, that is, $H(\mathbf{k})$ depends only on the square length $k^2(x) = k_x^2 + k_z^2$ of $\mathbf{k}(x)$. Equation (4.22) becomes

$$H(k_x, k_z, x) \equiv k_z - \left[k^2(x) - k_x^2 \right]^{1/2} = 0 \tag{4.25}$$

and the Hamilton equations (4.24a) and (4.24b) become, respectively,

$$\frac{dx}{dz} = \frac{k_x}{k_z} \tag{4.26a}$$

$$\frac{dk_x}{dz} = \frac{1}{2k_z} \frac{dk^2(x)}{dx} \tag{4.26b}$$

Equation (4.26a) is merely a statement that the **k**-vector is directed along the ray.

Square-Law Media

Let us specialize these equations further by assuming a square-law variation of k^2 as a function of the transverse coordinate:

$$k^2 = k_0^2(1 - \Omega^2 x^2) \tag{4.27}$$

Substituting in (4.26), we obtain

$$\frac{dx}{dz} = \frac{k_x}{k_z} \tag{4.28a}$$

$$\frac{dk_x}{dz} = -\Omega^2 k_0^2 k_z^{-1} x \tag{4.28b}$$

From (4.28a) and (4.28b), an equation for $x(z)$ is obtained (remember that k_z is a constant of motion):

$$\frac{d^2x}{dz^2} + \frac{\Omega^2 k_0^2}{k_z^2} x = 0 \tag{4.29}$$

whose solution is, for the initial condition $x(0) = 0$,

$$x(z) = \frac{k_{x0}}{k_0 \Omega} \sin\left(\frac{\Omega k_0}{k_z} z \right) \tag{4.30}$$

k_{x0} being the initial value of k_x. The amplitude factor in (4.30) is chosen to satisfy (4.28a) at the origin of coordinates. Note that for small amplitudes, $k_z \approx k_0$, and the period of ray oscillation is $2\pi/\Omega$. This period decreases as the ray amplitude increases.

Let us evaluate the ratio of the phase shift along a ray to its value on axis. In the absence of material dispersion, this ratio is indicative of pulse broadening in multimode waveguides. The ray period is from (4.30)

$$Z = \frac{2\pi k_z}{\Omega k_0} \tag{4.31}$$

and the phase shift on axis for that period is

$$S_0 \equiv k_0 Z = \frac{2\pi k_z}{\Omega} \tag{4.32}$$

The phase shift for a ray period is

$$S = \int_{\text{period}} k \, ds = k_z^{-1} \int_0^Z k^2 \, dz \tag{4.33}$$

because the elementary arc length ds is equal to $(k/k_z) \, dz$. Thus, using (4.27) with $x(z)$ from (4.30), we obtain

$$\frac{S}{S_0} = \frac{\Omega}{2\pi k_z^2} \int_0^Z \left(k_0^2 - k_{x0}^2 \sin^2 \frac{2\pi z}{Z} \right) dz$$

$$= \frac{1 - \frac{1}{2}\sin^2(\alpha_0)}{\cos(\alpha_0)} \approx 1 + \frac{\alpha_0^4}{8} + \cdots \tag{4.34}$$

where α_0 denotes the angle that the ray makes with the z-axis at the origin:

$$\cos(\alpha_0) = \frac{k_z}{k_0} \tag{4.35}$$

By comparison, if the medium is homogeneous and the ray is reflected by boundaries, as is the case for a clad slab or a waveguide, we obtain from simple geometric considerations

$$\frac{S}{S_0} = \frac{1}{\cos(\alpha_0)} \approx 1 + \frac{\alpha_0^2}{2} + \cdots \qquad (4.36)$$

The important point is that for square-law media S/S_0 is free of the second-order terms in α_0 that appear for step-index fibers.

Linear-Law Medium

As another example, consider a linear-law medium with

$$k^2(x) = k_0^2 - 2a|x| \qquad (4.37)$$

For the first half-period $0 < z < Z/2$, the rays are

$$x(z) = \tan(\alpha_0)z - \frac{a}{2k_0^2 \cos^2(\alpha_0)} z^2 \qquad (4.38)$$

where the period Z is

$$Z = \frac{4k_0^2 \sin(\alpha_0) \cos(\alpha_0)}{a} \qquad (4.39)$$

The ratio of the phase shift along a ray to the phase shift on axis is

$$\frac{S}{S_0} = \frac{1 - \frac{2}{3}\sin^2(\alpha_0)}{\cos(\alpha_0)} \approx 1 - \frac{\alpha_0^2}{6} + \cdots \qquad (4.40)$$

Here we have a second-order term but with a sign opposite to that in (4.36). In linear-law media, the phase shift is smaller for rays with large amplitudes than for rays with small amplitudes.

4.4 The Descartes–Snell Law of Refraction

The Hamilton equations show that if the medium is invariant along the x- and y-coordinates, the x- and y-components k_x, k_y of $\mathbf{k}$ are invariant along a ray trajectory. This is, in particular, the case at a plane interface between two homogeneous media perpendicular to the z-axis. Continuity of the transverse components of the wave vector is the general form of the Descartes–Snell law of refraction, originally established for isotropic media. Because rays are perpendicular to the surfaces of wave vectors, it is not difficult to trace the refraction of a light ray at the boundary between two anisotropic media. An example is shown in Fig. 4-3 for the refraction

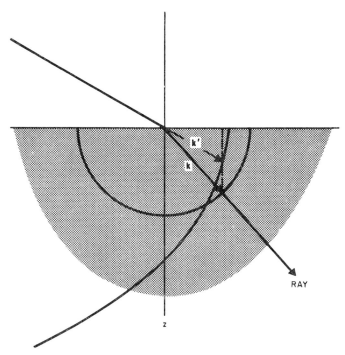

Fig. 4-3 Refraction at a plane interface between an isotropic medium (lower medium) and an anisotropic medium. Note that the incident ray does not have exactly the direction of **k'**.

between free space and the anisotropic medium described in (4.9). The case of two isotropic media (with wavenumbers k' and k, respectively) is shown in Fig. 4-4. For that case, we have clearly

$$k \cos(\alpha) = k' \cos(\alpha') \tag{4.41}$$

where α' and α denote the angles that the incident and refracted rays, respectively, make with the plane in the plane of incidence. Only the ratio k/k' enters in this relation. This ratio is the refractive index if the first medium is free space ($k' = \omega/c$).

The tangential x, y-components of **k**, as we have shown, are continuous at an interface. The normal z-component of **k**, on the other hand, suffers a discontinuity $k'_z - k_z$, which expresses the rate of change of the phase that results from a displacement of the boundary along the normal. More precisely, if ζ denotes a small displacement of the boundary as shown in Fig. 4-5, the phase shift ΔS experienced by the ray is

$$\Delta S = (k'_z - k_z)\zeta \tag{4.42}$$

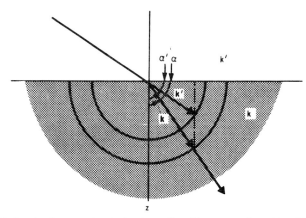

Fig. 4-4 Refraction between two isotropic media with wavenumbers k' and k, respectively. The law $k' \cos(\alpha') = k \cos(\alpha)$ is obeyed.

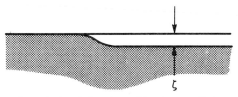

Fig. 4-5 The change in optical path length resulting from a displacement of the surface of discontinuity is the product of the displacement ζ and the variation of **k** along the normal to the surface.

For isotropic media, we have, in particular,

$$\Delta S = [k' \sin(\alpha') - k \sin(\alpha)]\zeta \tag{4.43}$$

For many systems (e.g., thin lenses, slightly curved mirrors), it is considerably simpler to evaluate the phase shift from (4.42) and (4.43), and subsequently the point-eikonal (see Section 4.5), than to apply directly the law of refraction (4.41). An example will be given in Section 4.21.

4.5 The Point-Eikonal

A fundamental concept of Hamiltonian optics is the point-eikonal, defined as the phase shift $S(\mathbf{x}; \mathbf{x}')$ along a ray from $\mathbf{x}'$ to $\mathbf{x}$. Let us again consider the function $H(\mathbf{k}, \mathbf{x})$ and define a Lagrangian function by

$$\mathcal{L}\left(\frac{d\mathbf{x}}{d\sigma} , \mathbf{x} \right) = \mathbf{k} \cdot \frac{d\mathbf{x}}{d\sigma} - H(\mathbf{k}, \mathbf{x}) \tag{4.44}$$

where **k** is expressed as a function of **x** and $dx/d\sigma$ through the first Hamilton equation

$$\frac{d\mathbf{x}}{d\sigma} = \frac{\partial H(\mathbf{k}, \mathbf{x})}{\partial \mathbf{k}} \tag{4.45}$$

The transformation in (4.44) is called a Legendre transformation. Note that if $\mathcal{L}$ in (4.44) were considered a function of the nine variables **x**, $dx/d\sigma$, and **k**, the partial derivative of $\mathcal{L}$ with respect to **k**, according to (4.45), would be equal to zero. This means that $\mathcal{L}$, in fact, does not depend explicitly on **k**.

Let us now evaluate $\partial\mathcal{L}/\partial(dx/d\sigma)$. We obtain

$$\frac{\partial\mathcal{L}}{\partial(d\mathbf{x}/d\sigma)} = \mathbf{k} + \frac{dx_\alpha}{d\sigma}\frac{\partial k_\alpha}{\partial(d\mathbf{x}/d\sigma)} - \frac{\partial H}{\partial k_\alpha}\frac{\partial k_\alpha}{\partial(d\mathbf{x}/d\sigma)} = \mathbf{k} \tag{4.46a}$$

Subscripts are used here to avoid a tensor notation. Summation over repeated indices is always implied. The last two terms in the expression between the equal signs cancel out because of (4.45).

Next, we evaluate

$$\frac{\partial\mathcal{L}}{\partial\mathbf{x}} = \frac{\partial k_\alpha}{\partial\mathbf{x}}\cdot\frac{dx_\alpha}{d\sigma} - \frac{\partial H}{\partial k_\alpha}\frac{\partial k_\alpha}{\partial\mathbf{x}} - \frac{\partial H}{\partial\mathbf{x}} = -\frac{\partial H}{\partial\mathbf{x}} = \frac{d\mathbf{k}}{d\sigma} \tag{4.46b}$$

using again (4.45) and (4.21). Equations (4.46a) and (4.46b) can be combined into a single equation

$$\frac{d}{d\sigma}\left(\frac{\partial\mathcal{L}}{\partial(d\mathbf{x}/d\sigma)}\right) = \frac{\partial\mathcal{L}}{\partial\mathbf{x}} \tag{4.47}$$

called the Euler equation. This equation can be solved for the ray trajectories $\mathbf{x}(\sigma)$. It is equivalent to the original Hamilton equations.

Let us now define the function

$$S(\mathbf{x}; \mathbf{x}') = \int_{\sigma'}^{\sigma}\mathcal{L}\left(\frac{d\mathbf{x}}{d\sigma}, \mathbf{x}\right)d\sigma \tag{4.48}$$

where the integral is, for the moment, along an arbitrary curve $\mathbf{x}(\sigma)$ from $\mathbf{x}'$ to **x**. Let us show that the variation of S is at most of second order with respect to a variation of the path of integration $\mathbf{x}(\sigma)$ from a ray path. It is important to remember that σ denotes an arbitrary parameter and not a specific function of the arc length. Thus the parameters σ, σ' at the endpoints can be assumed constant. We need vary only the integrand in (4.48):

$$\delta S(\mathbf{x}; \mathbf{x}') = \int_{\sigma'}^{\sigma}\left\{\frac{\partial\mathcal{L}}{\partial(d\mathbf{x}/d\sigma)}\cdot\delta\left(\frac{d\mathbf{x}}{d\sigma}\right) + \frac{\partial\mathcal{L}}{\partial\mathbf{x}}\cdot\delta\mathbf{x}\right\}d\sigma \tag{4.49}$$

We can exchange variation and total differentiation

$$\delta\left(\frac{d\mathbf{x}}{d\sigma} \right) = \frac{d}{d\sigma}\,\delta\mathbf{x} \qquad (4.50)$$

Thus, integrating by parts, we obtain

$$\delta S(\mathbf{x};\,\mathbf{x}') = \int_{\sigma'}^{\sigma}\left[-\frac{d}{d\sigma}\left(\frac{\partial\mathcal{L}}{\partial(d\mathbf{x}/d\sigma)} \right) + \frac{\partial\mathcal{L}}{\partial\mathbf{x}} \right] \cdot \delta\mathbf{x}\, d\sigma = 0 \quad (4.51)$$

if the Euler equation (4.47) is used. S is therefore stationary when the path of integration is a ray. In fact, it can be shown that S is then an extremum, that is, either an absolute maximum or an absolute minimum.

4.6 The Phase Space

The variation of $\mathbf{x}$ and $\mathbf{k}$ with σ can be represented as a curve in the six-dimensional phase space $\mathbf{x}$, $\mathbf{k}$. Let us consider a two-parameter manifold of rays $\mathbf{x}(u, v; \sigma)$ and $\mathbf{k}(u, v; \sigma)$ with parameters u, v. The quantity

$$I(u, v; \sigma) \equiv \frac{\partial\mathbf{x}}{\partial u} \cdot \frac{\partial\mathbf{k}}{\partial v} - \frac{\partial\mathbf{k}}{\partial u} \cdot \frac{\partial\mathbf{x}}{\partial v} \qquad (4.52)$$

known as the Lagrange ray invariant, is independent of σ.

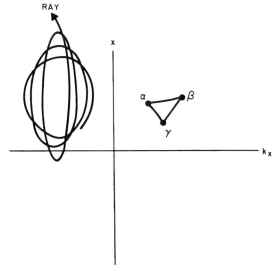

Fig. 4-6 Phase space k_x, x. The area enclosed by three neighboring rays (triangle) is a constant as z varies (Lagrange ray invariant). On the left of the figure, the trajectory in phase space of a ray in a lenslike medium with slowly varying parameters along the axis is shown. The area enclosed in phase space by the successive orbits is almost a constant.

Indeed, differentiating I with respect to σ and exchanging partial differentiations, we have

$$\frac{\partial I}{\partial \sigma} = \frac{\partial}{\partial u}\left(\frac{\partial \mathbf{x}}{\partial \sigma}\right) \cdot \frac{\partial \mathbf{k}}{\partial v} + \frac{\partial \mathbf{x}}{\partial u} \cdot \frac{\partial}{\partial v}\left(\frac{\partial \mathbf{k}}{\partial \sigma}\right) - \frac{\partial}{\partial u}\left(\frac{\partial \mathbf{k}}{\partial \sigma}\right) \cdot \frac{\partial \mathbf{x}}{\partial v}$$

$$- \frac{\partial \mathbf{k}}{\partial u} \cdot \frac{\partial}{\partial v}\left(\frac{\partial \mathbf{x}}{\partial \sigma}\right) \tag{4.53}$$

Using the Hamilton equation (4.5) and (4.21) and the identities

$$\frac{dH}{du} = \frac{\partial H}{\partial \mathbf{k}} \cdot \frac{\partial \mathbf{k}}{\partial u} + \frac{\partial H}{\partial \mathbf{x}} \cdot \frac{\partial \mathbf{x}}{\partial u}$$

$$\frac{dH}{dv} = \frac{\partial H}{\partial \mathbf{k}} \cdot \frac{\partial \mathbf{k}}{\partial v} + \frac{\partial H}{\partial \mathbf{x}} \cdot \frac{\partial \mathbf{x}}{\partial v} \tag{4.54a}$$

we obtain for the first term on the rhs of (4.53)

$$\frac{\partial}{\partial u}\left(\frac{\partial \mathbf{x}}{\partial \sigma}\right) \cdot \frac{\partial \mathbf{k}}{\partial v} = \frac{\partial}{\partial u}\left(\frac{\partial H}{\partial \mathbf{k}}\right) \cdot \frac{\partial \mathbf{k}}{\partial v} = \frac{\partial \mathbf{k}}{\partial v} \cdot \frac{\partial}{\partial \mathbf{k}}\left(\frac{dH}{du}\right)$$

$$= \frac{\partial k_\alpha}{\partial v} \frac{\partial}{\partial k_\alpha}\left(\frac{\partial H}{\partial k_\beta} \frac{\partial k_\beta}{\partial u} + \frac{\partial H}{\partial x_\beta} \frac{\partial x_\beta}{\partial u}\right)$$

$$= \frac{\partial k_\alpha}{\partial v} \frac{\partial k_\beta}{\partial u} \frac{\partial^2 H}{\partial k_\alpha \partial k_\beta} + \frac{\partial k_\alpha}{\partial v} \frac{\partial x_\beta}{\partial u} \frac{\partial^2 H}{\partial k_\alpha \partial x_\beta} \tag{4.54b}$$

Using similar expressions for the other terms in (4.53), it is easy to verify that all terms cancel out. It should be noted that the parameters u and v that define a ray do not uniquely define the vector $\mathbf{x}$ that goes from the origin of the coordinate system to a point on that ray because the endpoint is free to move along the ray. However, the quantity I is unaffected by a small displacement of that endpoint along the ray as we easily verify. We can therefore assume that the tip of the x-vector is constrained to remain in the $z = 0$ plane. Then $\partial \mathbf{x}/\partial u$ and $\partial \mathbf{x}/\partial v$ are 2-vectors in the $z = 0$ plane. Because their axial components are equal to zero, the components $\partial k_z/\partial u$, $\partial k_z/\partial v$ of $\partial \mathbf{k}/\partial u$ and $\partial \mathbf{k}/\partial v$ do not enter in I. That is, $\partial \mathbf{k}/\partial u$, $\partial \mathbf{k}/\partial v$ can also be considered 2-vectors instead of 3-vectors. This restricted form of I is the one most often given to Lagrange ray invariants.

Thus the phase-space representation is two dimensional for two-dimensional systems, as shown in Fig. 4-6. The invariance of I can be expressed in that case by saying that the area of a small triangle defined by three rays is invariant as some parameter, such as σ or z, varies. It follows that the area enclosed by any continuous set of rays is invariant. It also follows that for a dense set of rays with density f (number of points per unit area in k_x, x space), f is invariant

$$\frac{df}{d\sigma} = 0 \qquad (4.55)$$

Equation (4.55) is known as the Liouville theorem. If the medium varies slowly with z, a ray trajectory is almost periodic over a certain distance along the z-axis. Thus it follows that the area I enclosed in phase space by the (almost) closed trajectory is (almost) a constant. This constant area I enclosed by the ray trajectory is called an adiabatic invariant. It plays a fundamental role in wave optics and quantum mechanics because only adiabatic invariants can be quantized. In wave optics, the area I is required to be equal to $2\pi(m + \frac{1}{2})$, m being the mode number.

If we are dealing with a system that has two transverse dimensions (x, y) or more (x, y, t) the Liouville theorem (4.55) follows from the fact that the Jacobian of the transformation from the phase space at some transverse plane z' to the phase space at some other transverse plane z is unity and that the number of rays is invariant. Let us recall that the Jacobian of a transformation is the ratio of the elementary volumes $(dk'_x\, dk'_y\, dx'\, dy')$ and $(dk_x\, dk_y\, dx\, dy)$. Alternatively, we can say that the *ray matrix*, which relates small variations in k_x, k_y, x, y at the z' and z planes, respectively, has determinant unity. This property of ray matrices will be proved in a subsequent section of this chapter. The proof rests on the existence of eikonal functions. Liouville's theorem has exactly the same content as the theorem that says that paraxial ray matrices have determinant unity.

4.7 Time-Dependent Phenomena

The preceding discussion was applicable to time-harmonic sources and stationary media. Generalization of the Hamilton formalism to time-dependent phenomena is straightforward. We simply need to append time to the x-coordinates and ω to the wave vector. In order that the phase be

$$\mathbf{k} \cdot \mathbf{x} - \omega t \qquad (4.56)$$

we take ict as the fourth space coordinate and $i\omega/c$ as the fourth wave vector component. [An alternative choice is t and $-\omega$, respectively. The choice ict, $i\omega/c$ is convenient if Lorentz transformations are to be performed. Still another notation consists in introducing a metric tensor $\mathrm{diag}(1, 1, 1, -1)$ but one then must distinguish covariant and contravariant components.] Thus we define the two 4-vectors

$$\mathbf{X} \equiv \left\{ \begin{array}{c} \mathbf{x} \\ ict \end{array} \right., \qquad \mathbf{K} \equiv \left\{ \begin{array}{c} \mathbf{k} \\ i\omega/c \end{array} \right. \qquad (4.57)$$

The Hamilton equations are formally identical to (4.5) and (4.21) with 4-vectors replacing 3-vectors[5]

$$\frac{d\mathbf{X}}{d\sigma} = \frac{\partial H(\mathbf{K}, \mathbf{X})}{\partial \mathbf{K}} \tag{4.58a}$$

$$\frac{d\mathbf{K}}{d\sigma} = -\frac{\partial H(\mathbf{K}, \mathbf{X})}{\partial \mathbf{X}} \tag{4.58b}$$

It follows from these equations that if the medium is stationary, as assumed before, that is, if H does not depend explicitly on t, the angular frequency ω is a constant of motion. Conversely, if the medium is time dependent but homogeneous, the optical wave vector $\mathbf{k}$ is a constant. An optical pulse propagating in a homogeneous electrooptic crystal whose permittivity is made to vary in time has a varying frequency but a constant wavelength.

Because we are mostly interested in stationary media, we find it convenient to write the Hamiltonian in the form

$$H(\mathbf{K}, \mathbf{X}) \equiv \omega - \omega(\mathbf{k}, \mathbf{x}) = 0 \tag{4.59}$$

Because, from (4.58a) and (4.59), $dt/d\sigma = 1$, the parameter σ can be identified with time. If the medium is also independent of the z-coordinate, another convenient form is

$$H(\mathbf{K}, \mathbf{X}) \equiv k_z - k_z(\omega, k_x, k_y, x, y) = 0 \tag{4.60}$$

Thus, if the medium is isotropic with wavenumber $k(\omega, x, y)$,

$$H(\mathbf{K}, \mathbf{X}) \equiv k_z - \left[k^2(\omega, x, y) - k_x^2 - k_y^2\right]^{1/2} = 0 \tag{4.61}$$

The Hamilton equations (4.58) with H as in (4.61) are

$$\frac{dt}{dz} = \frac{\partial k_z}{\partial \omega} = \frac{1}{2} \frac{\partial k^2/\partial \omega}{k_z} \tag{4.62}$$

$$\frac{dx}{dz} = -\frac{\partial k_z}{\partial k_x} = \frac{k_x}{k_z} \tag{4.63a}$$

$$\frac{dk_x}{dz} = \frac{\partial k_z}{\partial x} = \frac{1}{2} \frac{\partial k^2/\partial x}{k_z} \tag{4.63b}$$

The time of flight of a pulse is obtained by integrating (4.62) along the z-axis,

$$t = \frac{1}{2k_z} \int_0^z \frac{\partial k^2}{\partial \omega} \, dz \tag{4.64}$$

or equivalently, by integrating ds/u along a ray trajectory, where ds denotes the elementary arc length and $u = \partial\omega/\partial k$ the local group velocity. Indeed, we have

$$\frac{ds}{u} = \frac{\partial k}{\partial \omega} \frac{k}{k_z} \, dz = \frac{1}{2k_z} \frac{\partial k^2}{\partial \omega} \, dz \tag{4.65}$$

Square-Law Medium

As an example of application of these equations, let us consider the propagation of a pulse of light in a square-law medium with wavenumber

$$k^2(x, \omega) = k_0^2(\omega)[1 - \Omega^2(\omega)x^2] \tag{4.66}$$

Substituting (4.66) in (4.64) and using the solution (4.30) for the ray $x(z)$, we obtain the time of flight per unit length

$$v_g^{-1} = \frac{1}{2k_z Z} \int_0^Z \left[\frac{\partial k_0^2}{\partial \omega} - \frac{\partial(k_0^2\Omega^2)}{\partial \omega} \left(\frac{k_{x0}}{k_0\Omega} \right)^2 \sin^2 \frac{2\pi z}{Z} \right] dz \tag{4.67}$$

where $Z = 2\pi k_z/\Omega k_0$ denotes as before the ray period. After integration, we obtain

$$v_g^{-1} = \frac{k_0}{k_z} \frac{dk_0}{d\omega} - \frac{1}{2} \frac{k_{x0}^2}{k_0\Omega k_z} \frac{d(k_0\Omega)}{d\omega} \tag{4.68}$$

In the absence of dispersion, $dk_0/d\omega = k_0/\omega$ and $d\Omega/d\omega = 0$. Then (4.68) reduces to

$$v_g^{-1} = \frac{k_0}{\omega} \frac{1 - \frac{1}{2}\sin^2(\alpha_0)}{\cos(\alpha_0)} \tag{4.69}$$

where $\sin(\alpha_0) = k_{x0}/k_0$, $\cos(\alpha_0) = k_z/k_0$, in agreement with (4.34).

Back to Time-Harmonic Sources

In Section 4.18 we shall consider in detail the approximation of Gauss for time-harmonic sources and stationary media. The Hamiltonian will be used in the form (4.60). For convenience, we may set

$$\mathbf{p} \equiv \begin{cases} k_x, \\ k_y \end{cases} \qquad \mathbf{q} \equiv \begin{cases} x \\ y \end{cases} \tag{4.70a}$$

and write the surface of wave vectors

$$k_z - k_z(\mathbf{p}, \mathbf{q}, z) = 0 \tag{4.70b}$$

The Hamilton equations are then

$$\frac{d\mathbf{q}}{dz} = \frac{-\partial k_z(\mathbf{p}, \mathbf{q}, z)}{\partial \mathbf{p}}, \qquad \frac{d\mathbf{p}}{dz} = \frac{\partial k_z(\mathbf{p}, \mathbf{q}, z)}{\partial \mathbf{q}} \qquad (4.71)$$

The dependence of k_z on ω is omitted for brevity. The Lagrangian is now defined from the Legendre transformation

$$L\left(\frac{d\mathbf{q}}{dz}, \mathbf{q}, z\right) = \mathbf{p} \cdot \frac{d\mathbf{q}}{dz} + k_z(\mathbf{p}, \mathbf{q}, z) \qquad (4.72)$$

with $\mathbf{p}$ substituted from (4.71). The point-eikonal, or phase shift along a ray, is

$$S(\mathbf{q}, z; \mathbf{q}', z') = \int_{z'}^{z} L\left(\frac{d\mathbf{q}}{dz}, \mathbf{q}, z\right) dz \qquad (4.73)$$

More advanced results in geometrical optics can be found in Stravroudis,[6] Brandstatter,[7] and Arnaud.[8] In cylindrical coordinates, the element of length is $ds^2 = dr^2 + r^2 d\varphi^2 + dz^2$. Thus the metric elements are $g_{11} = g_{33} = 1$, $g_{22} = r^2$, $g_{ij} = 0$, $i \neq j$. The Hamiltonian equations for rays in cylindrical coordinates can be obtained from the general form in Brandstatter,[7] applicable to a space with arbitrary metric. However, this method is rather formal for the application considered, and we shall not use it.

4.8   Time of Flight in Graded-Index Fibers without Material Dispersion

Here we evaluate the time of flight of pulses following rays of various amplitudes in graded-index fibers. Because the approximation of Gauss is not applicable, rays of different amplitudes have different optical lengths. On the other hand, material dispersion can sometimes be neglected. In that case, times of flights are proportional to optical lengths. If the source is spatially incoherent, many different rays are excited. Because these rays have different optical lengths, pulse broadening takes place. This problem is of paramount importance in optical communication.

The Hamilton equations for light rays $x(z)$, $y(z)$ in an isotropic z-invariant medium with wavenumber $k(x, y)$ are, from (4.26), restoring the y-coordinate

$$\frac{dx}{dz} = \frac{k_x}{k_z}, \qquad \frac{dy}{dz} = \frac{k_y}{k_z}$$

$$\frac{dk_x}{dz} = \frac{1}{2k_z} \frac{\partial k^2}{\partial x}, \qquad \frac{dk_y}{dz} = \frac{1}{2k_z} \frac{\partial k^2}{\partial y} \qquad (4.74)$$

The ratio of the optical length of a ray to the corresponding optical length on axis is

$$\tau = k_0^{-1} \int k \, ds = \frac{\langle k^2 \rangle}{k_0 k_z} \tag{4.75}$$

where $\langle \ \rangle$ denotes an average over a ray period[§] and $k_0 \equiv k(0, 0)$. The object of this section is to evaluate τ in (4.75) from (4.74) for various profiles $k^2(x, y)$ of practical interest.

From the ray equations in (4.74), we first obtain

$$\frac{1}{2} k_z^2 \frac{d^2 X}{dz^2} = k_x^2 + X \frac{\partial k^2}{\partial X}, \qquad \frac{1}{2} k_z^2 \frac{d^2 Y}{dz^2} = k_y^2 + Y \frac{\partial k^2}{\partial Y} \tag{4.76}$$

where we have set $X \equiv x^2$, $Y \equiv y^2$. Remember that k_z is a constant of motion.

Adding these two expressions and using the relation $k^2 = k_x^2 + k_y^2 + k_z^2$, we obtain

$$\frac{1}{2} k_z^2 \frac{d^2(X + Y)}{dz^2} = k^2 - k_z^2 + X \frac{\partial k^2}{\partial X} + Y \frac{\partial k^2}{\partial Y} \tag{4.77}$$

Let us now integrate (4.77) over a distance such that $d(X + Y)/dz$ assumes the same value at the ends of the integration interval. The lhs of (4.77) vanishes. Let us consider profiles such that $k^2(x, y)$ is equal to a constant k_0^2, plus a homogeneous function of degree κ in $X \equiv x^2$ and $Y \equiv y^2$. Then we have (see footnote, p. 204)

$$X \frac{\partial k^2}{\partial X} + Y \frac{\partial k^2}{\partial Y} = \kappa (k^2 - k_0^2) \tag{4.78}$$

It follows from (4.78), (4.77), and (4.75) that, in the absence of material dispersion,

$$\tau = (k_0 k_z)^{-1} \langle k^2 \rangle = (1 + \kappa)^{-1} \left(\frac{k_z}{k_0} + \kappa \frac{k_0}{k_z} \right) \tag{4.79}$$

It is interesting that τ, according to (4.79), depends only on k_z and not on the other parameters that define the ray. The variation of τ as a function of $k_z' \equiv k_z/k_0$ is shown in Fig. 4-7, for various values of κ. The relative time of flight τ in (4.79) can be expanded in power series of

$$\tfrac{1}{2} \alpha^2 \equiv 1 - k_z'^2 \ll 1 \tag{4.80}$$

we obtain

$$\tau = 1 + \frac{\alpha^2}{4} \frac{\kappa - 1}{1 + \kappa} + \frac{\alpha^4}{32} \frac{3\kappa - 1}{1 + \kappa} + \cdots \tag{4.81}$$

[§] If the ray trajectory is not periodic, $\langle a \rangle$ is to be understood as $\lim_{z \to \infty}$ of the integral of a from 0 to z, divided by z.

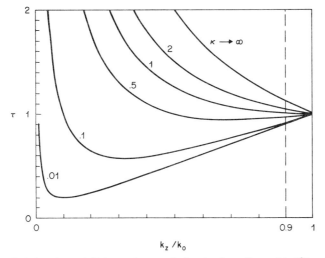

Fig. 4-7 Relative time of flight τ of an optical pulse in a fiber with $k^2(x, y) = k_0^2 +$ a homogeneous function of degree κ in x and y, as a function of the axial wavenumber k_z. In most fibers, k_z is bounded by k_s, the wavenumber of the surrounding medium (e.g., $k_s/k_0 = 0.9$) and $k_0 = k(0, 0)$. (From Arnaud.[8] Reprinted with permission from *The Bell System Technical Journal* copyright 1975, The American Telegraph and Telephone Co.)

Note that the second-order term in (4.81) vanishes when $\kappa = 1$ (square-law medium). For the special case where the fiber is circularly symmetric, the expression in (4.79) for $\tau(k_z)$ follows from a straightforward differentiation with respect to ω of the expression for the axial wavenumber k_z, given by Streifer and Kurtz.[9] (See also Section 4.24.) Impulse responses for circularly symmetric fibers free of material dispersion were given by Gloge and Marcatili.[9] Exact closed-form expressions for the impulse response will be given in Section 4.16.

4.9 Times of Flight in Circularly Symmetric Fibers without Material Dispersion

We now restrict ourselves to circularly symmetric fibers and assume that k^2 depends only on $X + Y \equiv R \equiv r^2$:

$$\frac{\partial k^2}{\partial X} = \frac{\partial k^2}{\partial Y} = \frac{\partial k^2}{\partial R} \tag{4.82}$$

The axial component of the ray canonical angular momentum

$$l_z \equiv xk_y - yk_x \tag{4.83}$$

is a constant of motion. This result is obtained by evaluating dl_z/dz, using (4.83) and the ray equations (4.74).

Equation (4.77) can be written as an equation for $R \equiv X + Y$:

$$\frac{1}{2} k_z^2 \frac{d^2 R}{dz^2} = \frac{d(k^2 R)}{dR} - k_z^2 \tag{4.84}$$

If we set $U \equiv dR/dz$, we have $d^2 R/dz^2 \equiv \frac{1}{2} dU^2/dR$. Thus (4.84) can be integrated:

$$\frac{U^2}{4} = R\left(\frac{k^2}{k_z^2} - 1 \right) - \frac{l_z^2}{k_z^2} \tag{4.85}$$

To verify the integration constant in (4.85), note that at the maximum or minimum R, $U = 0$, $k^2 = k_z^2 + k_\varphi^2$, where k_φ denotes the azimuthal wavenumber, and $l_z^2 = R k_\varphi^2$. Now (4.85) can be integrated in the form

$$z(R) = \frac{k_z}{2} \int_{R_{\min}}^{R} \left[R(k^2 - k_z^2) - l_z^2 \right]^{-1/2} dR \tag{4.86a}$$

and we have

$$z(R_{\max}) = \frac{Z_c}{2} \tag{4.86b}$$

where $R_{\min}$ and $R_{\max}$ denote the roots of the quantity in the bracket in (4.86a) and Z_c denotes the ray period. For meridional rays, Z_c is half the period Z of $x(z)$ defined in Section 4.3.

The integrand in (4.86a) is singular at $R_{\min}$ and $R_{\max}$, but the integral is well defined. Thus the variation of R with z can be obtained from a single integration once $k^2(R)$ is known. The ray is fully specified, except for arbitrary axial displacement and azimuthal rotation, when its two constants of motion k_z and l_z are given.

Square-Law Fibers

In particular, for square-law media $k^2 = 1 - R$, we obtain from (4.84) or (4.86a) a variation of the form

$$R(z) = R_0 + (R_0^2 - l_z^2)^{1/2} \cos \frac{2z}{k_z} \tag{4.87}$$

This result can be rewritten

$$R = \frac{1}{2} R_M (1 + \theta) + \frac{1}{2} R_M (1 - \theta) \cos \frac{2z}{k_z} \tag{4.88}$$

where $R_M \equiv R_{\max}$ and $\theta \equiv (l_z/R_M)^2$. For meridional rays, $\theta = 0$, and for helical rays, $\theta = 1$.

For later use, let us evaluate $\langle R^n \rangle$ using the binomial expansion and the result

$$\langle \cos^m \rangle = m! \, 2^{-m} [(\tfrac{1}{2} m)!]^{-2} \tag{4.89}$$

for m even, and 0 for m odd. We obtain

$$\langle R^n \rangle = n! \, 2^{-n} R_M^n \sum_{m=0, 2, \ldots}^{n} \frac{(1 + \theta)^{n-m}(1 - \theta)^m}{2^m (n - m)! \, [(\tfrac{1}{2} m)!]^2} \qquad (4.90)$$

In particular,

$$\langle R^2 \rangle = \tfrac{1}{8} R_M^2 (3\theta^2 + 2\theta + 3)$$

$$\langle R^3 \rangle = \tfrac{1}{16} R_M^3 (1 + \theta)(5\theta^2 - 2\theta + 5)$$

$$\langle R^4 \rangle = \tfrac{1}{128} R_M^4 (35\theta^4 + 20\theta^3 + 18\theta^2 + 20\theta + 35) \qquad (4.91)$$

Perturbed Square-Law Fibers

Let us consider next a perturbed square-law medium

$$k^2 = 1 - R + \sum_{n=2}^{N} \epsilon_n R^n \qquad (4.92)$$

where the $\epsilon_n R^{n-1}$ are of order ϵ. Integrating (4.84) over a ray period, we obtain

$$0 = \left\langle k^2 + R \frac{dk^2}{dR} - k_z^2 \right\rangle \qquad (4.93a)$$

Using (4.92), (4.93a) can be written

$$2\langle k^2 \rangle = k_z^2 + 1 + \sum_{n=2}^{N} (1 - n)\epsilon_n \langle R^n \rangle \qquad (4.93b)$$

Thus, the relative time of flight in (4.75) is

$$\tau = k_z^{-1} \langle k^2 \rangle = \tfrac{1}{2} k_z^{-1} \left[k_z^2 + 1 + \sum_{n=2}^{N} (1 - n)\epsilon_n \langle R^n \rangle \right] \qquad (4.94)$$

Equation (4.94) is exact. To first order in ϵ, R in (4.94) can be replaced by its zeroth-order approximation in (4.88), and we can set

$$k_z^2 = 1 - R_M(1 + \theta) \qquad (4.95)$$

A closed form for τ is therefore obtained for any small departure from square law that can be expanded in power series of R.

If we limit ourselves to a r^4 correction to the square-law profile, that is, if we set $\epsilon_3 = \epsilon_4 = \cdots = 0$, $\epsilon_2 \equiv \epsilon$, (4.94) becomes, using (4.95) and (4.91),

$$\tau = \tfrac{1}{2} [1 - R_M(1 + \theta)]^{-1/2} [2 - R_M(1 + \theta) - \tfrac{1}{8} \epsilon R_M^2 (3\theta^2 + 2\theta + 3)] \qquad (4.96)$$

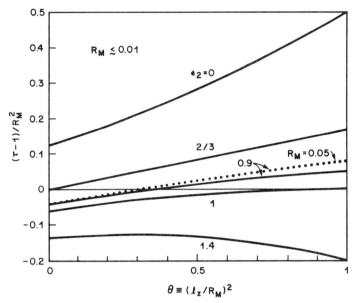

$$\theta \equiv (l_z/R_M)^2$$

Fig. 4-8 Relative time of flight τ of an optical pulse in a fiber with $k^2(r) = 1 - r^2 + \epsilon_2 r^4$ as a function of the axial component l_z of the angular momentum. For small ray radii, $\tau - 1$ is proportional to the fourth power of the maximum ray radius ($R_M \equiv r_{max}^2$). The rays are assumed to have the same maximum radius for any l_z. This condition is different from the condition $k_z > k_s$ in Fig. 4-7. (From Arnaud.[8] Reprinted with permission from *The Bell System Technical Journal*, copyright 1975, The American Telegraph and Telephone Co.)

Expanding τ in power series of R_M, we get

$$\tau \approx 1 + \tfrac{1}{16} R_M^2 [(2 - 3\epsilon) + (4 - 2\epsilon)\theta + (2 - 3\epsilon)\theta^2] \qquad (4.97)$$

This result is accurate when $R_M \lesssim 0.01$, that is, when the total change in refractive index ($\Delta n \approx R_M/2$) is less than about 0.005.

For meridional rays ($\theta = 0$), the fourth-order terms in (4.97) vanish when $\epsilon = \tfrac{2}{3}$. This is in agreement with the fact (Luneburg,[1] p. 180) that when

$$k^2(R) = \left[\cosh(R^{1/2})\right]^{-2} = 1 - R + \tfrac{2}{3}R^2 + \cdots \qquad (4.98)$$

$\tau = 1$, exactly, for meridional rays. For helical rays ($\theta = 1$), the fourth-order terms in (4.97) vanish when $\epsilon = 1$. This is in agreement with the fact[10] that when

$$k^2 = (1 \div R)^{-1} = 1 - R + R^2 + \cdots \qquad (4.99)$$

$\tau = 1$, exactly, for helical rays. $\tau - 1$ is plotted in Fig. 4-8 as a function of θ for various values of ϵ_2. Minimum pulse broadening is obtained (neglecting material dispersion) when $\epsilon_2 \approx 0.91$, that is, for a value close to the value that optimizes helical rays. For that value of ϵ_2, pulse broadening is eleven times smaller than for $\epsilon_2 = 0$, that is, for square-law media. In the above discussion we have specified the maximum ray radius $R_M^{1/2}$. If we specify instead the minimum value k_s of k_z, the minimum impulse width is obtained for $\epsilon_2 = \frac{2}{3} \approx 0.66$. This minimum is one-fourth the impulse width for square-law fibers.

It turns out that only slight improvement can be obtained by adding a r^6 correction term. This correction will therefore not be discussed here.

4.10 The Method of Strained Coordinates

To obtain terms of order ϵ^2 and higher, a more accurate expression for the ray trajectory than that given in (4.87) is needed. This corrected expression can be obtained by the method of strained coordinates.[11] Let us derive this result for the profile

$$k^2 = 1 - R + \epsilon R^2 \tag{4.100}$$

By setting, for convenience, $\alpha^2/2 \equiv 1 - k_z^2$, Eq. (4.84) for R becomes

$$k_z^2 \frac{d^2R}{dz^2} + 4R - \alpha^2 = 6\epsilon R^2 \tag{4.101}$$

For $\epsilon = 0$, the solution of (4.101) is, as we have seen in (4.87), of the form const $+ \cos(2z/k_z)$. For $\epsilon \neq 0$, let us set

$$t = \frac{2z}{k_z}, \qquad a = \frac{\alpha^2}{4}, \qquad \epsilon' = \frac{3\epsilon}{2} \tag{4.102}$$

Equation (4.101) simplifies slightly to

$$\frac{d^2R}{dt^2} + R - a = \epsilon'R^2 \tag{4.103}$$

Let now *both* R and t be expressed as power series of ϵ'

$$R = R_0 + \epsilon'R_1 + \cdots, \qquad t = s(1 + \epsilon'\omega_1 + \cdots) \tag{4.104}$$

Substituting (4.104) in (4.103) and identifying terms that are independent of ϵ' and proportional to ϵ', respectively, we obtain the differential equation

$$\frac{d^2R_0}{ds^2} + R_0 = a \tag{4.105}$$

whose solution is

$$R_0 = b^2 \cos(s) + a \tag{4.106}$$

where b denotes an integration constant, and

$$\frac{d^2 R_1}{ds^2} + R_1 = R_0^2 + 2\omega_1 \frac{d^2 R_0}{ds^2} \tag{4.107}$$

If we substitute in (4.107) the expression for R_0 in (4.106), we obtain

$$\frac{d^2 R_1}{ds^2} + R_1 = a^2 + \frac{b^4}{2} + \frac{b^4 \cos(2s)}{2} + 2b^2(a - \omega_1) \cos(s) \tag{4.108}$$

To avoid the secular terms that would make the expansion nonuniform, we select $\omega_1 = a$. The solution of (4.108) with $\omega_1 = a$ is simple

$$R_1 = -\frac{b^2}{6} \cos(2s) + a^2 + \frac{b^4}{2} \tag{4.109}$$

The complete solution of (4.103) is therefore

$$R = b^2 \cos(s) + a + \epsilon'\left[-\frac{b^4}{6} \cos(2s) + a^2 + \frac{b^4}{2} \right] + O(\epsilon'^2)$$

$$s = [1 - \epsilon'a + O(\epsilon'^2)]t \tag{4.110}$$

Going back to our original notation, we obtain the square R of the ray radius

$$R = b^2 \cos(s) + \frac{\alpha^2}{4} - \frac{1}{4} \epsilon b^4 \cos(2s) + \frac{3}{4} \epsilon\left(b^4 + \frac{\alpha^4}{8} \right) + O(\epsilon^2)$$

$$\tag{4.111a}$$

and

$$s = \left(1 - \frac{3\epsilon\alpha^2}{8} \right) \frac{2z}{k_z} \tag{4.111b}$$

The expression derived before for τ in (4.96) can be obtained alternatively by substituting expression (4.111a) for R in (4.75). A result of order ϵ^2 can be obtained by substituting this more accurate expression for R in (4.94). The details will not be given.

4.11 Time of Flight in Circularly Symmetric Fibers with Inhomogeneous Dispersion

The role of inhomogeneous dispersion[§] can be neglected in step-index fibers for quasi-monochromatic sources because of their large ray dispersion. For square-law fibers, however, inhomogeneous dispersion may be significant when the core is doped with high-dispersion materials such as TiO_2. The ratio τ of the time of flight of a pulse along a ray to the corresponding time on axis is, from (4.64),

$$\tau = \frac{k_0}{k_z} \frac{\langle \partial k^2 / \partial \omega^2 \rangle}{dk_0^2 / d\omega^2} \tag{4.112}$$

where, as before, $k_0 \equiv k(0)$ and $\langle\ \rangle$ denotes an average over a ray period.

Relative Permittivity a Power of Radius

Let us first consider a medium

$$k^2(R, \omega) = k_0^2(\omega) - k_\kappa^2(\omega)R^\kappa \tag{4.113}$$

where $R \equiv X + Y \equiv r^2$ denotes the square of the radius. The relative time of flight (4.112) is, for k^2 in (4.113),

$$\tau = \frac{k_0}{k_z} (1 - \epsilon_\kappa D_\kappa \langle R^\kappa \rangle) \tag{4.114}$$

where we have defined

$$\epsilon_\kappa \equiv \frac{k_\kappa^2}{k_0^2} \tag{4.115}$$

$$D_\kappa \equiv \frac{k_0^2(dk_\kappa^2/d\omega^2)}{k_\kappa^2(dk_0^2/d\omega^2)} \equiv \frac{k_0}{k_\kappa} \frac{dk_\kappa/d\omega}{dk_0/d\omega} \tag{4.116}$$

D_κ is a dispersion factor equal to unity in the absence of inhomogeneous dispersion. Thus we need to evaluate $\langle R^\kappa \rangle$. It is interesting that we can do so without solving the ray equations. The quantity $\langle R^\kappa \rangle$ is, of course, independent of dispersion. Thus we may omit the ω arguments.

[§] Inhomogeneous dispersion refers to the spatial (radial) variations of the ratio of the local phase to group velocities. Inhomogeneous dispersion should not be confused with the material dispersion $d^2k/d\omega^2$. See Section 5.18 for more details.

We have, directly from (4.113),

$$\langle k^2 \rangle = k_0^2 - k_\kappa^2 \langle R^\kappa \rangle \tag{4.117}$$

and, from (4.93a) and (4.113),

$$\langle k^2 \rangle = \frac{k_z^2 + \kappa k_0^2}{1 + \kappa} \tag{4.118}$$

Equating these two expressions for $\langle k^2 \rangle$, we obtain

$$\epsilon_\kappa \langle R^\kappa \rangle = \frac{1 - k_z'^2}{1 + \kappa} \tag{4.119}$$

where

$$k_z' \equiv \frac{k_z}{k_0} \tag{4.120}$$

Thus by substituting (4.119) in (4.114), the relative time of flight is

$$\tau = k_z'^{-1} - D_\kappa \frac{k_z'^{-1} - k_z'}{1 + \kappa} \tag{4.121}$$

This expression for τ reduces to (4.79) when $D_\kappa = 1$. The impulse response of a graded-index fiber, based on the above expression for τ, will be given in a later section.

Perturbed Square-Law Fibers

For square-law fibers,

$$k^2(R) = k_0^2 - k_1^2 R \tag{4.122}$$

the solution of (4.117) is

$$R(z) = R_0 + \left(R_0^2 - \frac{l_z^2}{k_1^2} \right)^{1/2} \cos \frac{2\Omega z}{k_z'} \tag{4.123}$$

where

$$R_0 \equiv \frac{1}{2} \frac{k_0^2 - k_z^2}{k_1^2} \equiv \frac{1}{2} \frac{1 - k_z'^2}{\Omega^2} \tag{4.124}$$

and $\Omega \equiv k_1/k_0$, a slight generalization of (4.87). $l_z \equiv xk_y - yk_x$ denotes as before the axial component of the angular momentum, which is the second constant of motion. Let us set

$$\theta \equiv \left(\frac{l_z}{k_1 R_M} \right)^2 \tag{4.125}$$

where R_M denotes the maximum radius squared. For meridional rays, $\theta = 0$ and for helical rays, $\theta = 1$. Equation (4.123) can be written in the convenient form (4.88) with the argument of the cosine now equal to $2\Omega z / k_z'$. Thus $\langle R^\kappa \rangle$ has the value in (4.90).

Let us now show that a closed-form expression can be obtained for the times of flight in fibers whose permittivity profiles depart slightly from a square-law, inhomogeneous dispersion being taken into account. Let

$$k^2(R) = k_0^2 - k_1^2 R + \sum_{n=2}^{N} k_n^2 R^n \tag{4.126}$$

We assume that $\epsilon_n R^{n-1}$, $n \geq 2$, is of order $\epsilon \ll 1$, where $\epsilon_n \equiv k_n^2 / k_0^2$.

Substituting (4.126) in (4.112), we obtain

$$\tau = k_z^{-1} \left(1 - D_1 \Omega^2 \langle R \rangle + \sum_{n=2}^{N} \epsilon_n D_n \langle R^n \rangle \right) \tag{4.127}$$

where we have defined the dispersion factors

$$D_n = \frac{k_0^2 (dk_n^2 / d\omega^2)}{k_n^2 (dk_0^2 / d\omega^2)} \tag{4.128}$$

The D_n are unity in the absence of material dispersion. Because the perturbation is small, $\langle R^n \rangle$ in the sum (4.127) can be replaced by the expression (4.90) applicable to square-law fibers. This approximation is not permissible, however, for the term $\langle R \rangle$ because this term is not small. We need an exact expression for $\langle R \rangle$. We shall proceed as in the previous paragraph. We first observe that, for k^2 in (4.126),

$$\frac{d(k^2 R)}{dR} = 2k^2 - k_0^2 + \sum_{n=2}^{N} (n-1)k_n^2 R^n \tag{4.129}$$

Integrating (4.84) over a ray period, the lhs vanishes and, using (4.129), we obtain an expression for $\langle k^2 \rangle$ that does not involve $\langle R \rangle$:

$$\langle k^2 \rangle = \tfrac{1}{2}(k_z^2 + k_0^2) + \tfrac{1}{2} \sum_{n=2}^{N} (1-n)k_n^2 \langle R^n \rangle \tag{4.130}$$

We also have, directly from (4.126),

$$\langle k^2 \rangle = k_0^2 - k_1^2 \langle R \rangle + \sum_{n=2}^{N} k_n^2 \langle R^n \rangle \qquad (4.131)$$

Thus by comparing (4.130) and (4.131),

$$k_1^2 \langle R \rangle = \tfrac{1}{2}(k_0^2 - k_z^2) + \tfrac{1}{2} \sum_{n=2}^{N} (n+1)k_n^2 \langle R^n \rangle \qquad (4.132)$$

Substituting this expression for $\langle R \rangle$ in (4.127), we obtain the relative time of flight for dispersive near square-law fibers:

$$\tau = k_z'^{-1} \left\{ 1 - \tfrac{1}{2}(1 - k_z'^2)D_1 + \sum_{n=2}^{N} [D_n - \tfrac{1}{2}(n+1)D_1]\epsilon_n \langle R^n \rangle \right\} \qquad (4.133)$$

where $\langle R^n \rangle$ is given in (4.90), $k_z' \equiv k_z/k_0$, and D_n is defined in (4.128). This expression for the time of flight will be used in subsequent sections to obtain the pulse broadening in near square-law fibers. It is written in Section 4.24 in terms of the mode numbers (α, μ).

4.12 Nonuniform Losses

Let us examine now the problem of loss. In graded-index fibers, the loss depends significantly on the radius. For example, in vapor-phase deposition, the central part of the fiber is richer in foreign material and is sometimes more lossy, either because of the impurities in the foreign material or because of composition fluctuations that introduce Rayleigh scattering. The core–cladding interface irregularities introduce some loss too. These losses, however, will not be considered here because they are of a different nature. Let us define a relative loss l as the ratio of the loss along a ray to the loss along the corresponding length on axis. If we set $k = k_r + ik_i$, the loss along an elementary path length ds is $k_i \, ds$, in nepers. It is convenient to consider the imaginary part of the square of the wavenumber $k^2 \equiv \kappa_r + i\kappa_i$. For small losses, we have the approximation

$$\kappa_i = 2k_r k_i \approx 2kk_i \qquad (4.134)$$

Thus the relative loss is simply

$$l = \frac{\int k_i \, ds}{\int k_{i0} \, dz} = \frac{k_0}{k_z \kappa_{i0}} \langle \kappa_i \rangle \qquad (4.135)$$

where we have again used $ds = k\, dz / k_z$ and the subscript 0 denotes values on axis. The average $\langle \kappa_i \rangle$ can be evaluated for an unperturbed ray trajectory. The integration is therefore simple.

4.13 Incoherent Sources

Blackbody radiation and light-emitting diodes are two important examples of spatially incoherent sources. They differ in their frequency spectrum and their radiance.

Blackbody Radiation

A blackbody is a collection of atoms in equilibrium with their radiation field. The modes of resonance of a cavity with sides x_0, y_0, z_0, along the x-, y-, and z-axes, respectively, satisfy the cyclic conditions $k_x x_0 = 2m\pi$, $k_y y_0 = 2n\pi$, $k_z z_0 = 2l\pi$, where m, n, l denote integers, positive and negative. The frequency of a mode m, n, l is obtained by substituting these values for k_x, k_y, k_z in the dispersion equation $\omega = \omega(k_x, k_y, k_z)$. According to Planck, each of these modes, in equilibrium with the cavity walls at a temperature T, has an energy

$$E(\omega) = \frac{\hbar\omega}{\exp(\hbar\omega / KT) - 1} \approx KT, \qquad \text{if} \quad \hbar\omega \ll KT \qquad (4.136)$$

In (4.136), $\hbar$ denotes the Planck constant divided by 2π, and K denotes the Boltzmann constant.

Because of the resonant conditions stated above, the number of modes between $k_x, k_x + dk_x, k_y, k_y + dk_y$, and $k_z, k_z + dk_z$ is

$$dk_x\, dk_y\, dk_z\, x_0 y_0 z_0 / (2\pi)^3.$$

Thus the number of modes between $k_x, k_x + dk_x$, $k_y, k_y + dk_y$, and $\omega, \omega + d\omega$ is

$$dk_x\, dk_y\, d\omega\, V / u_z (2\pi)^3$$

where $u_z \equiv \partial\omega / \partial k_z$ denotes the z-component of the group velocity $\mathbf{u}$, and $V \equiv x_0 y_0 z_0$ denotes the volume of the cavity. The power density, on the other hand, is the product of the group velocity $\mathbf{u}$ and the energy density $E(\omega)/V$. Thus the power per unit area radiated between $k_x, k_x + dk_x$, $k_y, k_y + dk_y$, and $\omega, \omega + d\omega$, divided by $dk_x\, dk_y\, d\omega$, is

$$\mathbf{P} = E(\omega)\mathbf{u} / u_z (2\pi)^3 \qquad (4.137)$$

The apparent cross section of an aperture in the xy plane, with area A, is $A(u_z/u)$, where u denotes the magnitude of $\mathbf{u}$. Thus the power radiated by the cavity through the aperture within $d\omega\, dk_x\, dk_y$ divided by $d\omega\, dk_x\, dk_y$ is

$$\mathbf{P}_A = \frac{AE(\omega)(\mathbf{u}/u)}{(2\pi)^3} \tag{4.138}$$

Note that $|\mathbf{P}_A|$ is independent of k_x, k_y. Another way of stating this result is to say that blackbody radiation is uniform in the phase space k_x, k_y, x, y. The variation of $\mathbf{P}_A$ with ω is that given in (4.136). At low frequencies, the spectral density increases in proportion to the square of the frequency. It drops to negligible values above a certain frequency proportional to the temperature (peak frequency $= 5.9 \times 10^{10} T^{\circ}\text{K}$). The peak wavelength λ_0 is of the order of 16 μm at room temperature $T = 300^{\circ}\text{K}$, and 0.85 μm at $T = 6000^{\circ}\text{K}$. In light-emitting diodes, electron–hole recombination gives rise to a spontaneous emission of light, whose frequency is related to the band gap. The spectral width is of the order of 0.04 μm; that is, it is considerably narrower than that of a blackbody radiator, but it is much broader than that of a single-mode laser. The above expressions are applicable to each of the two states of polarization of the electromagnetic field.

Liouville Equation

A monochromatic spatially incoherent source can be generated, for instance, by shining a monochromatic laser beam on a rotating diffusing plate. When a wave in a single transverse mode is transmitted through an irregular fiber, the coherence tends to be destroyed after certain length. To be precise, the concept of coherence is based on averagings over large numbers of similar fibers. However, the inevitable fluctuations in time that affect irregular fibers often simulate statistical averaging. Because we do not wish at this point to enter into any mathematical detail, we shall use the word "coherence" rather loosely. In any event, the broadening of pulses that results from the source being spatially incoherent can be studied independently from the broadening that results from the nonzero spectral width of the source.

The transmission of incoherent light through an optical system is easy to formulate. An incoherent source can be represented by a collection of optical pulses. Each pulse is characterized at the input plane of the optical system by its carrier angular frequency ω, the time t when it crosses the input plane, its position x, y, and the transverse components k_x, k_y of the wave vector. The latter define the direction in space of the light pulse. In short, the optical pulse is characterized by a 6-vector $\boldsymbol{\xi}$, with components k_x, k_y, ω, x, y, t. If the light pulse has a duration Δt, ω is defined only to within Δt^{-1}. Similarly, if the diameter of the light beam is Δx (respectively, Δy) the wavenumber Δk_x (respectively, Δk_y) is defined at best to within Δx^{-1} (respectively, Δy^{-1}). Thus the light pulse occupies a volume in the phase space k_x, k_y, ω, x, y, t at least equal to one. The basic assumption of geometrical optics is that the total volume in phase space is so large that the elementary volume defined above can be neglected. We deal with each pulse as if it were a mathematical point in phase space. Furthermore, the collection of points (or pulses) in phase space is assumed to be so dense that it is possible to define a density. This density, or distribution $f(\xi)$, is the number of points in the volume $\{k_x, k_x + dk_x; \ldots ; t, t + dt\}$ divided by $dk_x \cdots dt$. Each point can be given a weight $E(\xi)$ equal to the energy in the pulse. Thus the energy density in phase space is $E(\xi)f(\xi)$. As the pulses proceed along the z-axis, the distribution $f(\xi)$ evolves. The equations of Hamilton show that elementary volumes along any path remain constant. Because the number of points is invariant, the density f remains the same along a pulse trajectory. This is the Liouville theorem $df/dz = 0$ stated earlier. If the losses are not uniform, the energy E associated with each point may vary. In some cases, we can specify that pulses originating from some region of the phase space are unattenuated, while the others are completely absorbed. The evolution of $f(\xi)$ as a function of z is given by the Liouville equation, which is merely a restatement of $df/dz = 0$:

$$0 = \frac{df}{dz} = \frac{\partial f}{\partial z} + \frac{\partial f}{\partial x}\frac{dx}{dz} + \frac{\partial f}{\partial y}\frac{dy}{dz} + \frac{\partial f}{\partial t}\frac{dt}{dz}$$

$$+ \frac{\partial f}{\partial k_x}\frac{dk_x}{dz} + \frac{\partial f}{\partial k_y}\frac{dk_y}{dz} + \frac{\partial f}{\partial \omega}\frac{d\omega}{dz} \qquad (4.139)$$

The quantities dx/dz, dy/dz, dt/dz, dk_x/dz, dk_y/dz, and $d\omega/dz$ can be expressed as functions of x, y, t, k_x, k_y, ω, and z with the help of the Hamilton equations. Thus, if the initial source distribution $f(\xi)$ is known, and the medium parameters are known, it is possible to obtain the evolution of $f(\xi)$ by integrating the partial differential equation (4.139), perhaps with the help of a computer. Equation (4.139) provides the increment of f for an increment of z at a fixed ξ.

Assuming that this problem has been solved, the output pulse power $P(t)$ is obtained by integrating $E(\xi)f(\xi)$ over all variables except t. Detailed expressions for $P(t)$ will be given in subsequent sections.

Acceptance or Effective Number of Modes

A more global characteristic of the response of optical systems to incoherent sources is the *acceptance*, which is defined as follows: Let us first assume that the source is monochromatic and uniform in spatial phase space with a density $1/(2\pi)^2$. Note that, for isotropic media, $k_z = [k^2(\omega) - k_x^2 - k_y^2]^{1/2}$ is real only when $k_x^2 + k_y^2 < k^2(\omega)$. $f(k_x, k_y)$ is assumed to be a constant inside the circle defined by the above equation. The rays located outside that circle can be ignored. The power radiated through a small area dA in the xy plane is obtained by multiplying dA by the area πk^2 of the circle defined above and dividing by $(2\pi)^2$

$$P = \frac{\pi k^2 \, dA}{(2\pi)^2} \qquad (4.140)$$

Let us now evaluate the angular distribution of the radiation from dA. Let us consider a small solid angle $d\Omega$ originating from the center of the small area dA, in a direction that makes an angle α to the z-axis. The projection of the area $d\Omega$ (on the unit sphere) on the transverse plane xy is clearly $d\Omega \cos(\alpha)$. Assuming isotropy of the medium, the wave vectors have the direction of the rays. Thus the projection of the wave vectors within the solid angle $d\Omega$ on the transverse k_x, k_y plane covers an area $k^2 \, d\Omega \cos(\alpha)$. Because the density in the k_x, k_y plane is, by definition, $1/(2\pi)^2$, the power radiated within $d\Omega$, divided by $d\Omega$, is

$$R(\alpha) = \frac{k^2 \, dA \, \cos(\alpha)}{(2\pi)^2} \qquad (4.141)$$

The integral of $R(\alpha)2\pi \sin(\alpha) \, d\alpha$ from $\alpha = 0$ to $\alpha = \pi/2$ is the total power P radiated by the area A (irradiance), given in (4.140). If we define the radiance (or brightness)

$$B(\alpha) \equiv \frac{R(\alpha)}{dA \, \cos(\alpha)} \equiv \frac{k^2}{(2\pi)^2} = \frac{1}{\lambda^2} \qquad (4.142)$$

the radiation (4.141) takes the form of the Lambert law

$$R(\alpha) = B \, dA \, \cos(\alpha) \qquad (4.143)$$

where $B \equiv B(0)$.

A spatially uniform Lambertian source can be viewed as a surface whose elementary areas radiate power incoherently, the radiation pattern of each area being that given in (4.143), or as a spectrum of incoherent plane waves. The power per unit solid angle and unit area of these plane waves is the radiance B. It is independent of α. A Luneburg lens of large radius provides a concrete way of transforming a plane wave (within $d\Omega$) of the plane wave representation into a point source (within dA) of the point source representation. Locally, the transformation is based on the sine law, applicable to any aplanatic system (an optical system is called aplanatic if it is free of spherical aberration and coma).

We are now in a position to define the acceptance N^2 of an optical system. This is the power that the system transmits when the source has a constant distribution $1/(2\pi)^2$ in spatial phase space. It can be shown that N^2 represents the effective number of modes that can be transmitted by the optical system for a given polarization state. The relation between ray optics and wave optics is discussed further in Section 4.24.

The acceptance of an aperture with area A is simply the power given in (4.140) because all the rays within the area A are transmitted without loss. Thus for an aperture

$$N^2 = \frac{\pi k^2 A}{(2\pi)^2} = \frac{\pi A}{\lambda^2} \tag{4.144}$$

For two circular coaxial apertures with radii a_1 and a_2, respectively, and spacing d, we shall only give the result.[12] We have exactly

$$N^2 = \frac{k^2}{8}\left\{ d^2 + a_1^2 + a_2^2 - \left[d^4 + 2d^2(a_1^2 + a_2^2) + (a_1^2 - a_2^2)^2 \right]^{1/2} \right\} \tag{4.145}$$

This expression reduces to (4.144) when $d = 0$, if we set $A = \pi a_1^2$, $a_1 < a_2$, or $A = \pi a_2^2$, $a_2 < a_1$. In the limit of large spacing $d \gg a_1, a_2$, we obtain

$$N^2 = \frac{(\pi a_1^2)(\pi a_2^2)}{(\lambda d)^2} \tag{4.146}$$

Thus N^2 coincides for paraxial rays with the power coupling defined in (2.96).

Frequency filtering of spatially incoherent sources can be obtained with so-called degenerate optical cavities, which we shall discuss in more detail in Section 4.23. The concept of acceptance of an optical instrument is applicable to degenerate optical cavities. Here we shall give only an expression for the acceptance of a degenerate cavity incorporating circular apertures, but free of aberration.

A degenerate optical cavity is an optical system where all of the ray trajectories are closed curves of equal optical length. Because any wavefront reproduces itself at the resonant frequency after a round trip in the cavity, such cavities provide efficient frequency filtering of incident signals with arbitrary wavefronts. The degeneracy condition can be met exactly using continuously varying refractive-index media, such as Luneburg lenses. The configurations usually encountered, however, suffer from various types of aberrations, which severely limit their capability of accepting arbitrary wavefronts (see Section 4.23). This capability is expressed by

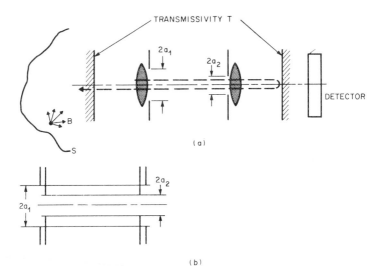

Fig. 4-9 (a) Degenerate optical cavity with stops. (b) Acceptance factor of a pair of stops. (From Arnaud,[20] by permission of the American Optical Society.)

the acceptance N^2, defined as earlier in this section. The acceptance factor of a lossless cavity resonating in a single mode is unity.

Let us consider a degenerate cavity free of aberrations but incorporating a number of stops along its optical axis as shown in Fig. 4-9a. Since any ray retraces its own path after a round trip in the cavity, it is sufficient to determine what rays originating from the source make a round trip without being intercepted by a stop. Following a procedure used in conventional optical instruments, we calculate the image of each stop back into the

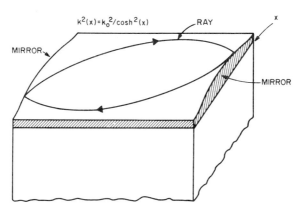

Fig. 4-10 Degenerate filter free of aberration. The fiber thickness is tapered in such a way that $k(x) = 1/\cosh(x)$ (see ref. 1). The acceptance of such a filter is the same as that of the fiber itself at the resonant frequency.

object space. For a linear cavity, however, stops whose location does not coincide with the end mirrors must be considered twice, as they are crossed twice by the optical axis in a round trip. Let us apply these conditions to the degenerate cavity shown in Fig. 4-9a, which incorporates two identical confocal lenses of focal length f between two plane end mirrors and two coaxial circular stops of radii a_1 and a_2, respectively, adjacent to the lenses. The images of these stops in the object space are shown in Fig. 4-9b. The acceptance is, therefore, that of a system of two coaxial circular apertures of radii $\min(a_1, a_2)$ separated by a distance $d = 2f$. This result was given in (4.145). An example of degenerate cavity free of aberration using thin films is shown in Fig. 4-10.

4.14 Acceptance of Optical Fibers

Let us consider first a conventional metallic waveguide with cross-sectional area A. Because all the rays within that area are transmitted, the acceptance is simply that in (4.144) of an aperture with area A.

Wave Optics Acceptance

The problem is slightly more complicated for a dielectric waveguide with wavenumber $k(x, y)$ in a medium with wavenumber k_s. If we assume that the rays whose k_z is less than k_s radiate away and may, therefore, be ignored, a simple and general result is obtained. Indeed, the condition $k_z > k_s$ is, for isotropic media,

$$k_x^2 + k_y^2 < k^2(x, y) - k_s^2 \qquad (4.147)$$

For each x, y point, the permitted area in k_x, k_y space is equal to $\pi[k^2(x, y) - k_s^2]$. Thus, integrating over the fiber cross section and multiplying by 2 to account for the two states of polarization, we obtain

$$N^2 = \frac{1}{2\pi} \int \int [k^2(x, y) - k_s^2]\, dx\, dy \qquad (4.148)$$

If we think of the $k^2 = k^2(x, y)$ surface as a well (upside down), we can say that the (wave optics) acceptance of a dielectric fiber is the volume enclosed in the $k^2(x, y)$ well, divided by 2π. The irradiance in the fiber is proportional to the r.h.s. of (4.147).

For a step-index fiber, with $k(x, y)$ equal to a constant k_0 within an area A and $k(x, y) = k_s < k_0$ outside that area, the acceptance is, from (4.148),

$$N^2 = \frac{A}{2\pi} (k_0^2 - k_s^2) \qquad (4.149)$$

As another example, let us assume that

$$k^2(r) = \begin{cases} k_0^2 - (k_0^2 - k_s^2)\left(\dfrac{r}{a}\right)^{2\kappa}, & r < a \\ k_s^2, & r \geqslant a \end{cases} \qquad (4.150)$$

We find from (4.148), after integration,

$$N^2 = \frac{\kappa}{2(\kappa + 1)}\,(k_0^2 - k_s^2)a^2 \qquad (4.151)$$

For square-law fibers $\kappa = 1$, the acceptance is half that of the step-index fiber having the same wavenumber on axis.

Ray Optics Acceptance

The Descartes–Snell law of refraction states that a ray ceases to be totally reflected if the tangential (rather than axial) component of $\mathbf{k}$ is less than k_s. Thus many rays are totally reflected whose k_z is, in fact, less than k_s. Because these rays are leaking out only slightly, the power that they carry needs to be taken into consideration, at least for highly oversized fibers. Expressions of the loss that they suffer will be given in Section 4.24. If we assume that these rays are unattenuated (as geometrical optics asserts), a new, larger acceptance may be defined. The derivation is more complicated than before. We shall give it only for circularly symmetric fibers.

In place of the condition $k_z > k_s$ used earlier, we now specify that the rays reaching the fiber–cladding interface must satisfy the less restrictive condition

$$k_z^2 + k_\varphi^2 > k_s^2, \qquad \text{at} \quad r = a \qquad (4.152)$$

where k_φ denotes the azimuthal wavenumber at the interface. Because of conservation of the axial component of the angular momentum l_z, we have

$$rk_y = ak_\varphi \qquad (4.153)$$

for a ray originating from $x = r$, $y = 0$, k_x, k_y at the input plane, that can reach the interface $r = a$. Thus condition (4.152) is

$$k_x^2 + \left(1 - \frac{r^2}{a^2}\right)k_y^2 < k^2(r) - k_s^2 \qquad (4.154)$$

Equation (4.154) defines an area in the k_x, k_y plane bounded by an ellipse. We need also specify that k_z is a positive quantity, that is,

$$k_x^2 + k_y^2 < k^2(r) \qquad (4.155)$$

This condition is not implied by (4.154). However, for weakly guiding fibers, the additional condition (4.155) can be ignored.

We have to make sure that rays outside the area defined in (4.154) do, in fact, reach the interface. This is not necessarily the case. The maximum radius r_M of a ray is defined implicitly by

$$k_x^2 + \left(1 - \frac{r^2}{r_M^2}\right)k_y^2 = k^2(r) - k^2(r_M) \tag{4.156}$$

r_M is the largest real number that satisfies (4.156). (The initial radius r is considered a constant in the present discussion.) Equation (4.156) shows that the k_x, k_y that correspond to some given r_M are contained in an ellipse with semiaxes squared

$$k_{x0}^2 = k^2(r) - k^2(r_M) \quad \text{and} \quad k_{y0}^2 = [k^2(r) - k^2(r_M)]/(1 - r^2/r_M^2)$$

respectively. If $k^2(r)$ is never increasing, we are sure that k_{x0} keeps increasing as r_M increases from r to a. We do not have any such assurance for k_{y0}, however. When r_M reaches a, there may be acceptable values of k_x, k_y that are located outside the ellipse defined in (4.154). For each profile, we need therefore to verify that $k_{y0}^2(r_M)$ never exceeds $k_{y0}^2(a)$. We easily verify that this is the case for square-law fibers, because

$$k_{y0}^2 = \frac{k_1^2(r_M^2 - r^2)}{1 - r^2/r_M^2} = k_1^2 r_M^2 \tag{4.157}$$

increases with r_M for any r.

Thus, for square-law fibers at least, we can proceed with the calculation of the area of the ellipse defined by (4.154), which is

$$\pi[k^2(r) - k_s^2]\left(1 - \frac{r^2}{a^2}\right)^{-1/2} \tag{4.158}$$

Substituting this result in the general expression for the acceptance factor we obtain, for two polarization states,

$$N^2 = \frac{1}{2}\int_0^{a^2}[k^2(r) - k_s^2]\left(1 - \frac{r^2}{a^2}\right)^{-1/2} dr^2 \tag{4.159}$$

This expression simplifies if we introduce the variable $u \equiv (1 - r^2/a^2)^{1/2}$. Equation (4.159) becomes

$$N^2 = a^2\int_0^1[k^2(u) - k_s^2]\,du \tag{4.160}$$

Thus the ray optics acceptance of most circularly symmetric fibers is half the area enclosed by the curve $k^2(u)a^2$. For a step-index fiber, we obtain from (4.160).

$$N^2 = (k_0^2 - k_s^2)a^2 \qquad \text{(step index, ray optics)} \qquad (4.161)$$

This is twice the wave optics acceptance. Thus, for step-index fibers, the slightly leaky rays carry half the power. For a square-law fiber, with $k(a) = k_s$, we obtain

$$N^2 = (k_0^2 - k_s^2)\,\frac{a^2}{3} \qquad \text{(square law, ray optics)} \qquad (4.162)$$

In square-law fibers, 25% of the total power is carried by slightly leaky rays. Alternative derivations of some of the above results and results applicable to other profiles are given in Ikeda,[15] Timmermann,[15] Cozannet et al.,[16] Ostermayer,[17] Miller,[17] Olshansky and Keck,[18] and Snyder and Mitchell.[19]

Concentration; Lens and Tapers[20]

We now examine the limitations that are imposed by the general theorems of ray optics in concentrating the power radiated by incoherent sources, such as light-emitting diodes, into optical fibers. Limitations concerning the concentration of rays from an extended source can be obtained from the second law of thermodynamics or from ray tracing, in the case of three-dimensional linear tapers (conical mirrors) and lenses (e.g., Luneburg lenses).

The (false) intuition that ray bundles can be concentrated into arbitrarily small areas by funnel-like devices stems from the common experience of funneling material particles into small bottlenecks. Funneling of material particles is successful because friction or collision with other particles sufficiently reduces their energy to prevent them from escaping from the funnel. The particles move in a nearly random fashion, and they eventually reach the outlet. In optical funnels, also called integration boxes, the photons can always escape by the opening through which they came in. If S denotes the area of this opening and s the area of the absorbing material (e.g., a detector) located inside the lossless integration box, only a proportion s/S of the incident photons is absorbed; the others are reflected back to the source.

Let us now see what conclusions can be drawn from the second law of thermodynamics. The second law of thermodynamics states that no work can be obtained from a body at a uniform temperature and that the temperature at any point inside a closed box at a uniform temperature eventually reaches the temperature of the box.

Let us consider a black box at a uniform temperature T corresponding to a radiance B and, inside the box, a small disk. For simplicity, we assume that one side of the disk is black, that the other side is perfectly reflecting, and that the refractive index is unity. According to the second law of thermodynamics, the disk is in equilibrium when its temperature is T. When this equilibrium is reached, it radiates with the same radiance B as the box, and it receives a power equal to the power that it radiates. This conclusion holds also in the case where the box contains lossless materials, such as lenses, mirrors, and total reflection prisms. But in that case, part of the power that the disk receives may originate from the disk itself rather than from the box. This is unlikely, however, if the disk is very small. Consequently, the conclusion is reached that in a closed box at a uniform temperature that contains lossless materials, the radiance is the same everywhere.

The acceptance of an optical system cannot be increased by external devices (such as lenses and mirrors); nor can it be reduced by these elements, provided they are lossless and do not reflect toward the instrument any of the light rays that could originate from it. This conclusion readily follows from the invariance of the radiance and the definition of the acceptance given earlier.

Let us verify that this principle is satisfied in the case of the linear taper shown in Fig. 4-11a. The absorber (detector) is assumed to be a section of sphere with center O and radius a. The convex side is assumed to be perfectly absorbent and the concave side perfectly reflecting. The area of this detector is equal to $a^2\Omega$, if Ω denotes the solid angle of the cone T defined by O and the contour of the detector. This detector is surrounded by a sphere of radius R acting as a source of blackbody radiation of radiance $1/\lambda^2$. Let us first evaluate the acceptance of the detector alone. Since the detector area is $a^2\Omega$, we obtain from (4.144) an acceptance

$$N^2 = \frac{\pi a^2 \Omega}{\lambda^2} \tag{4.163}$$

Let us now introduce in front of the detector a linear taper coincident with the cone T defined above; the surface of this cone is assumed to be perfectly reflecting and to extend up to the radiating sphere of radius R. It is easy to see that none of the rays originating from the detector are reflected back to it. The previous discussion indicates that in such conditions the acceptance of the detector and taper system should be just the same as the acceptance of the detector alone. Let us show that this result can be demonstrated solely on the basis of ray optics.

Consider an incident ray reflected from the cone surface at points $A, B, C, \ldots$. The cone formed by the planes $OAB, OBC, \ldots$ can be unfolded along the lines $OA, OB, OC, \ldots$ as shown in Fig. 4-11b and

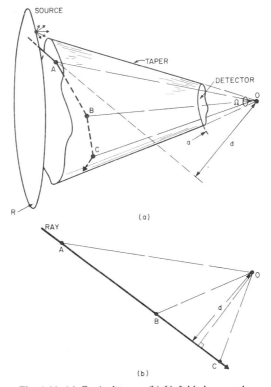

Fig. 4-11 (a) Conical taper. (b) Unfolded ray path.

made to coincide with a plane. The law of reflection then shows that the
segments AB, BC, ... of the trajectory are aligned in the unfolded representation. The point of closest approach of the ray trajectory to O is
consequently obtained by drawing a perpendicular from O to the straight
line ABC. Let us call d the minimum distance. It is clear that if this
minimum distance d between the incident ray (prolonged in space
beyond A if necessary) and O is larger than a (the radius of the detector)
to start with, this ray will never reach the detector. Conversely, if d is
smaller than a (or equal to a), the ray will certainly intersect the detector
(or be tangent to it) and be absorbed. Therefore, considering all of the
possible incident rays at the entrance of the taper, we see that the
acceptance area of the system is πa^2, the area of a circle of radius a. To
evaluate the accepted solid angle, let us assume for simplicity that $R \gg a$
(but Ω is not necessarily small compared with unity). In that case, the
accepted solid angle is simply Ω, and the acceptance is

$$N^2 = \frac{\pi a^2 \Omega}{\lambda^2} \qquad (4.164)$$

in exact agreement with the result obtained from the second law of thermodynamics. The role of a long taper with small Ω is to increase the accepted area of the system at the expense of the accepted angle. The total acceptance is invariant.

It has been assumed up to now that the radiating source completely surrounds the detecting system. Let us now consider a related problem formulated as follows: Given a thermal source of area $R^2\Omega$ and radiance B and a detector of area $a^2\Omega$, how much power can be transferred from the source to the detector? The previous discussion shows that the maximum power that can be transferred to the detector is equal to the power $\pi a^2\Omega B$ that the detector would receive if it were completely immersed in a thermal source of the same radiance B as the finite source. This upper limit is just reached in the configuration discussed above, using an ideal linear taper.

Instead of a taper, one may think of using a lens. If aberration-free lenses of unlimited transverse dimensions were available, it would be possible to transfer all the power radiated from a thermal source into a detector, in contradiction with the result established before. Such lenses, however, are physically unrealizable. Let us therefore consider a Luneburg lens, which is known to be free of aberration for a spherical object. Such a lens (see Fig. 4-12) is difficult to fabricate in practice, but its theoretical

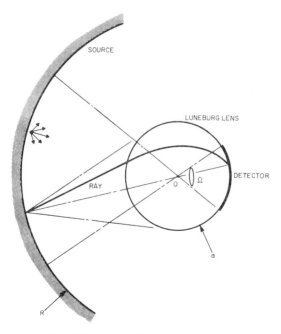

Fig. 4-12 Luneburg lens concentrator.

properties are well understood. A Luneburg lens of radius a is an inhomogeneous medium with spherical symmetry. Its refractive-index law need not be given here. It suffices to know that the refractive index at the boundary is unity and that a sphere of radius R can be imaged perfectly into a sphere of radius a. Every ray originating from a point of the sphere of radius R that intersects the lens is focused by the lens at the point of the sphere of radius a that is aligned with the lens center O. The limit rays, tangent to the Luneburg lens, arrive at the image point under grazing incidence. Knowing these properties of the Luneburg lens, we can evaluate the power received by the detector, taken as coincident with a portion of the sphere of radius a, which subtends a solid angle Ω. It is now assumed for convenience that the concave side of the detector is perfectly absorbent and that the convex side is perfectly reflective. The blackbody source coincides, as before, with a sphere of radius R. Consider first a small area dA of the radiating sphere. Only part of the power that this elementary area radiates reaches the detector. This power is obtained by integrating the power density $R(\alpha)$, given by (4.141), from 0 to the limit angle defined by the Luneburg lens $\sin^{-1}(a/R)$. Therefore we have

$$P = \int_0^{\sin^{-1}(a/R)} R(\alpha) 2\pi \sin(\alpha)\, d\alpha = \frac{\pi\, dA\, a^2}{R^2\lambda^2} \qquad (4.165)$$

The area of the portion of the radiating sphere, which is imaged into the detector, is equal to $R^2\Omega$. Therefore, the acceptance of the detector–Luneburg lens combination is

$$N^2 = \frac{\pi a^2 \Omega}{\lambda^2} \qquad (4.166)$$

as is the case for linear tapers.

Let us make a few additional remarks. The invariance of the radiance of extended Lambertian sources is easily demonstrated on the basis of the Hamilton theory of light $(df/dz) = 0$. The application of this theory, however, raises difficulties when some of the rays are reflected back with respect to the system axis. There are, therefore, some difficulties in applying the Hamilton theory of light to optical systems such as the linear tapers, which incorporate mirrors.

Second, the derivations given before are based on the consideration of blackbody radiation. From a ray optics point of view, the only thing that matters is the fact that the source is a Lambertian extended source. The geometric properties of ray trajectories and ray bundles in optical systems do not depend on the photon statistics or the intensity of the rays; the conclusions obtained are consequently applicable to the case of incident monochromatic Lambertian ray bundles of any intensity. The fabrication of perfect compressors of Lambertian ray bundles based on wave optics is not forbidden if the source is nearly monochromatic and intense. Indeed, arbitrary fields can be converted into single-mode patterns, in principle without losses, by using, for instance, bundles of fibers of adjustable length. The light can always be focused into areas of the order of a wavelength square. This fact does not contradict the second law of thermodynamics. To make the mode converter work with blackbody radiation, it would be necessary to filter out small bandwidths Δf from the broad-band radiation spectrum. This would give a time $t \ll \Delta f^{-1}$ to measure the phase with sufficient accuracy and correct for it. The energy contained in the small bandwidth Δf is not sufficient to perform this operation in the case of blackbody radiation, which has only one photon in each phase-space cell.

We shall now give an example of calculation of evolution of the distribution f, which is interesting in part because of the formal analogy that it exhibits with the propagation of gaussian beams.

4.15 Evolution of the Distribution in Spatial Phase Space

Let $f(k_x, k_y, x, y)$ denote the density of rays in spatial phase space at $z = 0$. As we have indicated above, f is a constant for a uniform Lambertian source. This distribution is approximately applicable to light-emitting diodes. However, not all sources are Lambertian. We wish to discuss more general incoherent sources. The evolution of f as z varies is given by the Liouville theorem $df/dz = 0$ in (4.55), where we have identified the arbitrary parameter σ with the axial coordinate z. Thus

$$0 = \frac{\partial f}{\partial z} + \frac{\partial f}{\partial x}\frac{dx}{dz} + \frac{\partial f}{\partial y}\frac{dy}{dz} + \frac{\partial f}{\partial k_x}\frac{dk_x}{dz} + \frac{\partial f}{\partial k_y}\frac{dk_y}{dz} \quad (4.167)$$

dx/dz, dy/dz, dk_x/dz, and dk_y/dz are given by the Hamilton equations (4.71).

For two dimensions x, z in free space, we have, within the paraxial approximation,

$$k_z = \left(k_0^2 - k_x^2\right)^{1/2} \approx k_0 - \tfrac{1}{2}k_x^2/k_0 \quad (4.168)$$

Thus

$$\frac{dx}{dz} = \frac{k_x}{k_0}, \qquad \frac{dk_x}{dz} = 0 \quad (4.169)$$

$f(k_x, x, z)$ in (4.167) therefore obeys the simple partial differential equation

$$k_0 \frac{\partial f}{\partial z} + k_x \frac{\partial f}{\partial x} = 0 \qquad (4.170)$$

whose general solution is

$$f(k_x, x, z) = g(k_x, k_0 x - k_x z) \qquad (4.171)$$

where g is an arbitrary differentiable function of two variables, as we readily verify by substituting (4.171) in (4.170). If the initial density of rays is gaussian at $z = 0$, that is, if

$$f(k_x, x, 0) = \frac{k_{x0} \xi_0}{\pi} \exp\left[-\left(\frac{x}{\xi_0} \right)^2 - \left(\frac{k_x}{k_{x0}} \right)^2 \right] \qquad (4.172)$$

the density at z is, in order that f be of the form in (4.171),

$$f(k_x, x, z) = \frac{k_{x0} \xi_0}{\pi} \exp\left[- \frac{(x - k_x z / k_0)^2}{\xi_0^2} - \left(\frac{k_x}{k_{x0}} \right)^2 \right] \qquad (4.173)$$

It therefore remains gaussian. The irradiance is obtained by integrating f in (4.173) over k_x from $-\infty$ to $+\infty$ (we assume that $k_{x0} \ll k_0$). The result is a gaussian irradiance

$$P(x) = \int_{-\infty}^{+\infty} f \, dk_x = \frac{\pi^{-1/2}}{\xi} \exp\left(\frac{-x^2}{\xi^2} \right) \qquad (4.174)$$

where we have set

$$\xi^2 = \xi_0^2 + \frac{k_{x0}^2 z^2}{k_0^2} \qquad (4.175)$$

This result coincides with that in (2.34) applicable to gaussian beams if $k_{x0} \xi_0 = 1$. Diffraction effects, however, play no role in the present problem. The analogy is purely formal. Optical beams cannot, in general, be represented by a (positive) probability distribution in phase space. Gaussian beams, as we just saw, are noteworthy exceptions to that rule.

4.16 Pulse Broadening in Multimode Optical Fibers

Light-emitting diodes supply their optical power in a time- and space-incoherent form. The linewidth is typically of the order of 0.04 μm, and the radiation is approximately Lambertian with an emissive area of the order of 50 μm $\times$ 50 μm. Time- and space-incoherent optical pulses can be

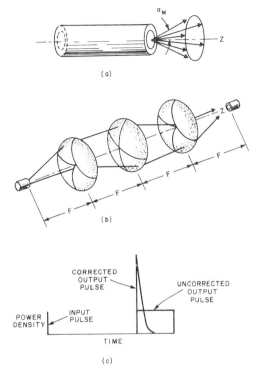

Fig. 4-13 (a) Radiation from a step-index fiber. $\sin(\alpha_M)$ is called the numerical aperture. (b) Ray equalizer using three confocal lenses. The first and last lenses are of an unconventional type with an 8-shaped profile. Fast rays (small α) and slow rays (large α) are exchanged.[14] (c) Calculated impulse response for an uncorrected step-index fiber and for a fiber corrected by the ray equalizer.[14]

transmitted by oversized optical fibers (core radius $\gg$ wavelength). However, optical pulses propagating in such fibers tend to broaden as they travel. This is in part due to the nonzero linewidth of the source and the dispersion $d^2k/d\omega^2$ of the fiber material. The other cause of pulse broadening is associated with the fact that the average group velocity in the fiber is a function of the ray trajectory. Pulses traveling along axial rays usually go faster than pulses traveling along rays of large amplitude. Because both types of rays are excited by spatially incoherent sources, this causes a broadening of the pulse. It was pointed out by Kompfner[13] that pulse broadening in fibers could be drastically reduced by introducing ray equalizers at various locations along the fiber. The role of ray equalizers is, in short, to exchange fast rays and slow rays. A possible implementation of this idea is shown in Fig. 4-13 together with the calculated impulse response for uncorrected and corrected step-index fibers.[14] Because natural mode mixing is very small in the most recently made optical fibers, ray converters may be useful. However, to our knowledge, these ideas have not been implemented. Thus we shall restrict ourselves to uniform uncorrected fibers.

Space–Time Phase Space Mapping

Let ξ' denote a point in phase space $(k'_x, k'_y, \omega', x', y', t')$ at the input plane and ξ a point in phase space at the output plane. The optical system maps the input phase space into the output phase space: $\xi' \to \xi$. The transformation is denoted

$$\xi = \xi(\xi') \qquad (4.176)$$

It follows from the Hamilton equations (4.58) that the Jacobian of the transformation (4.176) is unity, as we have seen before.

A source of light that is time and space incoherent is described by a distribution $S'(\xi')$ in phase space. Each small point in phase space can be pictured as an optical pulse, whose detailed structure is ignored.

The transmission T_1 of an optical pulse through the optical system is presumably a known function of ξ' that we denote

$$T_1 = T_1(\xi') \qquad (4.177)$$

For lossy optical systems, $T_1 < 1$. Because the Jacobian of the transformation $\xi' \to \xi$ is unity, the output distribution is simply

$$S(\xi) = S'(\xi')T_1(\xi') \qquad (4.178)$$

The power emitted by the source and the power that can be collected at the output of the optical system are obtained by integrating S (or S') over all variables, except t (or t'):

$$P(t) = \int S(\xi) \, (d\zeta) \qquad (4.179)$$

$$P'(t') = \int S'(\xi') \, (d\zeta') \qquad (4.180)$$

where $\zeta' \equiv (k'_x, k'_y, \omega', x', y')$ and ζ is similarly defined. $(d\zeta)$ and $(d\zeta')$ denote elementary volumes in ζ- and ζ'-spaces, respectively. The response of the detector could be described by a function $D(\xi)$. For simplicity we do not take the detector response into consideration. All subsequent results follow in a rather straightforward manner from the above expressions, through a succession of approximations.

Pulse Transformation in Time-Invariant Fibers

Let us assume that the fiber is time invariant and the source is time separable. These assumptions are applicable to most fibers and sources of interest. Let us first consider the assumption that the properties of the fiber do not vary with time. This means that the transmission T_1 and the mapping $\xi' \rightarrow \xi$ do not depend on time. In particular,

$$t = t' + t_1(\zeta') \tag{4.181}$$

Time-separable sources, on the other hand, have the property that

$$S'(\xi') = P'(t')F'(\zeta') \tag{4.182}$$

That is, the distribution in ζ'-space does not vary with time. As a counterexample, consider a hot tungsten wire whose temperature varies with time. The spatial phase-space distribution is almost Lambertian at all times, but the frequency spectrum (approximately given by the Planck law of radiation) varies with time. Thus (4.181) is not applicable to that source. For consistency with (4.156), we assume that $F'(\zeta')$ is normalized to unity.

For most sources, we can further assume that

$$F'(\zeta') = \Omega(\omega)f'(\mathbf{s}') \tag{4.183}$$

where $\mathbf{s}' \equiv (k_x', k_y', x', y')$ denotes a point in spatial phase space. That is, we assume that the spatial distribution does not depend on what part of the frequency spectrum we are considering. Both Ω and f' are assumed normalized to unity. This ensures that F' is normalized to unity. When the spectral width of the source is small (e.g., less than 1%, as is the case for light-emitting diodes) and the fiber material absorption does not exhibit sharp resonances in that band, we can assume that

$$T_1(\zeta') = T_0(\omega)T(\mathbf{s}') \tag{4.184}$$

and

$$t_1(\zeta') = t_0(\omega) + t(\mathbf{s}') \tag{4.185}$$

For definiteness, we assume that the maximum value of $T_0(\omega)$ is unity, and we define $t_0(\omega)$ as the delay experienced by axial pulses. We have evaluated in preceding sections the relative time $\tau \equiv t_1(\zeta')/t_0(\omega)$.

The pulse response is obtained from (4.178)–(4.185):

$$P(t) = \int P'[t - t_0(\omega) - t(\mathbf{s}')]\Omega(\omega)T_0(\omega)f'(\mathbf{s}')T(\mathbf{s}')\,(d\mathbf{s}')\,d\omega$$

$$= \int P''[t - t_0(\omega)]\Omega(\omega)T_0(\omega)\,d\omega \tag{4.186}$$

where

$$P''(t'') = \int P'[t'' - t(s')] f'(s') T(s') \, (ds') \qquad (4.187)$$

In writing (4.186) we have used the fact that the Jacobian of the transformation $\xi' \to \xi$ is unity and that $d\omega' = d\omega$. The pulse response is the convolution of the pulse response in (4.187), which we may call the quasi-monochromatic pulse response, and the spectral curve of the source, because in most cases $T_0(\omega)$ is a constant. For injection lasers, the quasi-monochromatic pulse response is the most important contribution. This response is considered in detail in subsequent sections for uniform fibers. In what follows, we assume that the fiber is uniform and long compared with the period of ray oscillation and, therefore, approximately z-invariant.

In almost any z-invariant focusing system, any initial distribution eventually reaches a steady state. This steady state, in general, differs from the initial distribution. A Lambertian distribution $f = \text{const}$ remains Lambertian because it is a (trivial) solution of the Liouville equation. Note that the distribution f represents a ray (or pulse) density. If the medium introduces a nonuniform attenuation on the rays, the power density in phase space must be distinguished from the distribution f.

A fiber is usually surrounded by a homogeneous material, called the cladding, with wavenumber k_s. For fibers that are not highly overmoded, the transmission law

$$T(x, y, k_x, k_y) = \begin{cases} 1 & \text{if } k_z > k_s \\ 0 & \text{if } k_z \leqslant k_s \end{cases} \qquad (4.188)$$

is often acceptable. Equation (4.188) says that rays whose axial wavenumber is less than the wavenumber of the surrounding medium are leaking sufficiently rapidly to be ignored. The distribution f of the Lambertian source is set equal to $1/(2\pi)^2$ in order that the radiance be unity.[20] In that case, the total power transmitted is the acceptance of the fiber.

Impulse Response for $k^2 - k_0^2$ a Homogeneous Function of x and y

In the case where $k^2(x, y)$ is equal to $k(0, 0)$ plus a homogeneous function of x and y, the relative time of flight τ is, as we have seen, solely a function of k_z. The upper and lower bounds on k_z are $k(x, y)$ and k_s, respectively. We need express the volume element $dk_x \, dk_y \, dx \, dy$ in (4.187) as a function of $dk_z \, dx \, dy$. For given x, y, a constant value of k_z corresponds to a circle of radius squared $k^2(x, y) - k_z^2$ in the k_x-, k_y-space. Thus

$$dk_x \, dk_y = \pi \, dk_z^2 \qquad (4.189)$$

By using (4.189), the pulse transformation in (4.187) becomes

$$P(t) = \frac{1}{4\pi} \int dx \, dy \int_{k_s^2}^{k^2(x,y)} P'[t - t(k_z)] \, dk_z^2 \qquad (4.190)$$

where we have omitted the double primes for brevity.

If the input pulse $P'(t)$ is a Dirac symbolic function (e.g., a rectangular pulse of width Δt and height Δt^{-1} in the limit $\Delta t \to 0$), the output pulse in (4.190) becomes

$$P(t) = \frac{1}{4\pi} \left| \frac{dk_z^2}{dt} \right| A(k_z), \qquad k_z > k_s \qquad (4.191)$$

where $A(k_z)$ denotes the cross-sectional area that satisfies $k(x, y) > k_z$. k_z can be expressed as a function of the delay t by inverting the relation $\tau(k_z)$ given earlier. We assume that the fiber length is such that the delay is unity for pulses on axis. We therefore do not need to distinguish t and τ. We obtain from (4.79)

$$\frac{dk_z'^2}{dt} = \frac{2(1 + \kappa)k_z'}{1 - \kappa/k_z'^2} \qquad (4.192)$$

where

$$k_z' = \frac{k_z}{k_0} = \frac{(1 + \kappa)t}{2} \pm \left\{ \left[\frac{(1 + \kappa)t}{2} \right]^2 - \kappa \right\}^{1/2} \qquad (4.193)$$

If $\kappa > 1$, there is only one value of k_z' between $k_s' \equiv k_s/k_0$ and 1, for any k_s'. If, however,

$$k_s'^2 < \kappa < 1 \qquad (4.194)$$

there are two values of k_z' that need to be considered. Their contributions to P should be added. If $\kappa < k_s'^2$, there is again only one relevant value of k_z'.

Impulse Response of Square-Law Fibers

Let us consider as an example a (noncircularly symmetric) square-law medium

$$k^2(x, y) = k_0^2(1 - \Omega_x^2 x^2 - \Omega_y^2 y^2) \qquad (4.195)$$

where Ω_x, Ω_y denote arbitrary constants. $2\pi/\Omega_x$ and $2\pi/\Omega_y$ are the periods of ray oscillation in the xz and yz planes, respectively. The area $A(k_z)$ defined earlier is the interior of an ellipse. We obtain

$$A(k_z) = \frac{\pi(1 - k_z'^2)}{\Omega_x \Omega_y} \qquad (4.196)$$

The impulse response is obtained from (4.191), (4.192) with $\kappa = 1$, and (4.196):

$$P(t) = \begin{cases} \dfrac{k_0^2 k_z^3}{\Omega_x \Omega_y}, & k_z > k_s \\[2mm] 0, & k_z \leqslant k_s \end{cases} \tag{4.197}$$

where, from (4.193),

$$k_z' = t - (t^2 - 1)^{1/2} \equiv k_z/k_0 \tag{4.198}$$

Because in most fibers k_z remains close to k_0, the variation of k_z' can be neglected, and the impulse response is almost rectangular.

Impulse Response of Step-Index Fibers

For a step-index fiber, the area A is the area of the core cross section, and $t = k_0/k_z$. Thus the impulse response of a step-index fiber with cross-sectional area A is simply[22]

$$P(t) = \frac{k_0^2 A}{2\pi t^3}, \qquad 1 < t < \frac{k_0}{k_s} \tag{4.199}$$

Because in most fibers t remains close to unity, the impulse response is almost rectangular.[14] The pulse width, however, is considerably larger than for square-law fibers, as we shall see in more detail later.

Impulse Response for $k^2 - k_0^2$ a Power of the Radius

Let now the wavenumber profile be of the form

$$k^2(R, \omega) = k_0^2(\omega) - k_\kappa^2(\omega)R^\kappa \tag{4.200}$$

where $R \equiv X + Y \equiv r^2$ denotes the square of the radius. The relative time of flight is given for that case in (4.121). Expressing k_z' as a function of t, we obtain

$$k_z' = \frac{t}{2D_\kappa'} \pm \left[\left(\frac{t}{2D_\kappa'} \right)^2 + 1 - D_\kappa'^{-1} \right]^{1/2} \tag{4.201}$$

and

$$\frac{dk_z'^2}{dt} = 2k_z' \left[D_\kappa' - \frac{1 - D_\kappa'}{k_z'^2} \right]^{-1} \tag{4.202}$$

where

$$D'_\kappa \equiv \frac{D_\kappa}{1 + \kappa} \tag{4.203}$$

To obtain the impulse response, we need the area $A(k_z)$ defined by $k_z < k(R)$. For $k(R)$ in (4.200), this area is

$$A(k_z) = \pi R(k_z) = \pi \left(\frac{1 - k_z'^2}{\epsilon_\kappa} \right)^{1/\kappa}, \qquad \epsilon_\kappa \equiv \frac{k_\kappa^2}{k_0^2} \tag{4.204}$$

If ϵ_κ were kept a constant as the parameter κ varies, the core radius a defined by $k(a) = k_s$ would vary. Thus it is preferable to express ϵ_κ as a function of the core radius a. We have

$$\epsilon_\kappa^{1/\kappa} = \frac{\left(1 - k_s'^2 \right)^{1/\kappa}}{a^2} \tag{4.205}$$

where $k_s' \equiv k_s/k_0$. The impulse response is finally obtained from (4.191) and (4.201)–(4.205):

$$P(t) = \frac{k_0^2 a^2}{2} k_z' \frac{\left[(1 - k_z'^2)/(1 - k_s'^2) \right]^{1/\kappa}}{D'_\kappa - (1 - D'_\kappa)k_z'^{-2}} \tag{4.206}$$

The (possibly doubled value) k_z' is expressed as a function of t in (4.201), and D'_κ is defined in (4.203) and (4.116). In the absense of dispersion, $D'_\kappa = 1/(1 + \kappa)$, and (4.206) reduces to

$$P(t) = \frac{k_0^2 a^2}{2} k_z' \left[\frac{(1 - k_z'^2)}{(1 - k_s'^2)} \right]^{1/\kappa} \frac{1 + \kappa}{1 - \kappa k_z'^{-2}} \tag{4.207}$$

As indicated in the previous section, there are in general two values of k_z' that contribute to P. Note that the shape of the impulse response does not depend on the core radius a.

The impulse response $P(t)$ in (4.207) is shown in Figs. 4-14 and 4-15 for various values of the parameter κ. We have assumed that $k_s/k_0 = 0.9$, that is, $\Delta n/n = 10\%$ (a rather large value). For $\kappa = 1$ (square-law fiber), the pulse width is 0.0054. For example, if $n = 1.45$ and the fiber length is 1 km, the pulse width is 26 nsec. For $\kappa = 0.9$, however, the corresponding pulse width is only 7 nsec. The minimum pulse width occurs when $\kappa = k_s'$. For a step-index fiber ($\kappa \to \infty$), the pulse width is as large as 630 nsec. Note the following detailed features on the curves in Figs. 4-14 and 4-15. For $(0.9)^2 < \kappa < 1$, the response starts from infinity because of the minimum in the $t(k_z)$ curve. For $\kappa = 0.85$, P drops suddenly for $t \approx 0.998$.

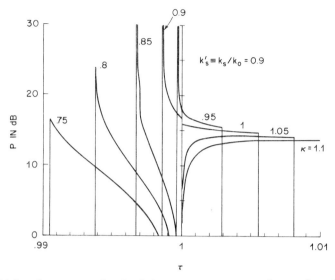

Fig. 4-14 Impulse response of a circularly symmetric fiber with $k^2(r) = k_0^2 - k_\kappa^2 r^{2\kappa}$ for a Lambertian source and various values of κ. We have assumed $k_s/k_0 = 0.9$. The optimum impulse response is for $\kappa \approx k_s/k_0$. (From Arnaud.[8] Reprinted with permission from *The Bell System Technical Journal*, copyright 1975, The American Telegraph and Telephone Co.)

This is because, at that time, the smaller of the two k_z' becomes less than 0.9 and is rejected. For $\kappa = 0.95$ the response extends from times smaller than unity to times larger than unity.

Figure 4-15 is applicable to larger values of κ. We note that for very large κ (step-index fiber), the response is almost rectangular.[14] The slow decay in power shown in Fig. 4-15 would be almost negligible for small $\Delta n/n$.

The effect of inhomogeneous dispersion is shown in Fig. 4-16. The parameter κ is kept equal to 0.9. This is the optimum value in the absence of material dispersion. The inhomogeneous dispersion parameter D_κ is made to vary in the neighborhood of unity. These curves have a striking resemblance to those in Fig. 4-14. This means that the effect of inhomogeneous dispersion merely consists in shifting the optimum value of κ. The impulse response remains essentially the same, at least for $\kappa \approx 1$. Olshansky and Keck[18] have shown that inhomogeneous dispersion is very significant for near square-law fibers doped with TiO_2. Results for germania doped fibers are given in Section 4.24.

Another quantity of interest is the power density in the fiber. It is obtained by integrating the phase-space distribution over k_x and k_y. The density in the k_x-, k_y-space is uniform within a circle of radius squared

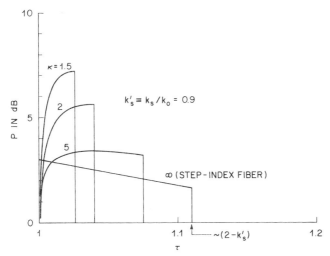

Fig. 4-15 Continuation of Fig. 4-14 for larger values of κ. $\kappa \to \infty$ corresponds to clad fibers. (From Arnaud,[8] by permission of The American Telegraph and Telephone Co.)

equal to $k^2(R) - k_s^2$. Thus the intensity P in the fiber varies radially as

$$P(R) \propto k_0^2 - k_s^2 - k_\kappa^2 R^\kappa \qquad (4.208)$$

A comparison between the ray technique considered here and the modal approach is made in Section 4.24 and concrete examples are given.

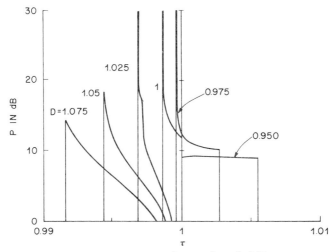

Fig. 4-16 Impulse response for a fiber with $k^2(r) = k_0^2 - k_\kappa^2(r^2)^{0.9}$ for various values of the parameter D that expresses the dispersion of the material. $D = 1$ corresponds to the absence of dispersion. $D \neq 1$ introduces a shift in the optimum value of κ. (From Arnaud,[8] by permission of The American Telegraph and Telephone Co.)

Summary

Let us summarize the most important results concerning the propagation of optical pulses in glass fibers. The acceptance of a fiber is the power transmitted for an incoherent source with density $1/(2\pi)^2$ in phase space x, y, k_x, k_y. For a step-index fiber with wavenumber k_0, radius a, in a medium with wavenumber k_s, the acceptance is

$$N^2 = (k_0^2 - k_s^2)a^2 \times \begin{cases} 1 & \text{(highly oversized)} \\ \frac{1}{2} & \text{(moderately oversized)} \end{cases} \qquad (4.209)$$

If the fiber is moderately oversized, the impulse response is almost rectangular, with a width

$$\Delta t = \left(\frac{k_0}{k_s} - 1 \right) \times \text{ time of flight of axial pulses} \qquad (4.210)$$

For example, at a wavelength of 1 μm, if $n_0 = 1.45$, $n_s = 1.4$, and $a = 50$ μm, we find $N^2 = 15,000$. Pulses broaden at a rate of 167 nsec/km.

For a near square-law fiber with $k(0) = k_0$ and $k(a) = k_s$, the coefficient in (4.209) is equal to $\frac{1}{2}$ for highly oversized fibers and to $\frac{1}{3}$ for moderately oversized fibers.

The impulse response of moderately oversized square-law fibers is almost rectangular with

$$\Delta t = \frac{1}{8} \left(1 - \frac{k_s^2}{k_0^2} \right)^2 \times \text{ times of flight of axial pulses} \qquad (4.211)$$

For the numerical values given earlier, $\Delta t = 2.7$ nsec/km. This pulse broadening can be reduced by a factor 4 by adding a r^4 term to the law $k^2(r)$. For highly oversized fibers, Δt is 4 times as large as given in (4.211). The reduction obtained by adding an r^4 term is in that case 11. If inhomogeneous dispersion is taken into account, the optimum profile is slightly modified, but the minimum pulse broadening is not drastically affected.

4.17 Experiments with Multimode Fibers

It has been observed in various laboratories[22] that fibers whose refractive index varies radially approximately as a square law exhibit much less pulse broadening than step-index fibers, for quasi-monochromatic sources, as we expect from the fact that in gaussian optics all the rays have the same optical length. It has also been observed that pulse broadening is considerably smaller when the source is a quasi-monochromatic injection laser than when the source is a rather broad-band light-emitting diode. Beyond these basic results, it seems fair to say that detailed agreement between theory and experiment has not been obtained. In one typical case, a reasonable agreement between theory and experiment is obtained for the 10-dB points of the impulse response, but the agreement is not obtained for the 3-dB points. The shape of the impulse response does not agree with the theoretical prediction. This is not surprising if we consider that neither the fiber, the source, nor the detector is fully characterized. In the following paragraphs, a number of difficulties that are encountered at present in characterizing the components of optical fiber transmission systems are pointed out. Partial answers are given in subsequent sections and at the end of Chapter 5.

Refractive-index profiles are measured by inserting a slice of the fiber in an interference microscope. The accuracy afforded by this technique is perhaps insufficient to measure the very small deviations from square law that are relevant to pulse broadening. Multiple-path interferometry is perhaps required.[23] An alternative technique is the measurement of Fresnel reflection from the fiber tip.[24]

The variation of the dispersion $(dn/d\lambda)$ with the distance from axis in graded-index fibers is not known accurately. There are uncertainties involved in estimating the inhomogeneous dispersion of the fiber from changes of composition.

The variation of the loss as a function of the distance from axis is likewise not known accurately. Rayleigh scattering depends on composition fluctuations. A theory of that effect is based on the assumption that Rayleigh scattering increases with dopant concentration (e.g., GeO_2). On that account, the loss of most of the recently made fibers would decrease away from axis. On the other hand, the polarizability of oxides originates mainly from the oxygen atoms, and significant composition fluctuations take place in pure silica (silicon atoms versus oxygen atoms). Dopants, in fact, occasionally reduce rather than increase the composition fluctuations.[25] In addition to the composition fluctuations, density fluctuations are significant. Measurements can be effected on the preform from which the fiber is pulled. For low-diffusivity dopants such as GeO_2, the composition law scales rather well. However, heterogeneities elongate when the fiber is pulled, and such elongated heterogeneities do not scatter light in the same way as the original defects. The attenuation due to impurities may be a function of the dopant concentration. In fact, dopants such as GeO_2 can be prepared in highly pure form, and impurities (HO radical, cobalt) are sometimes found in commercially available silica.

Axial variations of the fiber profile and bends of the fiber cause mode mixing and radiation. These effects, though not as severe as was originally

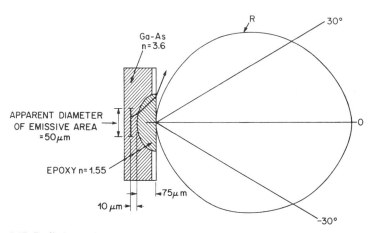

Fig. 4-17 Radiation R from a light-emitting diode (LED) with simple heterostructure. The radiance is $R/\cos(\alpha)$, where α is the angle to the axis. (From C. Burrus, private communication.)

thought, profoundly affect the characteristics of long fibers.[26] Most fibers are wound on an aluminum drum of large radius. In principle, the bending loss should be negligible. However, the mechanical tension exerted on the fiber sometimes introduces secondary bends that affect the fiber transmission. Plastic coatings also sometimes introduce bends and cause additional loss.[27]

In the theory given in the previous sections, the source is assumed for simplicity to be totally spatially incoherent, that is, uniform in phase space (k_x, k_y, x, y) and therefore Lambertian. Alternatively, from a wave optics point of view, all modes are assumed to be equally excited. This approximation is not always applicable. Commercial light-emitting diodes are often characterized only by their total radiated power. A more sophisticated characterization of light-emitting diodes sometimes available is in terms of the radiation pattern and the apparent area (e.g., 50 μm $\times$ 50 μm) as shown in Fig. 4-17. The latter is obtained by observing the diode with a microscope and an image converter. Although this characterization gives some indication of the phase-space occupation of the source, it falls short of providing the phase-space distribution. In light-emitting diodes, light is emitted at the junction almost isotropically. The radiance is therefore not Lambertian to start with. Further complication results from the absorption by the GaAs layer that the light must traverse and from refraction at a succession of curved interfaces, e.g., GaAs/epoxy and epoxy/air. Lensing effects are often introduced to make the radiation pattern more directive and the apparent area larger.

Light-emitting diodes cannot always be used as sources in fiber testing, because, in addition to being spatially incoherent, they are also temporally incoherent, and polychromatic pulse broadening tends to mask the effect of quasi-monochromatic pulse broadening. If a monochromatic spatially coherent source is used, spatial incoherence can be simulated by averaging the incident plane waves over all relevant angles of incidence. If the beam of light is focused, the averaging needs to be made over all relevant offsets and angles about a fixed point of rotation.[20]

It follows from the linearity of the pulse transformation given earlier that if a pulse $P'_1(t)$ gives a response $P_1(t)$ and a pulse $P'_2(t)$ gives a response $P_2(t)$, the pulse $P'(t) \equiv P'_1(t) + P'_2(t)$ gives a response $P(t) = P_1(t) + P_2(t)$. This simple result is tied up with the fact that the source is incoherent. Baseband linearity, however, does not hold if the source is partially coherent or if the relation optical power versus current is not linear at the source or at the detector.

It is sometimes possible to characterize the impulse response of a fiber by a single number. It has been suggested[28] that the pulse rms width σ is the most significant parameter with respect to error rate. It is defined as $\sigma \equiv (\langle t^2 \rangle - \langle t \rangle^2)^{1/2}$, where, for any function $f(t)$, $\langle f(t) \rangle$ is the integral of $f(t)P(t)\,dt$ over all times, divided by the integral of $P(t)\,dt$ over all times. In particular, the rms width of a rectangular pulse of width unity is $1/2\sqrt{3} \approx 0.288$. In previous sections, we have evaluated the output pulse power $P(t)$ for fibers excited by Dirac symbolic functions at the input. The rms pulse width σ can be obtained from $P(t)$ by integration according to the above definition. Note, however, that if the impulse response of the fiber consists of a sequence of pulses, as is the case for fibers carrying few modes, the rms pulse width σ may be insufficient to evaluate the error rate.

In the remainder of this section, we shall describe a few recent results obtained in the transmission of optical pulses through glass fibers. For the reasons discussed above, we shall not attempt to relate in detail these results to theoretical expressions.

Light-Emitting Diodes

Let us first give a few typical numbers for high-radiance light-emitting diodes (LED) such as those used in fiber optics. Figure 4-17 represents a typical high-radiance LED. It consists of a junction that radiates almost isotropically. In that particular geometry, the radiation pattern is rather broad. The highest radiances have been obtained with double heterojunctions. Radiances up to 1 W/sr $\times$ mm^2 have been measured. Table 4.1 illustrates the effect of area on radiance obtained by Burrus.[21]

The radiation is contained in a cone with angle $\alpha_0 = 65°$ from axis. From this angle, the solid angle is calculated $\Omega = 2\pi[1 - \cos(\alpha_0)] = 3.6$

Table 4.1   Effect of Area on Radiance

Contact diameter (μm)	Current (mA)	Current density (A/mm^2)	Radiance on axis (W/sr $\times$ mm^2)
13	100	750	1
25	150	300	0.4
100	300	37.5	0.06

sr. The total radiated power is therefore, for the 25-μm diameter junction in Table 4.1, $P = (\pi/4)(25 \times 10^{-3})^2 \times 0.4 \times 3.6 = 0.7$ mW. The acceptance (or number of modes) of the source is given by the product of the apparent radiating area $(\pi/4)(25 \times 10^{-3})^2$ mm^2 and radiation solid angle 3.6 sr, divided by the square of the wavelength $\lambda_0 = 0.9$ μm. Note that all these quantities are defined in free space. Thus $N^2 = 2200$. These 2200 modes can be transmitted, for example, by a fiber with core diameter 100 μm and $\Delta n/n \approx 3.2\%$. [We have used the relation $N^2 = (k_0^2 - k_s^2)a^2$

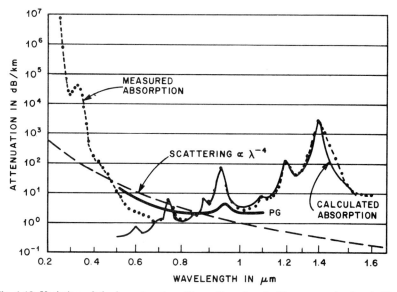

Fig. 4-18 Variation of the loss of a glass fiber with a doped silica core and a fused silica cladding. The numerical aperture sin $\alpha_M = 0.14$. The core radius is $a = 45$ μm. The measurement was made on a 0.55-km length. (Fron Keck and Maurer[18], by permission of the American Institute of Physics.) The solid curve marked PG is for a phosphosilicate fiber. (From Payne and Gambling,[22] by permission of the IEE.)

applicable to highly oversized step-index fibers transmitting two states of polarization.]

We shall report results obtained with two types of fibers. The first has a pure silica core and is most useful for short-distance communications. The second has a near square-law profile and is capable of high transmission capacity over long distances. A typical curve of material loss is shown in Fig. 4-18, as well as the loss curve of a phosphosilicate fiber.

Plastic-Clad Fused Silica Fiber

A homogeneous unclad silica fiber exhibits good transmission properties, but the distance between adjacent supports can hardly exceed 60 m. Furthermore, the fiber is soon contaminated by dust particles. A loosely

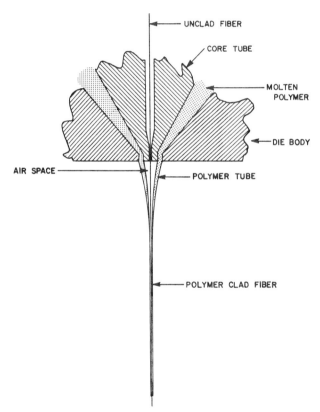

Fig. 4-19 The method of fabrication of FEP (Teflon) clad silica fibers. (From Kaiser *et al*,[29] by permission of the American Physical Society.)

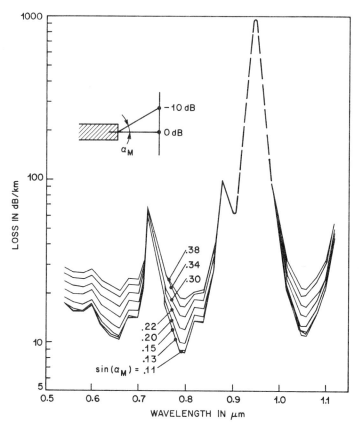

Fig. 4-20 Transmission loss as a function of wavelength of an FEP clad silica fiber for various divergence angles ($NA = \sin \alpha_M$) in air of the light beam. (From Kaiser *et al*,[29] by permission of the American Physical Society.)

fitting envelope with a refractive index lower than that of silica provides both support and protection. A suitable material is a plastic called per-fluoronated-ethylene-propylen (Teflon®-FEP 100) whose refractive index is 1.338, to be compared to that of quartz, which is 1.45 at $\lambda_0 = 1$ μm. The bulk loss of FEP is as high as 500,000 dB/km. Yet, for highly oversized fibers, this high cladding loss does not contribute significantly to the total loss. In the 0.8-μm range, the total loss is as low as 7 dB/km. This is in part due to the loose fitting between the core and the cladding achieved by a continuous extrusion process.[29] The fiber is drawn with an oxy-hydrogen torch from 7-mm diameter SiO_2 preform rods. The fiber immediately passes through the crosshead of a simple ram extruder (see Fig. 4-19). This type of SiO_2/FEP fiber is easy to fabricate and to handle. The large

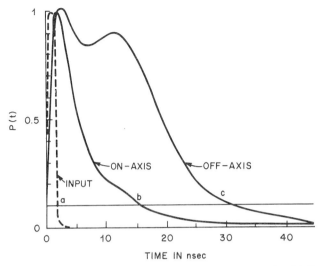

Fig. 4-21 Impulse response of an FEP clad silica fiber. (a) Input pulse. (b) Output pulse for on-axis excitation. (c) Output pulse for off-axis excitation. (From Kaiser *et al*,[29] by permission of the American Physical Society.)

acceptance of the fiber is well suited for short-distance transmission of the incoherent light emitted by light-emitting diodes. The total loss in decibels per kilometer is shown in Fig. 4-20 as a function of wavelength for various values of the angular divergence of the injected light (from a Xenon lamp) expressed as the sine of the cone half-angle in air (numerical aperture). A 11-dB/km loss is obtained at the HeNe wavelength: $\lambda_0 = 0.6328$ μm. Figure 4-21 shows the impulse response measured with a mode-locked GaAs laser at $\lambda_0 = 0.9$ μm. The input pulse (2 nsec wide) is shown in curve a. The impulse width defined at the 10-dB point is 16 nsec/230 m (curve b). It increases to 30 nsec/230 m (curve c) when the light is launched off axis to enhance the higher-order modes. Simple calculations show that the permissible data rate over the 230-m-long fiber is approximately 40 Mbit/sec, that is, about four television channels. Plastic materials with losses of the order of 1000 dB/km have been found recently that permit a tight fitting of the core and the plastic cladding.[30]

Near Square-Law Fiber

The second type of fiber that we shall describe is a graded-index fiber whose profile approaches the square-law profile. The fiber is made by a modified chemical vapor deposition technique (GeO_2 in a SiO_2 capillary). A borosilicate layer is deposited inside a commercial fused quartz tube to

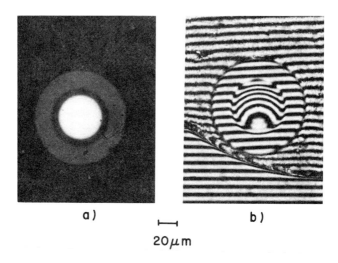

a)

b)

⊢——⊣
20μm

Fig. 4-22 (a) Photograph of an illuminated graded-index fiber (GeO$_2$ doping). (b) Interferogram with tilted wavefronts (From Cohen *et al.*,[22] by permission of the American Physical Society.)

prevent the diffusion of impurities into the core. The Ge halides are subsequently introduced. The preform is then collapsed and pulled into a fiber with an electric furnace. The transmission loss spectra has a minimum value of 3.8 dB/km at $\lambda_0 = 1.06$ μm. Figure 4-22 shows in (a) a photomicrograph of the fiber cross section and in (b) an interferogram of the fiber. If the plane of the wave transmitted through the sample and the plane of the reference wave were parallel, the interference pattern would consist of concentric circular rings. However, there would be very few rings because the sample is optically thin. It is therefore convenient to tilt one of the two interfering plane waves. This results in the pattern shown in Fig. 4-22b. The uniform lines in the cladding serve as reference lines. This technique allows, in principle, the measurement of the wavenumber k as a function of r and φ. The accuracy of the technique, however, is limited. Figure 4-23 shows the variation of k with x along a diametral line. A small lack of symmetry is noted. The solid and dashed lines correspond to samples cut near the two ends of the 1.12-km-long fiber. The impulse response is shown in Fig. 4-24, and the rms pulse width σ is shown in Fig. 4-25 as a function of the fiber length. For lengths exceeding the physical length of the fiber, a shuttle pulse technique was used. The measured pulse broadening is reduced by a factor of 11 from the values observed with a

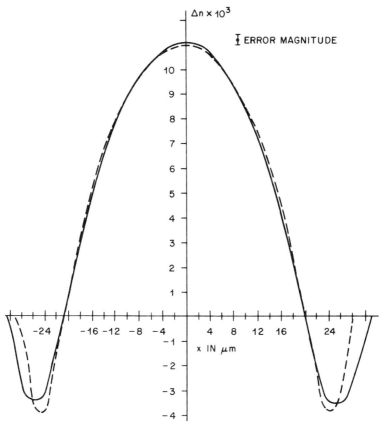

Fig. 4-23 Measured refractive-index profile along a diameter of the fiber shown in Fig. 4-22. Solid and dashed lines are for slices of the fiber located near the ends of a 1.1-km-long fiber. (From Cohen et al.,[22] by permission of the American Physical Society.)

step-index fiber having the same acceptance. This reduction is very significant for many applications. The observed pulse broadening, nevertheless, remains higher, by more than an order of magnitude, than that expected for the optimum profile. A considerable effort thus remains to be done to optimize the impulse response of graded-index fibers.

We now turn our attention to a rather different subject: the detailed analysis of optical systems in the approximation of Gauss. We are not interested in optical imaging (a subject discussed in many excellent textbooks) but in the evaluation of the point-eikonal. As we have seen in Chapter 2, the laws of propagation of gaussian beams follow readily from the knowledge of the point-eikonal.

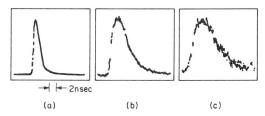

→| |←2nsec

(a)   (b)   (c)

Fig. 4-24 Impulse response of the fiber in Figs. 4-22 and 4-23. (a) L = fiber length = 1.2 km. (b) L = 3.36 km. (c) L = 5.6 km. (From Cohen *et al.*[22], by permission of the American Physical Society.)

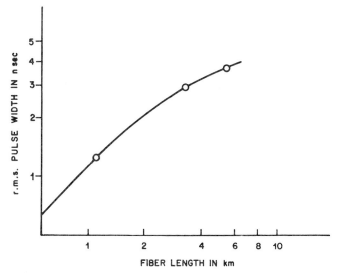

Fig. 4-25 Increase of the rms pulse width with fiber length. (From Cohen *et al.*,[22] by permission of the American Physical Society.)

4.18 General Results in Gaussian Optics

Most optical instruments have rotational symmetry with respect to some axis z, called the optical axis. However, in the field of optical communication, there is growing interest in arrangements of lenses or mirrors that do not possess such planes of symmetry. We therefore discuss the general case, which turns out to be formally identical to the case where only one transverse coordinate is considered,[31] provided a matrix notation is used.[32,33] The reader interested only in two-dimensional systems or in meridional rays in systems with circular symmetry may ignore signs such as ˜ indicating matrix transposition and may also ignore relations such as $\mathbf{A\tilde{B}} = \mathbf{B\tilde{A}}$ that are trivially true when $\mathbf{A}$ and $\mathbf{B}$ are scalar quantities.

The geometrical optics properties of an optical system are fully characterized by the point-eikonal. Note, however, that more than one eikonal must be defined if we consider different states of polarization and if more than one classical trajectory between two given points exist. Within the approximation of Gauss we have[§]

$$S(\mathbf{q}, z; \mathbf{q}', z') = \theta_0 + \mathbf{uq} + \mathbf{wq}' + \tfrac{1}{2}\mathbf{qUq} - \mathbf{qVq}' + \tfrac{1}{2}\mathbf{q'Wq'} \quad (4.212\mathrm{a})$$

where θ_0, $\mathbf{u}$, $\mathbf{w}$, $\mathbf{U}$, $\mathbf{V}$, and $\mathbf{W}$ are functions of z and z'. $\mathbf{U}$ and $\mathbf{W}$ are assumed symmetrical without loss of generality. The motivation for introducing the factors $\tfrac{1}{2}$ is to simplify the expressions for $\partial S/\partial\mathbf{q}$ and $\partial S/\partial\mathbf{q}'$ that we shall need later. The motivation for introducing a minus sign in front of $\mathbf{V}$ is to have, in free space, $\mathbf{U} = \mathbf{W} = \mathbf{V}$. In conventional geometrical optics, all these quantities are real. This assumption, however, is not essential, and we found it useful in Chapter 2 to consider complex point-eikonals. Whether the rays are real or complex need not be specified here. In a later section it will be shown how to evaluate explicitly the parameters θ_0, $\mathbf{u}$, $\mathbf{w}$, $\mathbf{U}$, $\mathbf{V}$, and $\mathbf{W}$ in (4.212) for a given optical system with input plane z' and output plane z. For the time being, let us assume that these parameters are known.

Let us consider now two adjacent optical systems, as shown in Fig. 4-26, characterized by point-eikonals S_1 and S_2, respectively:

$$S_1(\mathbf{q}''; \mathbf{q}') = \theta_1 + \mathbf{u}_1\mathbf{q}'' + \mathbf{w}_1\mathbf{q}' + \tfrac{1}{2}\mathbf{q}''\mathbf{U}_1\mathbf{q}'' - \mathbf{q}''\mathbf{V}_1\mathbf{q}' + \tfrac{1}{2}\mathbf{q}'\mathbf{W}_1\mathbf{q}'$$

$$S_2(\mathbf{q}; \mathbf{q}'') = \theta_2 + \mathbf{u}_2\mathbf{q} + \mathbf{w}_2\mathbf{q}'' + \tfrac{1}{2}\mathbf{qU}_2\mathbf{q} - \mathbf{qV}_2\mathbf{q}'' + \tfrac{1}{2}\mathbf{q}''\mathbf{W}_2\mathbf{q}'' \quad (4.212\mathrm{b})$$

The point-eikonal of the two optical systems is obtained by adding S_1 and S_2 and eliminating the intermediate coordinate $\mathbf{q}''$ by use of the Fermat principle, that is, by stating that the partial derivative of $S_1 + S_2$ with respect to $\mathbf{q}''$ vanishes. Thus

$$S(\mathbf{q}; \mathbf{q}') = S_2(\mathbf{q}; \mathbf{q}'') + S_1(\mathbf{q}''; \mathbf{q}') \quad (4.213)$$

where $\mathbf{q}''$ is obtained from the following equation, linear in $\mathbf{q}''$,

$$\frac{\partial}{\partial\mathbf{q}''}[S_2(\mathbf{q}; \mathbf{q}'') + S_1(\mathbf{q}''; \mathbf{q}')] = 0 \quad (4.214)$$

whose solution is

$$\mathbf{q}'' = -(\mathbf{U}_1 + \mathbf{W}_2)^{-1}(\mathbf{u}_1 + \mathbf{w}_2 - \mathbf{V}_1\mathbf{q}' - \tilde{\mathbf{V}}_2\mathbf{q}) \quad (4.215)$$

[§] For simplicity, we omit tildes on the first vector of a matrix product. Thus $\mathbf{ab}$ is equivalent to $\tilde{\mathbf{a}}\mathbf{b}$ (scalar product). It should not be confused with a tensor product.

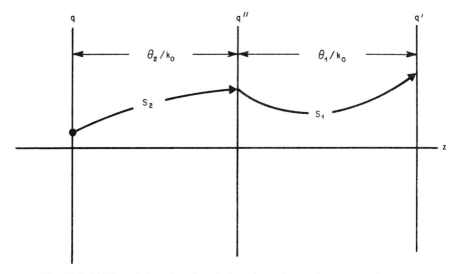

Fig. 4-26 Addition of the point-eikonals S_1 and S_2 of two adjacent optical systems.

Substituting in (4.213), we obtain after some rearranging

$$S(\mathbf{q}; \mathbf{q}') = \theta_0 + \mathbf{uq} + \mathbf{wq}' + \tfrac{1}{2}\mathbf{qUq} - \mathbf{qVq}' + \tfrac{1}{2}\mathbf{q}'\mathbf{Wq}' \qquad (4.216)$$

where

$$\theta_0 = \theta_1 + \theta_2 - \tfrac{1}{2}\mathbf{vK}^{-1}\mathbf{v}$$

$$\mathbf{u} = \mathbf{u}_2 + \mathbf{V}_2\mathbf{K}^{-1}\mathbf{v}, \qquad \mathbf{w} = \mathbf{w}_1 + \tilde{\mathbf{V}}_1\mathbf{K}^{-1}\mathbf{v}$$

$$\mathbf{U} = \mathbf{U}_2 - \mathbf{V}_2\mathbf{K}^{-1}\tilde{\mathbf{V}}_2, \qquad \mathbf{V} = \mathbf{V}_2\mathbf{K}^{-1}\mathbf{V}_1, \qquad \mathbf{W} = \mathbf{W}_1 - \tilde{\mathbf{V}}_1\mathbf{K}^{-1}\mathbf{V}_1$$

$$(4.217)$$

and the abbreviations

$$\mathbf{K} \equiv \mathbf{U}_1 + \mathbf{W}_2 \qquad (4.218)$$

$$\mathbf{v} \equiv \mathbf{u}_1 + \mathbf{w}_2 \qquad (4.219)$$

have been introduced.

It should be noted that $\mathbf{U}_1$ represents the optical thickness of an element located at the output plane of the first optical system and $\mathbf{W}_2$ the optical thickness of an element located at the input plane of the second system. Because these two elements coincide, only the sum $\mathbf{U}_1 + \mathbf{W}_2 \equiv \mathbf{K}$ enters in (4.217). The same observation applies to $\mathbf{u}_1$ and $\mathbf{w}_2$.

In a homogeneous medium with wavenumber k_0, S is the product of k_0 and the length of the ray, which is the straight line going from $\mathbf{q}'$, z' to $\mathbf{q}$, z,

$$S(\mathbf{q}; \mathbf{q}') = k_0\left[(z - z')^2 + |\mathbf{q} - \mathbf{q}'|^2\right]^{1/2}$$

$$\approx k_0(z - z') + \frac{1}{2}\,\frac{k_0|\mathbf{q} - \mathbf{q}'|^2}{z - z'} \tag{4.220}$$

Thus, for that system we have

$$\theta_0 = k_0 L, \quad L \equiv z - z' \tag{4.221}$$

$$\mathbf{u} = \mathbf{w} = \mathbf{0} \tag{4.222}$$

$$\mathbf{U} = \mathbf{W} = \mathbf{V} = \frac{k_0}{L} \tag{4.223}$$

On the rhs of (4.223) and in subsequent similar equations, we omit the unit matrix **1** for brevity.

As a simple example of application of (4.216), let us consider two adjacent sections of free space. Substituting

$$\mathbf{u}_1 = \mathbf{w}_1 = \mathbf{u}_2 = \mathbf{w}_2 = \mathbf{0} \tag{4.224}$$

$$\mathbf{U}_1 = \mathbf{W}_1 = \mathbf{V}_1 = k_0 L_1^{-1} \tag{4.225}$$

$$\mathbf{U}_2 = \mathbf{W}_2 = \mathbf{V}_2 = k_0 L_2^{-1} \tag{4.226}$$

in (4.217), we obtain after some rearranging

$$\theta_0 = \theta_1 + \theta_2 \tag{4.227}$$

$$\mathbf{u} = \mathbf{w} = \mathbf{0} \tag{4.228}$$

$$\mathbf{U} = \mathbf{W} = \mathbf{V} = \frac{k_0}{L_1 + L_2} \tag{4.229}$$

as expected from (4.221)–(4.223) since the total length of free space is $L_1 + L_2$.

If the second optical system happens to be the mirror image of the first with respect to the plane q'', we have

$$S_2(\mathbf{q}; \mathbf{q}') = S_1(\mathbf{q}'; \mathbf{q}) \tag{4.230}$$

or, explicitly,

$$\theta_2 = \theta_1, \quad \mathbf{u}_2 = \mathbf{w}_1, \quad \mathbf{w}_2 = \mathbf{u}_1$$
$$\mathbf{U}_2 = \mathbf{W}_1, \quad \mathbf{V}_2 = \tilde{\mathbf{V}}_1, \quad \mathbf{W}_2 = \mathbf{U}_1 \tag{4.231}$$

The parameters of the symmetrical system are, substituting (4.231) in (4.217),

$$\theta = 2\theta_1 - \mathbf{u}_1\mathbf{U}_1^{-1}\mathbf{v}_1, \qquad \mathbf{u} = \mathbf{w} = \mathbf{w}_1 + \tilde{\mathbf{V}}_1\mathbf{U}_1^{-1}\mathbf{u}_1$$

$$\mathbf{U} = \mathbf{W} = \mathbf{W}_1 - \tfrac{1}{2}\tilde{\mathbf{V}}_1\mathbf{U}_1^{-1}\mathbf{V}_1, \qquad \mathbf{V} = \tilde{\mathbf{V}} = \tfrac{1}{2}\tilde{\mathbf{V}}_1\mathbf{U}_1^{-1}\mathbf{V}_1 \tag{4.232}$$

Relation (4.217), giving the point-eikonal of a sequence of two optical systems, is the fundamental equation of gaussian optics. It is, however, sometimes convenient to introduce rays and ray matrices because the ray matrix of a sequence of optical systems is obtained by matrix multiplication. This is discussed in the next section.

4.19 General Properties of the Ray Matrix

The momentum $\mathbf{p}$ of a ray is, as we have seen, the transverse gradient of the eikonal. We have therefore, using for S the expression in (4.216),

$$\mathbf{p} = \frac{\partial S}{\partial \mathbf{q}} = \mathbf{u} + \mathbf{U}\mathbf{q} - \mathbf{V}\mathbf{q}' \tag{4.233}$$

and at the input plane

$$-\mathbf{p}' = \frac{\partial S}{\partial \mathbf{q}'} = \mathbf{w} - \tilde{\mathbf{V}}\mathbf{q} + \mathbf{W}\mathbf{q}' \tag{4.234}$$

These relations can be written in matrix form

$$\begin{bmatrix} \mathbf{q} \\ \mathbf{p} \\ 1 \end{bmatrix} = \begin{bmatrix} \mathbf{A} & \mathbf{B} & \mathbf{a} \\ \mathbf{C} & \mathbf{D} & \mathbf{c} \\ \mathbf{0} & \mathbf{0} & 1 \end{bmatrix} \begin{bmatrix} \mathbf{q}' \\ \mathbf{p}' \\ 1 \end{bmatrix} \equiv \mathfrak{M} \begin{bmatrix} \mathbf{q}' \\ \mathbf{p}' \\ 1 \end{bmatrix} \tag{4.235}$$

By comparing (4.233), (4.234), and (4.235), we readily find

$$\mathbf{U} = \mathbf{D}\mathbf{B}^{-1}, \qquad \mathbf{W} = \mathbf{B}^{-1}\mathbf{A}$$

$$\mathbf{V} = \mathbf{D}\mathbf{B}^{-1}\mathbf{A} - \mathbf{C}, \qquad \mathbf{u} = \mathbf{c} - \mathbf{D}\mathbf{B}^{-1}\mathbf{a} \tag{4.236}$$

$$\tilde{\mathbf{V}} = \mathbf{B}^{-1}, \qquad \mathbf{w} = \mathbf{B}^{-1}\mathbf{a}$$

or, inverting these relations (we assume that the matrices are nonsingular),

$$\mathbf{A} = \tilde{\mathbf{V}}^{-1}\mathbf{W}, \qquad \mathbf{B} = \tilde{\mathbf{V}}^{-1}$$

$$\mathbf{C} = \mathbf{U}\tilde{\mathbf{V}}^{-1}\mathbf{W} - \mathbf{V}, \qquad \mathbf{D} = \mathbf{U}\tilde{\mathbf{V}}^{-1} \tag{4.237}$$

$$\mathbf{a} = \tilde{\mathbf{V}}^{-1}\mathbf{w}, \qquad \mathbf{c} = \mathbf{u} + \mathbf{U}\tilde{\mathbf{V}}^{-1}\mathbf{w}$$

Because $\mathbf{U}$ and $\mathbf{W}$ are symmetrical, and comparing the two expressions given for $\mathbf{V}$ in (4.237), the following relation must hold between the elements of $\mathfrak{M}$:

$$\mathbf{A\tilde{B}} = \mathbf{B\tilde{A}}, \quad \mathbf{\tilde{B}D} = \mathbf{\tilde{D}B}, \quad \mathbf{D\tilde{C}} = \mathbf{C\tilde{D}}$$
$$\mathbf{\tilde{C}A} = \mathbf{\tilde{A}C}, \quad \mathbf{\tilde{D}A} - \mathbf{\tilde{B}C} = 1, \quad \mathbf{D\tilde{A}} - \mathbf{C\tilde{B}} = 1 \qquad (4.238)$$
$$\det(\mathfrak{M}) = 1$$

To obtain the last result, let us observe that, using (4.237),

$$\mathfrak{M} \equiv \begin{bmatrix} \mathbf{A} & \mathbf{B} \\ \mathbf{C} & \mathbf{D} \end{bmatrix} = \begin{bmatrix} 1 & 0 \\ 0 & \mathbf{U} \end{bmatrix} \begin{bmatrix} \mathbf{\tilde{V}}^{-1} & \mathbf{\tilde{V}}^{-1} \\ \mathbf{\tilde{V}}^{-1} - \mathbf{U}^{-1}\mathbf{VW}^{-1} & \mathbf{\tilde{V}}^{-1} \end{bmatrix} \begin{bmatrix} \mathbf{W} & 0 \\ 0 & 1 \end{bmatrix}$$
$$(4.239)$$

Thus

$$\det(\mathfrak{M}) = \det(\mathbf{U}) \det \begin{bmatrix} 0 & \mathbf{\tilde{V}}^{-1} \\ -\mathbf{U}^{-1}\mathbf{VW}^{-1} & 0 \end{bmatrix} \det(\mathbf{W}) = 1 \quad (4.240)$$

by subtracting lines and columns, and using the fact that the middle determinant in (4.240) is

$$\det(\mathbf{U}^{-1}\mathbf{VW}^{-1}) \det(\mathbf{\tilde{V}}^{-1}) = \left[\det(\mathbf{U}) \det(\mathbf{W}) \right]^{-1} \qquad (4.241)$$

We have also used the fact that the determinant of a product of matrices is the product of the determinant of the individual matrices, and that the determinant of the inverse of a matrix is the inverse of the determinant of the matrix. An alternative proof[34] of $\det(\mathfrak{M}) = 1$ consists of showing that the eigenvalues $\lambda_1, \lambda_2, \lambda_3, \lambda_4$ of $\mathfrak{M}$ are such that $\lambda_1\lambda_2 = \lambda_3\lambda_4 = 1$. Thus $\det(\mathfrak{M}) = \lambda_1\lambda_2\lambda_3\lambda_4 = 1$.

Because $\mathfrak{M}$ is never singular, $\mathfrak{M}^{-1}$ exists. We obtain

$$\mathfrak{M}^{-1} = \begin{bmatrix} \mathbf{\tilde{D}} & -\mathbf{\tilde{B}} & \mathbf{\tilde{B}c} - \mathbf{\tilde{D}a} \\ -\mathbf{\tilde{C}} & \mathbf{\tilde{A}} & \mathbf{\tilde{C}a} - \mathbf{\tilde{A}c} \\ 0 & 0 & 1 \end{bmatrix} \qquad (4.242)$$

In two dimensions, the relations (4.238) reduce to the well-known relation $AD - BC = 1$. Note that these results rest on the existence of a point-eikonal. They are unrelated to reciprocity (the medium may, in fact, be nonreciprocal).

Thus the elements of the ray matrix $\mathfrak{M}$ satisfy special relations (in fact, six independent relations). Such a matrix is called "symplectic" because it preserves the antisymmetric bilinear form[1]

$$q_a p_b - p_a q_b \qquad (4.243)$$

between any two rays q_a, p_a and q_b, p_b. The invariant scalar quantity in (4.243) is called a Lagrange ray invariant, a special form of the invariant given in (4.52).

An optical system can be characterized by the position and momentum at the output plane of five rays having specified position and momentum at the input plane. Let these rays be denoted $\bar{q}$, $\bar{p}$, Q, P, and $Q^\dagger$, $P^\dagger$. The pair of matrices (Q, P) and $(Q^\dagger, P^\dagger)$ actually represent two rays each because they are defined by the association of two ordinary rays ($[q_1, q_2]$, $[p_1, p_2]$) and ($[q_1^\dagger q_2^\dagger]$, $[p_1^\dagger p_2^\dagger]$), respectively. If the initial conditions for $(Q^\dagger, P^\dagger)$, (Q, P), and $\bar{q}$, $\bar{p}$ are, respectively,

$$\begin{bmatrix} 1 \\ 0 \\ 0 \end{bmatrix}, \quad \begin{bmatrix} 0 \\ 1 \\ 0 \end{bmatrix}, \quad \begin{bmatrix} 0 \\ 0 \\ 1 \end{bmatrix} \qquad (4.244)$$

the ray matrix can be written

$$\mathfrak{M} = \begin{bmatrix} A & B & a \\ C & D & c \\ 0 & 0 & 1 \end{bmatrix} = \begin{bmatrix} Q^\dagger & Q & q \\ P^\dagger & P & p \\ 0 & 0 & 1 \end{bmatrix} \qquad (4.245)$$

Note that, according to (4.238), PQ^{-1} and $Q^{-1}Q^\dagger$ are symmetrical. The point-eikonal can be expressed in terms of these five rays. Using (4.216) and (4.236), we obtain

$$S(q; q') = \theta_0 + (\bar{p} - PQ^{-1}\bar{q})q + \bar{q}\tilde{Q}^{-1}q' + \tfrac{1}{2} qPQ^{-1}q - q\tilde{Q}^{-1}q'$$

$$+ \tfrac{1}{2} q'Q^{-1}Q^\dagger q' \qquad (4.246)$$

It remains to express θ_0 in terms of $\bar{q}$. We have, setting $q = \bar{q}$, $q' = 0$, and using the definition of S,

$$S(\bar{q}; 0) = \theta_0 + \bar{p}\bar{q} - \tfrac{1}{2} \bar{q}PQ^{-1}\bar{q} = \int_{z'}^{z} L(\dot{\bar{q}}, \bar{q}, z)\, dz \equiv \bar{S} \qquad (4.247)$$

where L denotes the Lagrangian. We have used the fact that PQ^{-1} is symmetrical.

Substituting θ_0 from (4.247) in (4.246), we finally obtain

$$S(\mathbf{q}; \mathbf{q}') = \bar{S} + \bar{\mathbf{p}}(\mathbf{q} - \bar{\mathbf{q}}) + \tfrac{1}{2}(\mathbf{q} - \bar{\mathbf{q}})\mathbf{P}\mathbf{Q}^{-1}(\mathbf{q} - \bar{\mathbf{q}}) - (\mathbf{q} - \bar{\mathbf{q}})\tilde{\mathbf{Q}}^{-1}\mathbf{q}'$$

$$+ \tfrac{1}{2}\mathbf{q}'\mathbf{Q}^{-1}\mathbf{Q}^{\dagger}\mathbf{q}' \tag{4.248}$$

It is interesting that $\mathbf{q}$ enters only as $\mathbf{q} - \bar{\mathbf{q}}$, where $\bar{\mathbf{q}}$ denotes, as we recall, a ray with initial conditions $\bar{\mathbf{q}}' = \mathbf{0}$, $\bar{\mathbf{p}}' = \mathbf{0}$.

4.20 Evaluation of the Point-Eikonal in the Approximation of Gauss[32]

General relations that allow the evaluation of the point-eikonal in the approximation of Gauss for a continuous medium (e.g., a lenslike medium) or a sequence of homogeneous media (e.g., a sequence of lenses) are given in this section. Consideration is given to anisotropic media. The eikonal equation is written

$$\frac{\partial S}{\partial z} = k_z\left(\frac{\partial S}{\partial \mathbf{q}}, \mathbf{q}, z\right) \tag{4.249}$$

where k_z is assumed as before to be at most quadratic in $\mathbf{q}$ and $\mathbf{p}$. The dependence of k_z on z, however, is arbitrary. Thus let us set

$$k_z(\mathbf{p}, \mathbf{q}, z) = k_0 + \mathbf{n}\mathbf{q} + \tfrac{1}{2}\mathbf{q}\mathbf{N}\mathbf{q} - \tfrac{1}{2}(\mathbf{p} - \mathbf{g} - \mathbf{G}\mathbf{q})\mathbf{F}(\mathbf{p} - \mathbf{g} - \mathbf{G}\mathbf{q}) \tag{4.250}$$

This form for the Hamiltonian is chosen in order that the Lagrangian, to be derived next, be simple. In this expression, the parameters k_0, $\mathbf{n}$, $\mathbf{N}$, $\mathbf{g}$, $\mathbf{G}$, and $\mathbf{F}$ are functions of z that can be obtained from knowledge of the material parameters from a power series expansion of (3.115) or (3.147). Without loss of generality, we can assume that $\mathbf{N}$ and $\mathbf{F}$ are symmetrical. The Hamilton ray equations (4.71) are

$$\dot{\mathbf{q}} = -\frac{\partial k_z}{\partial \mathbf{p}} = \mathbf{F}(\mathbf{p} - \mathbf{g} - \mathbf{G}\mathbf{p}) \tag{4.251}$$

$$\dot{\mathbf{p}} = \frac{\partial k_z}{\partial \mathbf{q}} = \mathbf{n} + \mathbf{N}\mathbf{q} + \tilde{\mathbf{G}}\mathbf{F}(\mathbf{p} - \mathbf{q} - \mathbf{G}\mathbf{q}) \tag{4.252}$$

where overdots denote differentiation with respect to z. Using for $\mathbf{p}$ the expression from (4.251), we find that the Lagrangian, defined in (4.72), is

$$L(\dot{\mathbf{q}}, \mathbf{q}, z) \equiv \mathbf{p}\dot{\mathbf{q}} + k_z = k_0 + \mathbf{n}\mathbf{q} + \tfrac{1}{2}\dot{\mathbf{q}}\mathbf{N}\mathbf{q} + \dot{\mathbf{q}}(\mathbf{g} + \mathbf{G}\mathbf{q}) + \tfrac{1}{2}\dot{\mathbf{q}}\mathbf{F}^{-1}\dot{\mathbf{q}}$$

$$\tag{4.253a}$$

and

$$\int_{z'}^{z} L(\dot{\mathbf{q}}, \mathbf{q}, z)\, dz = \int_{z'}^{z} k_0\, dz + \tfrac{1}{2}\left[\mathbf{qp} - \mathbf{q'p'} + \int_{z'}^{z}(\mathbf{nq} + \mathbf{g}\dot{\mathbf{q}})\, dz\right] \quad (4.253\mathrm{b})$$

The first term on the rhs of (4.253a) is the geometrical phase shift θ_0. If the system is aligned, $\mathbf{n} = \mathbf{g} = \mathbf{0}$, and the last integral vanishes. The term $\mathbf{qp} - \mathbf{q'p'}$ is then simply interpreted as originating from the extra path lengths at the ends of the ray, resulting from their tilt.

Let us now derive the differential equations obeyed by the parameters θ_0, $\mathbf{u}$, $\mathbf{w}$, $\mathbf{U}$, $\mathbf{V}$, and $\mathbf{W}$ in expression (4.212a) for $S(\mathbf{q}, z; \mathbf{q'}, z')$. Introducing this expression in (4.249) with k_z given in (4.250) and comparing terms that have the same power in $\mathbf{q}$ and $\mathbf{q'}$, we obtain

$$\dot{\mathbf{U}} + (\mathbf{U} - \mathbf{G})\mathbf{F}(\mathbf{U} - \mathbf{G}) = \mathbf{N}, \qquad \dot{\mathbf{u}} + (\mathbf{U} - \mathbf{G})\mathbf{F}(\mathbf{u} - \mathbf{g}) = \mathbf{n}$$

$$\dot{\mathbf{V}} + (\mathbf{U} - \mathbf{G})\mathbf{F}\mathbf{V} = \mathbf{0}, \qquad \dot{\mathbf{w}}\tilde{\mathbf{V}}\mathbf{F}(\mathbf{u} - \mathbf{g}) = \mathbf{0} \qquad (4.254)$$

$$\dot{\mathbf{W}} + \tilde{\mathbf{V}}\mathbf{F}\mathbf{V} = \mathbf{0}, \qquad \theta_0 + \tfrac{1}{2}(\mathbf{u} - \mathbf{g})\mathbf{F}(\mathbf{u} - \mathbf{g}) = k_0$$

where the overdots denote differentiation with respect to z. Because these equations are of first order, specification of θ_0, $\mathbf{u}$, $\mathbf{w}$, $\mathbf{U}$, $\mathbf{V}$, and $\mathbf{W}$ at some plane (e.g., at the plane $z = z'$) uniquely defines their values at any z.

Equations (4.254) are nonlinear. The first equation in (4.254), for instance, is a matrix Riccati equation. It is therefore sometimes easier to use form (4.248) of the point-eikonal. In this equation, $\bar{\mathbf{q}}$, $\bar{\mathbf{p}}$ is a solution of the ray equations (4.251) and (4.252) with the initial condition $\bar{\mathbf{q}}' = \bar{\mathbf{p}}' = \mathbf{0}$. The matrices $\mathbf{Q}$, $\mathbf{P}$ formally obey the ray equation without a misalignment term, namely,

$$\dot{\mathbf{Q}} = \mathbf{F}(\mathbf{P} - \mathbf{GQ}), \qquad \dot{\mathbf{P}} = \mathbf{NQ} + \tilde{\mathbf{G}}\mathbf{F}(\mathbf{P} - \mathbf{GQ}) \qquad (4.255)$$

with the initial conditions $\mathbf{Q}' = \mathbf{0}$, $\mathbf{P}' = \mathbf{1}$. $\mathbf{Q}^{\dagger}$, $\mathbf{P}^{\dagger}$ are also solutions of (4.255) with the initial conditions $\mathbf{Q}^{\dagger'} = \mathbf{1}$, $\mathbf{P}^{\dagger'} = \mathbf{0}$. In order to verify that this is indeed the case, it suffices to substitute in (4.254) expressions (4.236) for $\mathbf{u}$, $\mathbf{w}$, $\mathbf{U}$, $\mathbf{V}$, $\mathbf{W}$ in terms of $\bar{\mathbf{q}}$, $\bar{\mathbf{p}}$, $\mathbf{Q}$, $\mathbf{P}$, $\mathbf{Q}^{\dagger}$, $\mathbf{P}^{\dagger}$ from (4.245).

When the medium is stepwise homogeneous, we need an expression for the point-eikonal of a section of homogeneous medium and an expression for the effective optical thickness of a curved surface. These two problems are now considered.

When the medium is homogeneous, k_z in (4.250) does not depend on $\mathbf{q}$ or z, that is, $\mathbf{G} = \mathbf{0}$, $\mathbf{N} = \mathbf{0}$, $\mathbf{n} = \mathbf{0}$, and the other parameters are independent of z. Thus (4.252) reduces to $\ddot{\mathbf{q}} = \mathbf{0}$, the rays being straight lines. The point-eikonal is obtained by multiplying the right-hand side of (4.253a) by $z - z'$,

$$S(\mathbf{q}, z; \mathbf{q'}, z') = (z - z')(k_0 + \dot{\mathbf{q}}\mathbf{g} + \tfrac{1}{2}\dot{\mathbf{q}}\mathbf{F}^{-1}\dot{\mathbf{q}}) \qquad (4.256)$$

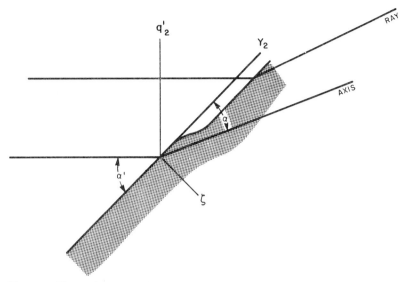

Fig. 4-27 Change in optical length resulting from a displacement ζ of the surface of discontinuity for oblique rays.

where

$$\dot{\mathbf{q}} = \frac{\mathbf{q} - \mathbf{q}'}{z - z'} \tag{4.257}$$

It remains to obtain the effective optical thickness associated with curved surfaces of discontinuity. We assume that the discontinuity is tapered in such a way that the initial state of polarization is preserved, yet sufficiently abrupt from a ray optics point of view.

If the incident ray is defined by its position q_1', q_2' in the plane perpendicular to the z-axis, it reaches the reference plane at $y_1 = q_1'$, $y_2 = q_2'/\sin(\alpha')$, where α' denotes the angle of the ray to the plane, and y_1, y_2 are coordinates in the reference plane as shown in Fig. 4-27 with y_1 perpendicular to the plane of incidence. The displacement ζ is to be evaluated at the point q_1', $q_2'/\sin(\alpha')$. The expression for the effective optical thickness was given in (4.43). This expression is not exact, but it is sufficiently accurate when calculations are limited to the approximation of Gauss (and somewhat beyond that approximation).

4.21 Focusing and Deflection of Optical Beams by Cylindrical Mirrors

An application of the method described in Section 4.4, which consists in evaluating the effective optical thickness of curved surfaces, is discussed in this section. Optical waveguides have been investigated as means of transmitting communication from city to city or within cities. They usually incorporate sequences of focusers (lenses or pairs of mirrors) whose separation is close, though not equal, to the confocal spacing. In most practical circumstances, it is necessary to bend or tilt in three dimensions the waveguide axis in order to circumvent natural obstacles or right-of-way restrictions. Although small tilts of less than, say, 1°, can be obtained by offsetting the focusers, large tilts are more easily obtained by rotating the mirrors in a periscopic configuration. It is consequently desirable to have periscopic systems providing variable deflections of the optical axes in three dimensions (to follow irregular paths) and variable effective focal lengths (to accommodate changes in focuser separation). In order to preserve the circular cross section of the beams, it is further required that the focusers behave as ordinary lenses with rotational symmetry, i.e., that they be free of astigmatism, to first order. It is the purpose of this section to show that these various conditions are met in a periscopic configuration that incorporates two cylindrical mirrors of equal (and fixed) curvature. Changes in direction and focal length can both be obtained through proper angular orientation of the two mirrors. The flexibility of the configuration discussed in this section results from the introduction of nonplanar optical axes.[33] Geometrical optics aberrations are relatively unimportant because the effect of such aberrations is masked by diffraction effects. As we have seen in Chapter 2, the transformation of beams propagating in optical waveguides at a sufficient distance from the edges of the apertures can be obtained from the results of a paraxial ray theory. We therefore restrict ourselves to a paraxial ray theory of curved mirrors under oblique incidence. General conditions for sharp focusing by two cylindrical mirrors are discussed, under the assumption that the separation between the two mirrors is small compared with the focal length. Experiments made at visible wavelengths are also described. To avoid confusion between subscripts, we change the notation slightly with $q_1 \rightarrow x$, $q_2 \rightarrow z$. Subscripts 1 and 2 now refer to the first and second mirrors, respectively. Primes still refer to incident quantities.

Consider a curved mirror such as R_1 shown in Fig. 4-28 and an incident optical axis. Let $x_1 y_1 \zeta_1$ be a rectangular coordinate system, with ζ_1 directed along the inner normal to the mirror and x_1 perpendicular to the plane of incidence. The x_1-axis is oriented in such a way that the y_1-component of the ray direction vector is positive. The position of an incident ray, offset with respect to the incident optical axis, is defined by coordinates x_1 and z_1', the z_1'-axis being perpendicular to both the x_1-axis and the incident optical axis. If $\zeta_1 = \zeta_1(x_1 y_1)$ denotes the equation of the mirror surface and i_1 the angle of incidence, (4.43) shows that the extra path length experienced by a ray x_1, z_1', as a result of the mirror deforma-

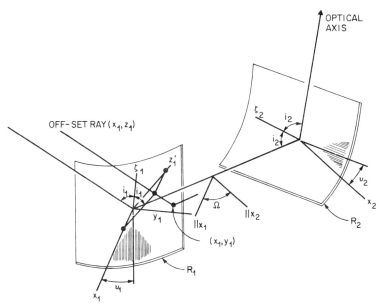

Fig. 4-28 Schematic representation of the two cylindrical mirror system. Notice that the optical axis, in general, does not lie in a plane; the angle between the planes of incidence at the two mirrors (or between their normals x_1, x_2) is denoted Ω. The angles ν_1, ν_2, and Ω as shown in the figure are positive. (From Arnaud and Ruscio,[33] by permission of the American Optical Society.)

tion, is

$$\Delta_1 (x_1, z_1') = 2 \cos(i_1)\zeta_1\left(x_1, \ \frac{z_1'}{\cos i_1}\right) \qquad (4.258)$$

This change in path length is evaluated by considering rays parallel to the optical axis, incident on an area of the mirror that has been displaced along the mirror normal. To be sure, the rays can be oblique with respect to the optical axis, but a small tilt introduces a negligible change in path length at the mirror. For the case of a cylindrical mirror of equation

$$\zeta_1(x_1, y_1) = \frac{1}{2R_1} (x_1 \sin \nu_1 - y_1 \cos \nu_1)^2 \qquad (4.259)$$

where R_1 denotes the mirror radius and ν_1 an angle between the mirror generatrix and the x_1-axis, we obtain, from (4.258) and (4.259),

$$\Delta_1 = \frac{\cos(i_1)}{R_1} \left[x_1 \sin(\nu_1) - \frac{z_1'}{\cos i_1} \cos(\nu_1) \right]^2 \qquad (4.260)$$

By rotating the coordinate system x_1, z'_1 by an angle ϵ_1 given by

$$\tan(\epsilon_1) = \cos(i_1) \tan(\nu_1) \tag{4.261}$$

we find that Δ_1 is equal to the path length that would be introduced by a cylindrical lens of focal length

$$f_1 = \frac{R_1}{2} \left(\cos i_1 \sin^2 \nu_1 + \frac{\cos^2 \nu_1}{\cos i_1} \right)^{-1} \tag{4.262}$$

The orientation of this equivalent cylindrical lens with respect to the z'_1-axis is defined by the angle ϵ_1 given in (4.261). Similar results hold, of course, for the second mirror, to which the index 2 henceforth refers. The coordinate system, however, must be rotated by an angle Ω (corresponding to the angle between the planes of incidence at the two mirrors) when going from the first to the second mirror.

As is well known, two cylindrical lenses provide a sharp focusing of incident homocentric ray pencils when they have equal focal lengths and are oriented 90° to each other. These two conditions, when applied to the lenses equivalent to the two mirrors, can be written, from (4.262) and (4.261),

$$\frac{R_1}{2} \left(\cos(i_1) \sin^2(\nu_1) + \frac{\cos^2(\nu_1)}{\cos(i_1)} \right)^{-1}$$

$$= \frac{R_2}{2} \left(\cos(i_2) \sin^2(\nu_2) + \frac{\cos^2(\nu_2)}{\cos(i_2)} \right)^{-1} = f \tag{4.263}$$

where f is the equivalent focal length of the system, and

$$\Omega = \tan^{-1}[\tan(\nu_1) \cos(i_1)] + \tan^{-1}[\tan(\nu_2) \cos(i_2)] \pm (\tfrac{1}{2}\pi) \tag{4.264}$$

The total deflection angle α experienced by the optical axis as a result of the reflections on the two mirrors is given, from elementary trigonometry, by

$$\cos(\alpha) = \cos(2i_1) \cos(2i_2) - \cos(\Omega) \sin(2i_1) \sin(2i_2) \tag{4.265}$$

Notice now that the angle $\alpha = 0$ (no deflection) can be achieved only if $i_1 = i_2 = i$. This condition is henceforth assumed to hold. It is also assumed, for simplicity, that $R_1 = R_2 = R$. Equation (4.263) then shows that we must have

$$\nu_1 = \pm \nu_2 \tag{4.266}$$

Let us consider first the solution $\nu_1 = \nu_2 = \nu$ and rewrite (4.263), (4.264),

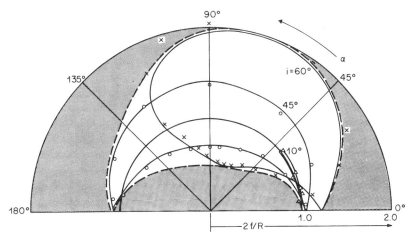

Fig. 4-29 The clear area is the region in space where an incident collimated beam can be focused by changing the orientation of the mirrors. (From Arnaud and Ruscio,[33] by permission of the American Optical Society.)

and (4.266), taking into account the above assumptions. We obtain

$$f = \frac{R}{2}\left(\cos(i)\,\sin^2(\nu) + \frac{\cos^2(\nu)}{\cos(i)}\right)^{-1} \tag{4.267}$$

$$\Omega = \frac{\pi}{2} + 2\,\tan^{-1}[\tan(\nu)\,\cos(i)] \tag{4.268}$$

$$\cos(\alpha) = \cos^2(2i) - \cos(\Omega)\,\sin^2(2i) \tag{4.269}$$

For a given value of R, we may consequently choose the effective focal length f of the system and the total deflection angle α and find the necessary value for i, ν, and Ω from (4.267)–(4.269). Figure 4-29 shows the range of values of f and α that can be obtained in that way, i being restricted to angles between 10° and 60° to avoid interferences between the two mirrors. This figure also shows that for $\alpha = 0$ (no deflection), the system focal length f can be varied between $0.5R$ and $0.6R$. For larger deflection angles such as 75°, f can be varied in a larger range ($0.25R$ to R).

The solution $\nu_1 = -\nu_2$ of (4.266) corresponds, from (4.264) and (4.265), to $\Omega = \pi/2$ and $\cos(\alpha) = \cos^2(2i)$. For a given value of α, f can consequently be varied by varying ν, the incident angles staying constant. This solution covers, in particular, the area enclosed by the curve $i = 10°$ in Fig. 4-29, which is not covered by the previous solution. Both solutions are consequently of interest.

Fig. 4-30 Photograph of the experimental setup incorporating two cylindrical mirrors. (From Arnaud and Ruscio,[33] by permission of the American Optical Society.)

A mechanical setup, shown in Fig. 4-30, was constructed to verify the above results. Adjustment and direct dial reading with precision better than $1°$ are provided for all relevant angles (i_1, i_2, ν_1, ν_2, and Ω). The variations of i_1 and i_2, however, are restricted to a 7–60° range. A deformable parallelogram orients the first mirror in such a way that the reflected beam is always directed to the second mirror. The whole system can rotate about the incident optical axis to provide beam deflection in two dimensions. The radii of curvature of the 50 mm × 50 mm × 13 mm mirrors are 6 m, and their accuracy is better than 2 μm. The focal length of the system is measured by sending a collimated ray pencil onto the system and measuring the position of the focus with respect to the point halfway between the two mirrors. A small astigmatic difference of 300 mm, which is of the order of the mirror separation (260 mm), is observed. This astigmatic difference would be masked by diffraction effects for long focal length systems such as those considered for communication purposes. For shorter focal lengths and fixed object planes, perfect focusing can be achieved by introducing a small correction of the orientation angles given before. The experimental values obtained for f and α, shown in Fig. 4-29 for $i = 10°$, $45°$, and $60°$, are in good agreement with the calculated values.

Thus a system of two cylindrical mirrors properly oriented in space can provide optimum focusing of incident beams, with variable focal lengths and deflection angles. Cylindrical optical mirrors, although somewhat more difficult to manufacture than spherical mirrors with the techniques presently available, can be made with acceptable accuracy. At millimeter wavelengths, cylindrical mirrors of circular cross section are easy to make by bending aluminum plates with bending moments exerted along the plate edges (rms errors of less than 0.1 mm have been measured on 1 m $\times$ 1 m plates, bent with 40-m radii; see Section 2.22). As in any periscopic system, the two mirrors, once properly oriented, must be held rigidly with respect to each other. The positioning of the periscopic system as a whole is just as critical as the positioning of lenses in conventional optical waveguides and must be controlled by similar devices.

4.22 Transformation of the Polarization

Let us consider first the change in polarization of a wave reflected by a sequence of perfectly conducting plane mirrors. Two general methods may be used.

Method I.   Because the mirrors are perfect conductors, the total electric field has no component in the mirror plane. In addition, the reflected field $\mathbf{E}_r$ and the incident field $\mathbf{E}_i$ have the same intensity. Consequently, the vectors $\mathbf{E}_r$ and $\mathbf{E}_i$ are symmetrical with respect to the normal $\mathbf{n}$ to the reflecting mirror, as shown in Fig. 4-31a. This relation can be expressed by the vectorial equality

$$\mathbf{E}_r = -\mathbf{E}_i + 2\mathbf{n}(\mathbf{n} \cdot \mathbf{E}_i) \qquad (4.270)$$

where $\mathbf{n}$, $\mathbf{E}_i$, and $\mathbf{E}_r$ are assumed to have a length of unity.

Let $\mathbf{I}$ and $\mathbf{R}$ be unit vectors directed, respectively, along the incident and reflected rays in the sense of propagation. $\mathbf{n}$ is given by (see Fig. 4-31b)

$$\mathbf{n} = \frac{\mathbf{R} - \mathbf{I}}{|\mathbf{R} - \mathbf{I}|} \qquad (4.271)$$

$\mathbf{E}_r$ as given by (4.270) and (4.271) is easily expressed in terms of the components of $\mathbf{E}_i$, $\mathbf{R}$, and $\mathbf{I}$ in a rectangular coordinate system. These equations are applicable to the successive reflecting mirrors. It can be checked at each step that $\mathbf{E}_r$ is perpendicular to $\mathbf{R}$. This obvious property results from the substitution of (4.271) in (4.270), multiplication of both sides by $\mathbf{R}$, and the fact that $\mathbf{I} \cdot \mathbf{E}_i = 0$.

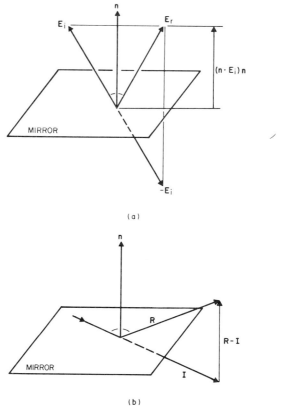

Fig. 4-31 Reflection of a perfectly conducting mirror. (a) Electric field. (b) Ray direction.

Method II. Consider the ray iO_1O_2r reflected at O_1 and O_2 by two mirrors M_1 and M_2 as shown in Fig. 4-32. Let the indices 1 and 2 refer, respectively, to the first and second mirrors. We define $\mathbf{n}_1$ and $\mathbf{n}_2$, as before, as the inner normals to the mirrors [see (4.271)]. We also define the "binormals" $\mathbf{B}_1$ and $\mathbf{B}_2$ as unit vectors perpendicular to both the incident and reflected rays. $\mathbf{B}_1$ and $\mathbf{B}_2$ are in the mirror planes. We use, as a reference system before a reflection, the triplet of vectors $\mathbf{n}_i$, $\mathbf{B}$, $\mathbf{I}$, where $\mathbf{I}$ denotes as before a unit vector directed along the incident ray. $\mathbf{n}_i$ is a unit vector perpendicular to both $\mathbf{B}$ and $\mathbf{I}$ such that $(\mathbf{n}_i \cdot \mathbf{n})$ is a positive quantity. The sense of $\mathbf{B}$ is now defined by stating that $(\mathbf{n}_i, \mathbf{B}, \mathbf{I})$ forms a direct trihedral. Just after reflection on the mirror, the reference system is defined by the vectors $\mathbf{n}_r$, $\mathbf{B}$, $\mathbf{R}$ that are similarly defined.

Because the polarization vector is perpendicular to the ray, it can be represented in the plane defined by the vectors $\mathbf{n}_i$, $\mathbf{B}$ before reflection and $\mathbf{n}_r$, $\mathbf{B}$ after reflection. This coordinate system (moving along the ray path) is

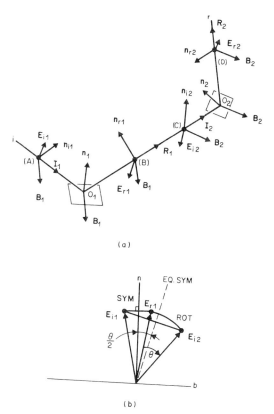

(a)

(b)

Fig. 4-32 Transformation of the polarization upon reflection on a pair of mirrors.

denoted henceforth **n**, **b**. $\mathbf{E}_r$ can be derived from $\mathbf{E}_i$ through a symmetry with respect to the **n**-axis. We have now to take into consideration the rotation θ_{12} of the coordinate system[§] between $(\mathbf{n}_{r1}, \mathbf{B}_1, \mathbf{R}_1)$ just after (M_1) and $(\mathbf{n}_{i2}, \mathbf{B}_2, \mathbf{I}_2)$ just before (M_2). Notice that $\mathbf{I}_2 \equiv \mathbf{R}_1$. Because of this rotation, the polarization vector experiences a rotation $-\theta_{12}$ with respect to the **n**, **b** coordinate system as shown in Fig. 4-32b. $\mathbf{E}_{r1}$ is, of course, identical to $\mathbf{E}_{i2}$; the rotation shown in that figure results only from the rotation of the coordinate system. We must now combine the symmetry with respect to the **n**-axis with a rotation $-\theta_{12}$. These two transformations are equivalent to a single symmetry with respect to an axis that makes an angle $-\theta_{12}/2$ with the *n*-axis. To prove that statement, just notice that a rotation θ is equivalent to two symmetries with respect to two axes that make an angle $\theta/2$ between them and take the first of these two axes as coincident with the **n**-axis. Since two symmetries with respect to the same

---

[§] This rotation becomes an "integrated torsion" in the case of continuous media.

axis cancel each other, a single symmetry remains. Let us not forget that
these transformations do not commute and must be applied in the proper
order.

In conclusion, the net effect of the reflection at the first mirror M_1 is, in
the $\mathbf{n}$, $\mathbf{b}$ coordinate system, a symmetry with respect to an axis that makes
an angle $-\theta_{12}/2$ with respect to the $\mathbf{n}$-axis as shown in Fig. 4-32b. θ_{12} is
the angle between $\mathbf{B}_1$ and $\mathbf{B}_2$ or the angle between the incidence planes at
M_1 and M_2. The vector $\mathbf{B}_i$ can be expressed by

$$\mathbf{B}_i = \frac{\mathbf{I}_i \times \mathbf{R}_i}{|\mathbf{I}_i \times \mathbf{R}_i|} \tag{4.272}$$

and we may express θ_{12} by

$$\cos(\theta_{12}) = \mathbf{B}_1 \cdot \mathbf{B}_2 = \frac{\mathbf{I}_1 \times \mathbf{R}_1}{|\mathbf{I}_1 \times \mathbf{R}_1|} \cdot \frac{\mathbf{R}_1 \times \mathbf{R}_2}{|\mathbf{R}_1 \times \mathbf{R}_2|} \tag{4.273}$$

For a sequence of mirrors $M_1, M_2, M_3, \ldots, M_N$, we have to combine a
number of symmetries with respect to axes that make angles $-\theta_{12}/2$,
$-\theta_{23}/2$, $-\theta_{34}/2, \ldots, -\theta_{N-1\,N}/2$, and 0, respectively, with the $\mathbf{n}$-axis.

From the above discussion, it is clear that if the number of mirrors is
odd, the total transformation is a symmetry. If the number of mirrors is
even, the total transformation is a rotation Ω equal to

$$\Omega = -2\left(\frac{\theta_{23}}{2} - \frac{\theta_{12}}{2}\right) - 2\left(\frac{\theta_{45}}{2} - \frac{\theta_{34}}{2}\right) - \cdots + 2\,\frac{\theta_{N-1\,N}}{2}$$

$$= \theta_{12} - \theta_{23} + \theta_{34} - \theta_{45} + \cdots + \theta_{N-1\,N} \tag{4.274}$$

For two mirrors the rotation given by (4.274) reduces to $\Omega = \theta_{12}$.

In any case, it is sufficent to consider the transformation of a single
$\mathbf{E}$-vector. The transformation of the other polarization vectors readily
follows from the knowledge of the number of mirrors (odd or even).

Method I described above, using a fixed coordinate system, appears to
be formally simpler. The second method, however, using a coordinate
system that moves along the ray path, has a more direct physical significance. In particular, it can be applied to continuous media. Both methods
are exact and give, of course, the same final result.

The optical cavity represented in Fig. 2-22d incorporates four mirrors
that define a nonplanar path. The rotation experienced by linearly
polarized axial rays can be obtained by the methods discussed above.
Because this angle is not, in general, a multiple of π, linearly polarized
fields do not reproduce themselves after a round trip. The eigenstates of
polarization are necessarily circular (clockwise and counterclockwise). The
rotation discussed above affects the *phase shift* of these circularly polarized

fields. For clockwise waves the phase shift is equal to the rotation angle, while for counterclockwise waves, the phase shift is opposite to the rotation angle. These phase shifts entail opposite shifts in the frequencies of resonance. Thus, a twist of the path of an optical cavity has the effect of lifting the degeneracy that otherwise exists (in the absence of other polarization-sensitive elements along the path) between clockwise and counterclockwise states of polarization.

Polarization Effects in Optical Fibers

These results can be generalized to continuous isotropic media. The rotation angle of the field is opposite to the integrated torsion of the ray path. As an example, let us consider rays in isotropic, z-uniform, and circularly symmetric optical fibers. The free wavenumber is denoted $k(r)$ $\equiv (\omega/c)n(r)$. To evaluate the torsion of the ray path, we first need the direction of the binormal **b** of the ray, which is perpendicular to both the ray and the gradient ∇n of the refractive index. The torsion of the ray is the ratio $d\theta/ds$ of the infinitesimal rotation $d\theta$ of **b** and the infinitesimal ray length ds. We shall evaluate the rotation $d\theta/d\varphi$ per infinitesimal azimuthal angle $d\varphi$. Using the Hamilton equations, it is not difficult to derive the following exact result:

$$\frac{d\theta}{d\varphi} = \frac{k_z k(r)}{k_z^2 + l_z^2/r^2}$$

where k_z is the axial wavenumber and l_z the axial angular momentum of the ray, two constants of motion. (According to the WKB approximation, $l_z \equiv \mu$ is the azimuthal mode number.) Within the weakly guiding approximation $k_z \gg k_\varphi \equiv l_z/r$, $k(r) \approx k_z$, the quantity $d\theta/d\varphi$ is unity. Because the field direction is referred to the principal normal of the ray, which coincides with the radial direction at the maxima of r, this result means that the direction of the field is, in fact, invariant, to first order.

Because $d\theta/d\varphi$ departs slightly from unity, however, waves whose polarization vector rotates in the same direction as the ray ($HE_{\mu+1}$ modes) and modes whose polarization vector rotates in the opposite direction ($EH_{\mu-1}$ modes) have slightly different axial wavenumbers, which can be obtained by integrating $\pm(d\varphi - d\theta)$ over a ray period, and dividing the result by the ray period.

The maximum difference in time of flight between corresponding HE and EH modes can be obtained on the basis of the previous expression for $d\theta/d\varphi$. It is of the order of $10{,}000(\Delta n/n)^2/F$ nsec/km, where F $\equiv (2\Delta n/n)^{1/2}(k_0 a)$ is the normalized frequency of the fiber. This spread in time is small compared to the minimum impulse width predicted by scalar ray optics provided $F \gg 20$.

When a ray is totally reflected at a discontinuity of the refractive index, E and H waves suffer different phase shifts. For a step-index fiber, the depolarization due to this effect is of the same order of magnitude as the effect discussed above. Thus, the above expressions are applicable only to smooth profiles.

4.23 Aberrated Degenerate Optical Resonators[20]

An optical resonator is called degenerate when all the modes corresponding to the same axial-mode number resonate at the same frequency. In other words, the frequency of resonance is independent of the transverse-mode number. Any incident beam at that frequency is transmitted. The frequency response is essentially that of conventional resonators with matched excitation field.

In degenerate resonators, all the rays recycle, that is, reproduce their own path after a round trip in the resonator. The Fermat principle shows that when this condition is satisfied, all the rays have the same optical length or phase shift. This round-trip length defines the frequency of resonance. An example of degenerate resonator free of aberration is the Maxwell fish-eye medium. The free-wavenumber law has spherical symmetry, and $k(r) = k_0/[1 + (r/r_0)^2]$, where k_0, r_0 denote constants, and r the radial distance. For that medium, all the rays are circles with round-trip phase shift $\theta_0 = \pi k_0 r_0$. This result follows from the properties of the spherical configuration lens, illustrated in Fig. 1-14. Because the Maxwell fish-eye cavity is free of aberration, its acceptance is the same as that of an aperture having the same radius, given in (4.144). Most degenerate optical cavities suffer, however, from geometrical optics aberration. Incident plane waves get distorted after successive round trips and do not add up exactly in phase. To preserve the frequency resolution, the transverse extent and the angular spread of these waves must be restricted by apertures. Indeed, if the total phase distortion is allowed to become large compared to 2π, the total power is almost the sum of the powers of the successive passes, rather than the modulus square of the sum of their fields. All phase information, therefore, is lost. For the resonator frequency response to be preserved, the apertures in the resonator must have radii and locations such that the phase distortion per pass in the resonator is much smaller than $2\pi/F$, where F denotes the cavity finesse, defined as the ratio of the free spectral range to the 3-dB bandwidth of a resonance. The finesse F is a measure of the effective number of passes in the resonator.

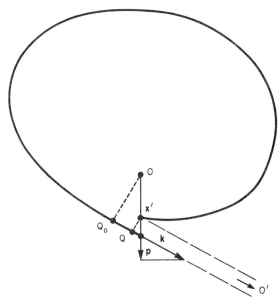

Fig. 4-33 Schematic of an aberrated ring-type degenerate cavity showing a ray trajectory originating from the reference plane x'. The ray direction after a round trip is defined by $\mathbf{p}$, the transverse component of the wave vector. The aberration is expressed by the variation of the length from x' to point Q, a function of x' and $\mathbf{p}$. (From Arnaud,[20] by permission of the American Optical Society.)

Let us now consider the ring-type resonator in Fig. 4-33 with a reference plane x at $z = 0$. The transverse coordinates at plane x are denoted $\mathbf{x} \equiv (x_1, x_2)$. We assume for simplicity that the free wavenumber at the reference plane, and in the neighborhood of that plane, is a constant k_0. The direction of a ray is defined by the transverse component $\mathbf{p}$ of the wave vector $\mathbf{k}$ (with $|\mathbf{k}| = k_0$), with components p_1, p_2. The resonator is assumed to be free of dissipation losses and to incorporate two identical mirrors with real reflectivity $R^{1/2}$. The round-trip field loss is equal to the power reflectivity $R \equiv 1 - T$ of one mirror. The cavity finesse defined above is $F = \pi/T$.

Consider a Lambertian source at the reference plane. The radiance at that plane is taken equal to T (instead of unity) to account for the transmissivity of the coupling mirror. Let us consider a small area $(d\mathbf{x}')$ $\equiv dx_1' \, dx_2'$ at $\mathbf{x}'$, as shown in Fig. 4-33. This small area radiates an almost spherical wavefront. Let us evaluate the total field radiated by this elementary area at some point O', a large distance from the area, in a direction defined by $\mathbf{p} \equiv (p_1, p_2)$. If we omit a complex factor common to all the successive passes that does not enter in the final expression, the field at the observation point is, for the first pass,

$$\psi(\mathbf{p}, \mathbf{x}') = T^{1/2}(d\mathbf{x}')^{1/2}(d\mathbf{p})^{1/2} \tag{4.275}$$

After a round trip in the resonator, the field is multiplied by $R \exp[i\phi(\mathbf{p}, \mathbf{x}')]$, where ϕ denotes the phase shift from $\mathbf{x}'$ to Q in Fig. 4-33, along the ray whose initial location is $\mathbf{x}'$, and the direction, after a round trip, is defined by $\mathbf{p}$. We cannot use in this problem the point-eikonal, as we did earlier in this chapter, because in a degenerate cavity the reference plane is imaged into itself. Thus the point-eikonal $S(\mathbf{x}; \mathbf{x}')$ is undefined for $\mathbf{x} \neq \mathbf{x}'$. The function to be used is the mixed eikonal $W(\mathbf{p}; \mathbf{x}')$ defined as the phase shift from $\mathbf{x}'$ to the point obtained by projecting orthogonally the origin on the image ray. To obtain ϕ we need to add $\mathbf{p}\mathbf{x}'$ to the mixed eikonal W. The phase shift ϕ from $\mathbf{x}'$ to Q, as well as W, is considered a function of $\mathbf{p}$ and $\mathbf{x}'$ (rather than $\mathbf{x}$ and $\mathbf{x}'$). The total field is obtained by summing over the successive passes

$$\psi_T(\mathbf{p}, \mathbf{x}') = T^{1/2}(d\mathbf{x}')^{1/2}(d\mathbf{p})^{1/2}\{1 - R \exp[i\phi(\mathbf{p}, \mathbf{x}')]\}^{-1} \tag{4.276}$$

and the acceptance of the resonator is obtained by integrating $\psi_T\psi_T^*$ over all $\mathbf{x}'$, $\mathbf{p}$ permitted by the apertures (as yet undefined), and dividing by $(2\pi)^2$. The resulting expression is multiplied by T to account for the transmissivity of the output mirror. After rearranging, we obtain

$$N^2 = \left(\frac{1}{2\pi}\right)^2 \int \frac{(d\mathbf{x}')(d\mathbf{p})}{1 + (2F/\pi)^2 \sin^2(\phi/2)} \tag{4.277}$$

for high cavity finesses $F \approx \pi/T$. The range of integration in (4.277) will be defined later. For two states of polarization, (4.277) should be multiplied by 2.

Let us now assume that the resonator is excited near one of its resonances and that the apertures are sufficiently small that the wave aberration is small compared with 2π. We set

$$\phi(\mathbf{p}, \mathbf{x}') = 2l\pi + \sum_{m=0}^{\infty} \Delta_m(\mathbf{p}, \mathbf{x}') \tag{4.278}$$

where l denotes the axial-mode number and $\Delta_m(\mathbf{p}, \mathbf{x}')$ is a polynomial of degree m in $\mathbf{p}, \mathbf{x}'$. Δ_0 represents a frequency offset that we can adjust for optimum transmission, Δ_1 a lack of alignment of the resonator, and Δ_2 a lack of first-order degeneracy (e.g., Δ_2 accounts for the fact that one end mirror may not be perfectly imaged into the opposite end mirror for the case of a symmetrical resonator). Δ_3 is absent in resonators with circular symmetry. We are concerned mainly with Δ_4, a polynomial of degree 4 in

p_1, p_2, x_1', and x_2'. This polynomial expresses the third-order (or Seidel) aberrations, most often considered in conventional imaging. Assuming for simplicity that only Δ_4 is present, the acceptance is

$$N^2 = \left(\frac{1}{2\pi} \right)^2 \int \frac{(d\mathbf{p})(d\mathbf{x}')}{1 + (F\Delta_4/\pi)^2} \qquad (4.279)$$

Let us evaluate Δ_4 for the confocal resonator, which is the most commonly used degenerate resonator. A confocal resonator incorporates two identical concave mirrors, whose radii of curvature are equal to their spacing. Thus the focal points of these mirrors (located halfway between the mirror center and the mirror vertex) are both at the center of the cavity. A confocal resonator is only half-degenerate, in the sense that it takes a ray two round trips (instead of one) to retrace its own path. For that reason, the input beam must be restricted to one-half the input mirror area. The confocal resonator has the great advantage over other types of degenerate resonators that it is rather insensitive to misalignments. A further advantage is that the confocal resonator does not incorporate lenses. Unless special precautions are taken, lenses exhibit troublesome Fresnel reflection.

The confocal resonator is most commonly used to analyze the spectral output of lasers. Because of its degeneracy, precise matching to the incident beam is unnecessary. The confocal resonator nevertheless suffers from limitations that result from aberrations that we now evaluate.

In order to evaluate the third-order aberration Δ_4, it is sufficient to consider a closed path defined by the law of paraxial ray optics. Let this closed path be defined by four points located on the mirror surfaces with polar coordinates $(r_1, 0)$, $(r_2, 0)$, (r_1, φ), $(r_2, \varphi + \pi)$, respectively, where r_1 and r_2 denote distances from axis, and φ denotes polar angles. This closed path does not obey the laws of reflection at the mirror surfaces exactly, but the accuracy is sufficient for our purpose. We now evaluate the exact length of that path (which is not a ray).

The exact distance from A_1 to A_2 (see Fig. 4-34) is obtained from elementary geometry:

$$A_1A_2 = \left(R'^2 + r_1^2 + r_2^2 + 2r_1r_2 \cos \varphi \right)^{1/2} \qquad (4.280)$$

where R' denotes the axial distance between A_1 and A_2. We have

$$R' = \left(R^2 - r_1^2 \right)^{1/2} + \left(R^2 - r_2^2 \right)^{1/2} - R \qquad (4.281)$$

because the mirrors are spherical. The distance between A_2 and A_3 is given by (4.280) with φ changed to $\varphi + \pi$, and $\overline{A_3A_4} = \overline{A_1A_2}$, $\overline{A_4A_1} = \overline{A_2A_3}$.

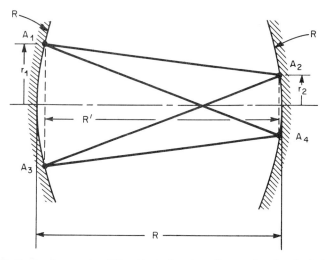

Fig. 4-34 Confocal resonator. To evaluate the aberration, a closed path $A_1A_2A_3A_4A_1$ is defined that follows the laws of paraxial optics but is not exactly a ray. The length of that path can be evaluated exactly.

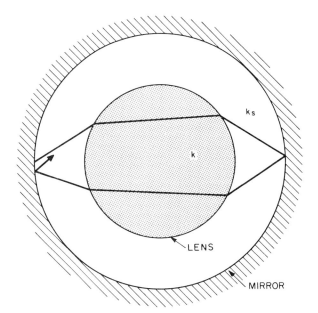

Fig. 4-35 Pole's concentric degenerate resonator.[35] The homogeneous spherical lens images the opposite points of the mirror surface. The acceptance of such spherically symmetric resonators is very large. A further improvement consists in replacing the homogeneous lens by a Luneburg lens. (From Arnaud,[20] by permission of the IEEE.)

Thus the aberration is

$$\Delta \equiv \phi - 4k_0 R = 2k_0 \left(R'^2 + r_1^2 + r_2^2 + 2r_1 r_2 \cos \varphi \right)^{1/2}$$

$$+ 2k_0 \left(R'^2 + r_1^2 + r_2^2 - 2r_1 r_2 \cos \varphi \right)^{1/2} - 4k_0 R$$

If we expand Δ in powers of r_1^2 and r_2^2, we find that the first nonvanishing term is

$$\Delta_4 = \frac{-k_0 r_1^2 r_2^2 \cos(2\varphi)}{R^3} \tag{4.282}$$

In the language of the theory of Seidel aberrations, (4.282) indicates that the resonator suffers from astigmatism. A numerical calculation shows that the actual aberration differs from Δ_4 by less than 5% if r_1/R and r_2/R are less than 0.1.

In most cases, the apertures are located at the mirrors. That is, we can now specify that $r_1 < a$, $r_2 < a$, where a denotes the aperture radii. We can take the reference plane x' at the left-hand mirror location. Because

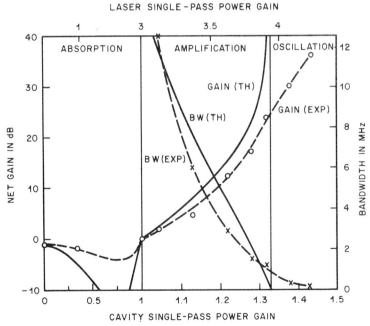

LASER SINGLE-PASS POWER GAIN

Fig. 4-36 Gain of a regenerative mode-degenerate laser amplifier as a function of discharge current. (From Arnaud,[36] by permission of the IEEE.)

only rather small angles are considered, and because of the rotational symmetry, we can easily change the integration over x_1', x_2', p_1', p_2' to an integration over r_1 and r_2. We shall omit the details of the integration. The acceptance of the confocal resonator is found to be approximately

$$N^2 = (R/\lambda)F^{-1} \tag{4.283}$$

where λ denotes the wavelength. For example, if $R = 10$ mm, $\lambda = 1\,\mu$m, and the finesse is $F = 100$, the acceptance is $N^2 = 100$ per state of polarization. In other words, 200 modes can be transmitted through the resonator without significant degradation of the frequency response.

The acceptance of resonators with spherical symmetry is larger than that of resonators with only rotational symmetry by orders of magnitudes. For example, the peak acceptance of the concentric resonator proposed by Pole (Fig. 4-35), which incorporates a spherical homogeneous lens centered on a

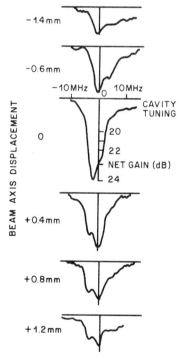

Fig. 4-37 The relative insensitivity of the gain to an offset of the incident beam demonstrates that the degeneracy of the resonator is preserved. (From Arnaud,[36] by permission of the IEEE.)

spherical resonator, is[20]

$$N^2 \approx 2\left(\frac{R}{\lambda} \right)^{3/2} F^{-1/2}(n-1)n^{-1/2} \qquad (4.284)$$

where n denotes the lens refractive index. If $n = 1.4$, $R = 10$ mm, $\lambda = 1$ μm, and $F = 100$ as before, we find $N^2 = 60,000$. In many practical cases, however, it is possible to use only part of the very large acceptance solid angle. If the homogeneous lens is replaced by a Luneburg lens, the resonator is free of aberration and has a much larger acceptance. This Luneburg lens resonator has the property of giving a perfect real image of a homogeneous volume.[20] That such a system may even exist had been questioned in the past.

When the medium in the cavity has gain, a regenerative gain can be obtained, which enhances the finesse. A new regenerative mode-degenerate amplifier, which could amplify with high-gain distorted wavefronts, was reported by Arnaud.[36] The degenerate resonator used was a triangular path with two confocal lenses. Figure 4-36 shows the gain as a function of the current of excitation of the HeXe discharge ($\lambda_0 = 3.5$ μm), and Fig. 4-37 shows that a moderate offset of the incident beam can be tolerated. This proves that the resonator degeneracy is not affected very much by the nonuniformities of the medium gain.

4.24 The WKB Approximation in Graded-Index Fibers

The results obtained earlier from the Hamilton equations of ray optics are related here to solutions obtained from wave optics. For simplicity, we shall consider the scalar Helmholtz equation rather than the exact Maxwell equations. The former is applicable when the variations of refractive index are small (but not necessarily continuous). The (WKB) approximation that leads from wave optics to ray optics has already been discussed in Section 2.21. However, it was restricted there to square-law two-dimensional non-dispersive media. The results given here are more general.

Let us consider a z-invariant isotropic fiber with free wavenumber $k(x, y, \omega)$. The time of flight per unit length obtained from ray optics was given in (4.64). We rewrite the result as

$$v_g^{-1} = \lim_{z \to \infty} \frac{1}{2zk_z} \int_0^z \frac{\partial k^2(x, y, \omega)}{\partial \omega} \, dz \qquad (4.285)$$

where k_z denotes the axial component of the wave vector $\mathbf{k}$, a vector that has magnitude k and the direction of the ray. In (4.285), $x = x(z)$, $y = y(z)$ represent a ray trajectory. Equation (4.285) is essentially the integral of ds/u along the ray, where $ds = (dx^2 + dy^2 + dz^2)^{1/2}$ is the elementary ray length, and $u \equiv (\partial k/\partial \omega)^{-1}$ the local group velocity.

Let us show that this result can be obtained alternatively from the scalar Helmholtz equation, which follows from the substitution $ik_x \to \partial/\partial x$, $ik_y \to \partial/\partial y$, in the relation $k^2 = k_x^2 + k_y^2 + k_z^2$:

$$\left[\frac{\partial^2}{\partial x^2} + \frac{\partial^2}{\partial y^2} + k^2(x, y, \omega) \right] \psi_m(x, y, \omega) = k_{zm}^2(\omega)\psi_m(x, y, \omega) \quad (4.286)$$

where $m = 0, 1, 2, \ldots$, for trapped modes. Given $k(x, y, \omega)$, we look for solutions of (4.286) that are square integrable. The time of flight of a pulse in a mode m is subsequently obtained by differentiating $k_{zm}(\omega)$ with respect to ω. Instead of solving (4.286) for k_z, it is often more convenient to use the Hellmann–Feynman theorem derived in Section 3.17. We have

$$v_{gm}^{-1} = \frac{1}{2k_{zm}} \frac{\langle \psi_m^2(x, y)\partial k^2(x, y, \omega)/\partial \omega \rangle}{\langle \psi_m^2(x, y) \rangle} \quad (4.287)$$

where $\langle a \rangle$ denotes the integral of a over the cross section of the fiber.

For simplicity, we shall compare (4.285) and (4.287) only for the two-dimensional case and omit the y argument. In two dimensions, bound trajectories are always periodic. Thus, we can take for z in (4.285) the ray period Z. Let us recall that within the WKB approximation, a mode m is represented by a manifold of rays $x(z + \zeta)$, where the translation parameter ζ varies from 0 to Z. A typical ray manifold is represented in Fig. 2-34. The ray amplitude is quantized by the Bohr–Sommerfeld condition

$$\oint k_x \, dx = (m + \tfrac{1}{2})2\pi \quad (4.288)$$

where $m = 0, 1, 2, \ldots$. The meaning of this integral has been explained in Section 2.21.

Let us now consider two closely spaced rays, namely, $x(z)$ and $x(z + dz)$. The spatial variation of $|\psi^2|$ is defined by the condition that the energy flowing between these two rays is invariant. Because $|\psi^2|$ represents the z component of the density of energy flow, the power between the two rays is $(k/k_z)|\psi^2|\epsilon$, where $\epsilon = dz(k_x/k)$ is the distance between two rays. Thus, $|\psi^2|$ is proportional to $k_z/k_x = [dx(z)/dz]^{-1}$. The first factor on the rhs of (2.255), in particular, follows from this observation. The integral of $a(x)$ $|\psi_m^2| \, dx$ over x is consequently proportional to the integral of $a[x(z)] \, dz$

over a ray period, for any function $a(x)$. Setting $a(x) \equiv \partial k^2(x, \omega)/\partial \omega$, the equivalence of (4.285) and (4.287) is established.

Let us consider as an example a square-law medium

$$k^2(x, \omega) = k_0^2(\omega)[1 - \Omega^2(\omega)x^2] \qquad (4.289)$$

as in (4.66). The ray trajectory given in (4.30),

$$x(z) = \frac{k_{x0}}{k_0 \Omega} \sin\left(\frac{\Omega k_0}{k_z} z\right) \qquad (4.290)$$

is now quantized by condition (4.288). From this condition it follows that

$$k_{x0m}^2 = 2(m + \tfrac{1}{2})k_0\Omega \qquad (4.291a)$$

$$k_z^2 = k_0^2 - 2(m + \tfrac{1}{2})k_0\Omega \qquad (4.291b)$$

For the case considered, (4.291b) turns out to be the exact solution of the scalar Helmholtz equation (4.286). The time of flight per unit length v_g^{-1} can be obtained by direct differentiation of (4.291b). It is easy to see that the result coincides with (4.68). Alternatively, we may use the Hellmann–Feynman theorem (4.287) with

$$|\psi_m^2| = \left[1 - \left(x/\bar{\xi}_m\right)^2\right]^{-1/2}, \qquad \bar{\xi}_m^2 = 2(m + \tfrac{1}{2})/k_0\Omega$$

as given in (2.296). The fast variations of $|\psi_m^2|$ can be ignored in the present case.

Circularly Symmetric Fibers

Let us now consider circularly symmetric fibers, which are of greater practical importance than slabs. If a term $\exp[i(k_z z + \mu\varphi - \omega t)]$ is factored out, the equation for $\psi(r)$ is

$$r^{-1}\frac{d(r\, d\psi/dr)}{dr} + \left[k^2(r, \omega) - k_z^2 - \frac{\mu^2}{r^2}\right]\psi = 0 \qquad (4.292)$$

The integer μ is called the azimuthal mode number. Inaccurate results would be obtained if the WKB approximation were applied directly to (4.292) because of the singularity at $r = 0$. However, if we make the change of independent variable $r = e^u$, we obtain readily from (4.292)

$$\frac{d^2\psi}{du^2} + \{[k^2(e^u, \omega) - k_z^2]e^{2u} - \mu^2\}\psi = 0 \qquad (4.293)$$

which is nonsingular. Assuming an $\exp(ik_u u)$ dependence of ψ on u, we find from (4.293) that

$$k_u^2 = [k^2(e^u, \omega) - k_z^2]e^{2u} - \mu^2 \qquad (4.294)$$

The quantum condition (4.288) can now be applied. Reverting to the r (radius) notation, we have

$$I(\omega, k_z, \mu) = \oint k_u \, du = 2\int_{r_1}^{r_2} k_r \, dr = 2\pi(\alpha + \tfrac{1}{2}), \qquad \alpha = 0, 1, 2 \dots$$

$$(4.295a)$$

where k_r is the radial component of the wave vector

$$k_r = \left[k^2(r, \omega) - k_z^2 - \frac{\mu^2}{r^2} \right]^{1/2} \qquad (4.295b)$$

and r_1, r_2 are adjacent roots of k_r, between which k_r is real.

Note the important relations

$$-\frac{\partial I}{\partial k_z} = Z_c = 2k_z \int_{r_1}^{r_2} k_r^{-1} \, dr$$

$$-\frac{\partial I}{\partial \mu} = \phi_c = 2\mu \int_{r_1}^{r_2} (k_r r^2)^{-1} \, dr \qquad (4.296)$$

$$\frac{\partial I}{\partial \omega} = T_c = \int_{r_1}^{r_2} \frac{\partial k^2(r, \omega)}{\partial \omega} k_r^{-1} \, dr$$

where Z_c, ϕ_c, T_c denote the variations of z, φ, t, respectively, over a ray period. Thus Z_c is the axial distance between two adjacent maxima of r. For meridional rays, $Z_c = Z/2$, where Z is the full two-dimensional period used earlier.

The integrals over r in (4.296) are easily changed to integrals over φ, z, t, or s, whenever this is convenient, with the help of the relations

$$\frac{ds}{k} = \frac{dz}{k_z} = \frac{dr}{k_r} = \frac{r \, d\varphi}{k_\varphi} = \frac{r^2 \, d\varphi}{\mu} \qquad (4.297a)$$

which say that the ray has the direction of $\mathbf{k}$. We also have

$$k^2(r, \omega) = k_r^2 + k_\varphi^2 + k_z^2 \qquad (4.297b)$$

$$rk_\varphi \equiv l_z \equiv \mu \qquad (4.297c)$$

where l_z is the axial component of the ray angular momentum.

From (4.296), it follows that the spacings in axial wavenumber between adjacent azimuthal modes having the same radial mode number and between adjacent radial modes having the same azimuthal mode number are, respectively,

$$k_{z\mu} - k_{z\mu+1} \approx \frac{\phi_c}{Z_c} \tag{4.298a}$$

$$k_{z\alpha} - k_{z\alpha+1} \approx \frac{2\pi}{Z_c} \tag{4.298b}$$

There is therefore degeneracy when the rotation angle per period ϕ_c is equal to π. This is the case, exactly, for square-law media and, approximately, when $\alpha \ll \mu$, for any continuous profile $k(r)$. Indeed, for $\alpha \ll \mu$, r does not vary much. Over a small range of r, any continuous profile $k(r)$ can be approximated by a square law.

The behavior of the field, within the WKB approximation, is

$$\psi(r) = (k_r r)^{-1/2} \exp\left(i \int^r k_r \, dr \right) \tag{4.299}$$

The prefactor follows from the requirement that the radial energy flow $2\pi r |\psi^2|(k_r/k_z)$ be independent of r. For sufficiently large values of k_z, k_r is imaginary. In that case, (4.299) describes radially decaying waves.

Power Law Profiles

Let us now assume that the profile is a power of the radius

$$k^2(r) = k_0^2 - k_\kappa^2 r^{2\kappa} \equiv K_0 - K_\kappa R^\kappa \tag{4.300}$$

Streifer and Kurtz[9] found an expression for the axial wavenumber that is exact for square-law fibers ($\kappa = 1$) and for any κ for circularly symmetric modes ($\mu = 0$) within the WKB approximation. This expression is sufficiently accurate for any large-capacity optical fiber whose profile has the form in (4.300). We have

$$B \equiv 1 - \frac{k_z^2}{K_0} = \frac{A(\kappa)(2g)^{2\kappa/(\kappa+1)} K_\kappa^{1/(\kappa+1)}}{K_0} \tag{4.301}$$

where

$$g \equiv 2\alpha + |\mu| + 1 \tag{4.302}$$

and

$$A(\kappa) \equiv \left\{ \frac{(\pi^{1/2}/2)\Gamma[3/2 + 1/(2\kappa)]}{\Gamma[1 + 1/(2\kappa)]} \right\}^{2\kappa/(\kappa+1)} \tag{4.303}$$

where Γ denotes the gamma function. The coefficient $A(\kappa)$ is unity for square-law fibers ($\kappa = 1$) and equal to $(\pi/4)^2$ for step-index fibers ($\kappa \to \infty$). For $0.8 < \kappa < 1.2$, a good approximation for $A(\kappa)$ is

$$A(\kappa) = 1 - 0.2(\kappa - 1) \tag{4.304}$$

The above result is used in the procedure discussed below.

Procedure for Evaluating the Impulse Width of a Multimode Fiber

We now describe a procedure for evaluating the impulse width of large-capacity multimode fibers, which is based on the analytic results previously derived in this chapter. For the reader's convenience, the notation is repeated.

The basic information that one needs is the refractive index profile $n(r)$ of the fiber measured at the operating wavelength (λ_0) and at two closely spaced wavelengths, one above (λ_0') and one below (λ_0''), and the source distribution in phase space. The latter includes both the angular distribution of the rays radiated by the source and the power spectrum $\Omega(\omega)$. The output of the program is the root-mean-square (rms) output pulse width σ, often expressed in nsec/km. In some cases σ may not be a satisfactory representation of the pulse. The output pulse $P(t)$ itself then needs to be fed into an error rate computer program. We shall discuss here only the evaluation of σ.

Let us assume that the refractive index n of the fiber material has been measured as a function of radius r, at the operating wavelength (λ_0) and at a slightly different wavelength (λ_0'). The refractive index profiles $n(r, \lambda_0)$ and $n'(r, \lambda_0')$ are written

$$\left(\frac{2\pi}{\lambda_0} \right)^2 n^2(r, \lambda_0) \equiv K_0 - K_\kappa r^{2\kappa} + K_2 r^4 + K_3 r^6 + \cdots \tag{4.305}$$

$$\left(\frac{2\pi}{\lambda_0'} \right)^2 n'^2(r, \lambda_0') \equiv K_0' - K_\kappa' r^{2\kappa} + K_2' r^4 + K_3' r^6 + \cdots$$

The parameters K_0, K_κ and the exponent κ are selected to best fit the given profile at the wavelength considered, and the difference is best fitted to a polynomial in r^2. For the choice $\kappa = 1$, (4.305) is a conventional expansion in powers of r^2. From the coefficients K_γ, K_γ', $\gamma = 0, \kappa, 2, 3 \ldots$, of (4.305), we define the inhomogeneous dispersion parameters

$$D_\gamma \equiv \frac{K_0(K_\gamma' - K_\gamma)}{K_\gamma(K_0' - K_0)} \tag{4.306}$$

The relative time of flight τ of a pulse, defined as the ratio of the group velocity of plane waves on axis to the group velocity of a mode with radial mode number $\alpha \geqslant 0$ and azimuthal mode number μ, is

$$\tau(\alpha, \mu) = (1 - B)^{-1/2}\left(1 - BD'_\kappa + \sum_{\gamma=2}^{\infty} B^\gamma F_\gamma N_\gamma \epsilon_\gamma\right) \qquad (4.307)$$

where

$$D'_\kappa \equiv \frac{D_\kappa}{1 + \kappa}$$

$$B \equiv \frac{[1 - 0.2(\kappa - 1)](2g)^{2\kappa/(\kappa+1)}K_\kappa^{1/(\kappa+1)}}{K_0}$$

$$g \equiv 2\alpha + |\mu| + 1, \qquad |\mu| \equiv \text{absolute value of } \mu$$

$$F_\gamma \equiv \gamma! \, 2^{-\gamma}[D_\gamma - (\gamma + 1)D'_\kappa] \qquad (4.308)$$

$$N_\gamma \equiv \sum_{m=0,2}^{\gamma} \left\{2^m(\gamma - m)!\left[\left(\frac{m}{2}\right)!\right]^2\right\}^{-1}\left[1 - \left(\frac{\mu}{g}\right)^2\right]^{m/2}$$

$$\epsilon_\gamma \equiv \frac{K_\gamma K_0^{\gamma-1}}{K_\kappa^{2\gamma/(1+\kappa)}}$$

All the parameters of the lhs of (4.308) are dimensionless.

For preliminary designs, we can assume that there is only one dopant material and that n^2 varies linearly with the dopant concentration. In that case, the $D_\gamma \equiv D_1$ are all equal. This is assumed in the two special cases treated below.

Let us first set $\kappa = 1$ in the previous expressions and assume that the series in (4.305) terminates at the r^4 term. The total impulse width for long fibers is proportional to the difference between the maximum and minimum values of τ in (4.307), with α, μ restricted by the condition that the mode wavenumber k_z be larger than the cladding wavenumber $k_s \equiv (\omega/c)n_s$, that is, with the notation used in (4.307), $B < 1 - K_s/K_0$ where $K_s \equiv k_s^2$. The particular value of ϵ_2 that minimizes the total impulse width is

$$(\epsilon_2)_{\text{opt}} \approx \frac{2}{3}\left[1 - \frac{2(D_1 - 1)}{1 - n_s/n_0}\right] \qquad (4.309)$$

where n_s and n_0 denote the cladding and axial refractive indices, respectively. The relative time of flight predicted by (4.307) with $\kappa = 1$ can be compared to the exact value obtained by numerical integration of the space–time Hamilton equations. The expression in (4.307) is found very accurate for near optimum profiles provided $|D_1 - 1| \lesssim 0.02$. This condition is satisfied for many fiber materials. Nevertheless, in some cases, it is useful to let κ be different from unity.

If we keep κ arbitrary in (4.305) but omit the series, the value of κ that minimizes the total impulse width for long fibers is exactly (within the scalar ray optics approximation)

$$\kappa_{opt} = D_1\left(1 + \frac{n_s}{n_0}\right) - 1 \qquad (4.310)$$

This result follows from the fact that D_1 and κ enter in the expression of τ in (4.307) only as $D_1/(1 + \kappa)$. Values for D_1 are given on p. 420.

Let us now go back to the general case. Usually we are interested, not in the impulse response itself, but in the rms impulse width, defined as

$$\sigma = 5000[\langle\tau^2\rangle - \langle\tau\rangle^2]^{1/2} \text{ nsec/km}$$

$$\equiv 5000\left[\langle(\tau - 1)^2\rangle - \langle\tau - 1\rangle^2\right]^{1/2} \text{ nsec/km} \qquad (4.311)$$

where, for any quantity $q(\alpha, \mu)$, such as τ or τ^2, we have defined

$$\langle q\rangle \equiv \frac{\displaystyle\sum_{\alpha,\mu} T(\alpha, \mu)\, q(\alpha, \mu)}{\displaystyle\sum_{\alpha,\mu} T(\alpha, \mu)} \qquad (4.312)$$

Assuming that the material loss is independent of radius, the power transmission $T(\alpha, \mu)$ is, to within an unimportant constant factor,

$$T(\alpha, \mu) \equiv \exp(-2k_{zi}L) = \begin{cases} 1, & 0 < B < 2\Delta \\[2mm] T_l(\alpha, \mu), & 2\Delta < B < 2\Delta + \dfrac{\mu^2}{K_0A} \\[2mm] 0, & 2\Delta + \dfrac{\mu^2}{K_0A} < B \end{cases}$$

$$(4.313a)$$

where $A \equiv a^2$ is the square of the core radius, k_{zi} the imaginary part of the axial wavenumber, L the fiber length, and

$$2\Delta \equiv 1 - \frac{K_s}{K_0} \approx \frac{2\,\Delta n}{n} \qquad (4.313b)$$

If the fiber is long and only moderately overmoded, we can set $T_l = 0$ in (4.313a). This is done in the example that follows. It is more accurate, however, to use the expressions that we shall give in the next subsection.

To illustrate the method, let us consider a germania doped silica fiber, with a dopant concentration (in percent)[8]

$$d = 10 - 10\left(\frac{r_{\mu m}}{40}\right)^2 - E\left(\frac{r_{\mu m}}{40}\right)^4 \qquad (4.314)$$

where E is a small adjustable parameter. The dopant concentration d is equal to 10% on axis and 0% at the core radius $a \approx 40$ μm. The refractive index of germania doped silica is given in Chapter 5, Table 5.7, for 0% and 13.5% GeO_2 concentration. Linear interpolation and use of (4.314) gives the refractive index laws at $\lambda_0 = 0.88$ and 0.9 μm, respectively.

For the special value $E = 0.051$, for example, we obtain

$$n^2(r) = 2.15439 - 2.87281 \times 10^{-5}r^2 - 9.193 \times 10^{-11}r^4$$

$$n'^2(r) = 2.15352 - 2.87075 \times 10^{-5}r^2 - 9.186 \times 10^{-11}r^4 \qquad (4.315)$$

where r is in micrometers. The coefficients K_0, K_1, K_2, D_1, and D_2 are obtained from (4.315) and the definitions in (4.305) and (4.306). Substituting the values in (4.311) and (4.307), we obtain, with $1 - K_s/K_0 \approx 2 \Delta n/n = 0.017$, an rms impulse width $\sigma = 0.016$ nsec/km. The value of E selected above corresponds to the minimum value of σ. The σ becomes at least twice as large if $|E - 0.051| > 0.15$. This example shows how critical is the selection of the profile.

The impulse response could be optimized further by adding r^6 terms to the profile. However, there is some uncertainty as to whether measurements made on bulk samples are applicable accurately to fibers. Thus, very detailed expressions for $d(r)$ derived from measurements on bulk samples may not be of practical interest. Measurements can be made directly on the fiber at various wavelengths. Numerical calculations, however, show that the wavelengths λ_0 and λ_0' in (4.305) should not be separated by more than 0.05 μm if the error on σ is to be less than 50%. In spite of recent progress, the accuracy of presently available measurements seems to be insufficient for the application considered. Thus we have to rely, at least partly, on measurements made on bulk samples.

In the foregoing, the optical carrier was assumed monochromatic. If this is not the case, the sign $\langle \ \rangle$ in (4.311) should be replaced by

$$\langle\!\langle \ \rangle\!\rangle \equiv \int \langle \ \rangle_{\lambda_0} \Omega(\lambda_0) \, d\lambda_0 \qquad (4.316)$$

where $\langle \ \rangle_{\lambda_0}$ is evaluated by the procedure outlined above at various wavelengths λ_0 within the (normalized) power spectrum $\Omega(\lambda_0)$ of the source. This conclusion is based on space–time ray optics. In order to evaluate (4.316) we need to know the variation of the dispersion coefficients D_γ with wavelength (optical frequency). This variation can be obtained by parabolic interpolation, if the refractive index profiles are known at three wavelengths (λ_0, λ_0', λ_0'').

Tunneling Loss

The WKB method also permits the evaluation of the power transmission $T_l(\alpha, \mu)$ of slightly leaky modes. A rigorous theory of the WKB method can be found in a book by Froman et al.[37] The results given below, which are directly applicable to optical fibers, were first derived by Petermann.[38] The reader is referred to his paper for further details.

Let r_1 and r_2 denote the turning points between which the rays are trapped. That is, the square k_r^2 of the radial component of the wave vector

$$k_r^2(r) = k^2(r) - k_z^2 - \frac{\mu^2}{r^2} \tag{4.317}$$

is assumed to be positive between r_1 and r_2 for some given k_z, μ, ω, and to change sign at the endpoints r_1 and r_2. A new phenomenon takes place if, beyond some radius $r_3 > r_2$, k_r^2 becomes positive again. Between r_2 and r_3, k_r^2 is negative and the wave is radially decaying. This intermediate region is called a *barrier*. Modes that are, in first approximation, trapped between r_1 and r_2, in fact leak slightly through this barrier and slowly lose power. This power loss is expressed by the imaginary part k_{zi} of k_z. It is given in Nepers/unit length by the following formula:

$$k_{zi} = Z^{-1}\exp\left(-2\int_{r_2}^{r_3}|k_r|\,dr\right) \tag{4.318a}$$

where

$$Z = 4k_z\int_{r_1}^{r_2}\frac{dr}{k_r} \tag{4.318b}$$

denotes the full ray period (defined as twice the axial distance between adjacent maxima of r). This expression for k_{zi} can be substituted in (4.313a) to obtain the power transmission T_l. For truncated square-law profiles, the integrals in (4.318) can be evaluated in closed form. The expressions, however, will not be given. In most practical cases it is easier to perform the numerical integration of $|k_r|$.

Strictly speaking, (4.318) is applicable only when the profile $n(r)$ is continuous. However, this expression gives the correct order of magnitude even if there is a discontinuity at the core–cladding interface. The transmission of leaky modes for step-index fibers is shown in Fig. 5.32.

Perturbation Caused by the Cladding

Let us now assume that the field decays into the cladding at all radii but that the cladding has dissipation loss. That is, the cladding wavenumber squared k_s^2 has a small imaginary part that we denote Δk_s^2. This dissipation loss of the cladding introduces a modal loss, which depends essentially on the strength of the field at the core–cladding interface.

We shall give only the result based on perturbation theory. The perturbation of Δk_z of the axial wavenumber is approximately, from (3.201),

$$\Delta k_z = \frac{1}{2Z} \frac{\Delta k_s^2}{|k_r(a)|^2} \exp\left(-2 \int_{r_2}^{a} |k_r| \, dr\right) \tag{4.319a}$$

If the perturbation Δk_s^2 of the cladding is purely imaginary, the perturbation $\Delta k_z \equiv i k_{zi}$ is also imaginary. The period Z is defined in (4.318b) and

$$|k_r(r)| = \left[k_z^2 + \frac{\mu^2}{r^2} - k^2(r)\right]^{1/2} \tag{4.319b}$$

The expressions given in this final section of Chapter 4 allow the accurate evaluation of the impulse response of any large-capacity uniform multimode graded-index fiber. Moderate departures from the square-law profile, the effect of material dispersion, and cladding loss, either through tunneling or dissipation, have been accounted for. However, the effect of strain-induced anisotropy of the material, for which there is little direct evidence available, and that of nonuniformities either in azimuth or in the axial direction, were not considered.

References

1. R. K. Luneburg, "The Mathematical Theory of Light." Univ. of California Press, Berkeley, California, 1964.
2. M. Kline and I. M. Kay, "Electromagnetic Theory and Geometrical Optics." Wiley (Interscience), New York, 1965.
3. H. A. Buchdahl, "Introduction to Hamiltonian Optics." Cambridge Univ. Press, London and New York, 1970.
4. R. Descartes, "La Dioptrique," 1637. English summary by J. F. Scott, *in* "The Scientific Work of René Descartes," p. 32. Taylor and Francis, London, 1952.
5. M. J. Lighthill, *J. Inst. Math. Appl.* **1**, 1-28 (1965).
6. O. N. Stravroudis, "The Optics of Rays, Wavefronts and Caustics." Academic Press, New York, 1972.
7. J. J. Brandstatter, "Waves, Rays and Radiation in Plasma Media." McGraw-Hill, New York, 1963.

8. J. A. Arnaud, *Bell Syst. Tech. J.* **53**, 1599 (1974), **54**, 1179 (1975); *Electron. Lett.* **11** (1) 8, (18) 447 (1975).
9. W. Streifer and C. N. Kurtz, *J. Opt. Soc. Amer.* **57**, 779, 1967. D. Gloge and E. A. J. Marcatili, *Bell Syst. Tech. J.* **52**, 1563 (1973).
10. S. Kawakami and J. Nishizawa, *IEEE Trans. Microwave Theory Tech.* MTT **16**, 814 (1968).
11. A. H. Nayfeh, "Perturbation Methods." Wiley, New York, 1973.
12. H. Kogelnik and A. Yariv, *Proc. IEEE* **52**, 165 (1964).
13. R. Kompfner, 1971 (unpublished).
14. J. A. Arnaud, "Delay Equalizer for Multimode Fibers Using Ordinary Lenses," September 1971 (unpublished). U.S. Patent 3,759,590.
15. M. Ikeda, *IEEE Trans. Quant. Elec.* QE10, 362 (1974). C. C. Timmermann, *AEU* **28**, 183, 344 (1974).
16. A. Cozannet, M. Tréheux and R. Bouillie, *CNET Ann. Telecommun.* **29**, 219 (1974). K. H. Steiner, *Nachrichtentech. Z* **27**, 250 (1974).
17. S. E. Miller, *Bell Syst. Tech. J.* **53**, 177 (1974). F. W. Ostermayer, Topical meeting on fiber optics, Williamsburg, Virginia, Jan. 1975.
18. D. B. Keck, *Appl. Opt.* **13**, 1882 (1974). R. Olshansky and D. B. Keck *Appl. Opt.* (to appear). D. B. Keck, R. D. Maurer, and P. C. Schultz, *Appl. Phys. Lett.* **22**, 307 (1973).
19. A. W. Snyder and D. J. Mitchell, *Electron. Lett.* **9** (19), 437 (1973). A. W. Snyder, *Appl. Phys.* **4**, 273 (1974).
20. J. A. Arnaud, *Appl. Opt.* **9**, 1192 (1970), "Note on Optical Funnels," 1971 (unpublished).
21. C. B. Burrus, *Proc. IEEE Lett.* **60**, 231 (1972).
22. D. N. Payne and W. A. Gambling, *Electron. Lett.* **10**, 335 (1974). *Opt. Commun.* **13**, 422 (1975). L. G. Cohen, P. Kaiser, J. B. MacChesney, P. B. O'Connor, and H. Presby, *Appl. Phys. Lett.* **26**, 472 (1975). L. G. Cohen, *Appl. Opt.* **14**, 1351 (1975). L. G. Cohen and S. D. Personick, *Appl. Opt.* **14**, 1357 (1975). L. G. Cohen and H. M. Presby, *Appl. Opt.* **14**, 1361 (1975).
23. E. G. Rawson and R. G. Murray, *IEEE Trans. Quant. Electron.* QE-9, 1114 (1973).
24. W. Eickoff and E. Weidel, *Opt. Quant. Electron.* QE7, 109 (1975).
25. A. Tynes, private communication. G. W. Morey, "Properties of Glass," p. 446. Reinhold, New York, 1954. J. Schroeder, F. Mohr, P. B. Macedo, and C. J. Montrose, *J. Amer. Ceram. Soc.* **56**, 510 (1973).
26. H. E. Rowe and D. T. Young, *IEEE Trans. Microwave Theory Tech.* MTT **20**, 349 (1972). S. D. Personick, *Bell Syst. Tech. J.* **52**, 843 (1973), **50**, 3075 (1971).
27. W. B. Garnder, *Bell Syst. Tech. J.* **54**, 457 (1975).
28. S. D. Personick, *Bell Syst. Tech. J.* **52**, 1175 (1973).
29. P. Kaiser, A. C. Hart, Jr., and L. L. Blyler, Jr., *Appl. Opt.* **14**, 156 (1975).
30. S. Tanaka, K. Inada, T. Akimoto, and M. Kozima, *Electron. Lett.* **11**, 153 (1975).
31. J. A. Arnaud, *Appl. Opt.* **8**, 189 (1969).
32. J. A. Arnaud, in "Progress in Optics" (E. Wolf, ed.), Vol. 11. North-Holland Publ., Amsterdam, 1973.
33. J. A. Arnaud and J. T. Ruscio, *Appl. Opt.* **9**, 2377 (1970).
34. J. A. Arnaud, *Bell Syst. Tech. J.* **49**, 2311 (1970).
35. R. V. Pole, *J. Opt. Soc. Amer.* **55**, 254 (1965).
36. J. A. Arnaud, *IEEE Trans. Quant. Electron.* QE4, 893 (1968).
37. N. Fröman and P. O. Fröman, "JWKB Approximation." North-Holland Publ., Amsterdam, 1965.
38. K. Petermann, *A.E.U.* **29**, 345 (1975).

Piecewise Homogeneous Media

This chapter is devoted to waves guided by piecewise homogeneous media. A typical example of such a guide is the dielectric slab. We shall consider first a plane boundary between two dielectrics, perhaps a dielectric and free space. The mathematical formalism is simple because the waves separate into H-waves (electric field perpendicular to the plane of incidence) and E-waves (electric field in the plane of incidence). Many important effects, such as the Goos–Hänchen shift, show up in that simple configuration.

When the permittivities of the materials are all real positive, as is usually the case, at least two boundaries are needed to guide waves. The symmetrical slab, supported by identical upper and lower media, is a good model for many problems of wave propagation in fiber optics. The unsymmetrical slab is most often encountered in integrated optics. In that case, the slab is called a "film" and the lower medium a "substrate." The upper medium is usually free space. Exact solutions for the propagation of electromagnetic waves along round dielectric rods are known. These solutions, however, are difficult to use in practice. The scalar approximation, which is applicable when the dielectric rod and the surrounding medium have almost the same refractive index, leads to simplified expressions.

Various modifications of the two basic structures—the dielectric slab and the dielectric rod—are investigated. For example, the tapered dielectric slab and the slab-coupled rod. Simple expressions for the coupling between slabs and between rods and for the bending losses are derived, and the application of these results to fiber optics is discussed. The effect of axial irregularities of the guide (tapers or random defects) is briefly considered in Section 5.18. The most important results applicable to communication by glass fibers are summarized in Section 5.19.

5.1 Stratified Media

Let us consider first a medium whose scalar permittivity $\epsilon(x)$ and scalar permittivity $\mu(x)$ have arbitrary x-variations, but are independent of y and z, as shown in Fig. 5-1. Such a medium is called "stratified." Because of the translational invariance of the medium in the y, z plane, plane wave solutions of the form

$$E(x, y, z) = E(x) \exp[i(k_y y + k_z z)] \tag{5.1}$$

can be found. We shall consider only solutions with $k_y = 0$. The reader is cautioned that these solutions are not the most general, because, unless k_y and k_z are known to be real, it is not possible to rotate the yz coordinate system to make one of the two terms in parentheses in (5.1) vanish. Nevertheless, we shall proceed with the assumption that $k_y = 0$. With the change[§] $(x, y, z) \to (y, z, x)$, the Maxwell equations in (3.170) become

$$\partial_x \begin{bmatrix} E_y \\ E_z \end{bmatrix} = \begin{bmatrix} i\omega\mu - \partial_y(i\omega\epsilon)^{-1}\partial_y & -\partial_y(i\omega\epsilon)^{-1}\partial_z \\ -\partial_z(i\omega\epsilon)^{-1}\partial_y & i\omega\mu - \partial_z(i\omega\epsilon)^{-1}\partial_z \end{bmatrix} \begin{bmatrix} H_z \\ -H_y \end{bmatrix}$$

$$\partial_x \begin{bmatrix} H_z \\ -H_y \end{bmatrix} = \begin{bmatrix} i\omega\epsilon - \partial_z(i\omega\mu)^{-1}\partial_z & \partial_z(i\omega\mu)^{-1}\partial_y \\ \partial_y(i\omega\mu)^{-1}\partial_z & i\omega\epsilon - \partial_y(i\omega\mu)^{-1}\partial_y \end{bmatrix} \begin{bmatrix} E_y \\ E_z \end{bmatrix} \tag{5.2}$$

where $\partial_x \equiv \partial/\partial x$, $\partial_y \equiv \partial/\partial y$, $\partial_z \equiv \partial/\partial z$.
Setting $\partial_y = 0$ in (5.2) and $\partial_z = ik_z$, we obtain

$$\partial_x E_y = i\omega\mu H_z, \qquad \partial_x H_z = -\frac{k_x^2}{i\omega\mu} E_y$$

$$\partial_x H_y = -i\omega\epsilon E_z, \qquad \partial_x E_z = \frac{k_x^2}{i\omega\epsilon} H_y \tag{5.3}$$

[§] In the expressions given in Section 3.9, (x, y, z) should be changed to (y, z, x). The reason for this change of notation is that we are now mostly interested in waves guided by the stratifications rather than in waves propagating through them, and we want to maintain k_z as the most important wavenumber. The slab or rod wavenumber is now denoted k instead of k_0.

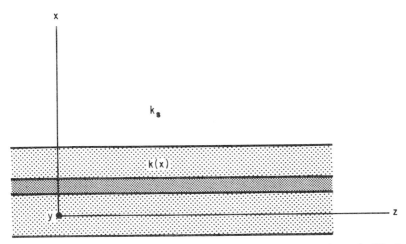

Fig. 5-1 Stratified medium. The free wavenumber k is a function of x only. The free wavenumber in the cladding is denoted k_s.

where

$$k_x^2 \equiv \omega^2 \epsilon \mu - k_z^2 \equiv k^2 - k_z^2 \qquad (5.4)$$

Equation (5.3) shows that two independent sets of solutions exist, one with E directed along the y-axis (H-waves) and one with H directed along the y-axis (E-waves). If we set for convenience

$$V \equiv \begin{cases} E_y & (H\text{-waves}), \\ H_y & (E\text{-waves}), \end{cases} \qquad \kappa \equiv \begin{cases} \mu & (H\text{-waves}) \\ \epsilon & (E\text{-waves}) \end{cases}$$

$$I \equiv (i\omega\kappa)^{-1} \partial_x V = \begin{cases} H_z & (H\text{-waves}) \\ -E_z & (E\text{-waves}) \end{cases} \qquad (5.5)$$

the H- and E-waves can be discussed with the same formalism. For both types of waves

$$\frac{dV}{dx} = i\omega\kappa I, \qquad \frac{dI}{dx} = i\frac{k_x^2}{\omega\kappa} V \qquad (5.6)$$

These equations coincide with those applicable to a transmission line with inductance per unit length $L \equiv \kappa$ and capacitance per unit length $C \equiv k_x^2/\omega^2\kappa$:

$$\frac{dV}{dx} = iL\omega I, \qquad \frac{dI}{dx} = iC\omega V \qquad (5.7)$$

This is not coincidental. The transmission along a conventional transmission line is a special case of the propagation considered above. We have seen that for H-waves the electric field is everywhere directed along the y-axis. Thus two perfectly conducting planes can be introduced, perpendicular to the y-axis, without perturbing the propagation. Let us assume, for simplicity, that the spacing between these two planes is unity. The voltage then is precisely equal to E_y. If we further assume that the width of the conducting strips, in the z-direction, is unity, and that $k_z = 0$ ($\Rightarrow k_x^2 = k^2 = \omega^2 \epsilon \mu$), the inductance per unit length is $L = \mu$, and the capacitance per unit length is $C = \epsilon$, in agreement with the notation introduced earlier. For E-waves, similar conclusions can be reached by introducing magnetic walls, instead of electric walls, perpendicular to the y-axis. A characteristic admittance can be defined by

$$Y = \left(\frac{C}{L} \right)^{1/2} = \frac{k_x}{\omega \kappa} \tag{5.8}$$

In defining k_x, we select the root of $(k^2 - k_z^2)^{1/2}$ that has a positive real part: $\mathrm{Re}(k_x) > 0$. If ϵ or μ is discontinuous, V and I remain continuous.

The x-component of the power density $\frac{1}{4}(\mathbf{E} \times \mathbf{H}^* + \mathbf{E}^* \times \mathbf{H})$ is, using the V, I-notation,

$$S_x = \frac{1}{4}(V^*I + VI^*) \tag{5.9}$$

To evaluate the z-component of the power density, we need the x-components of E and H. They are obtained directly from the Maxwell equations

$$H_x = -\frac{k_z}{\omega \kappa} V \quad (H\text{-waves}) \tag{5.10a}$$

$$E_x = \frac{k_z}{\omega \kappa} V \quad (E\text{-waves}) \tag{5.10b}$$

In some problems we need, instead of the real component S_z, the complex component

$$S_z' = \frac{1}{2} \mathbf{E} \times \mathbf{H}\big|_z = \frac{1}{2}(E_x H_y - E_y H_x) = \frac{1}{2} \frac{k_z}{\omega \kappa} V^2 \tag{5.10c}$$

for both H- and E-waves. S_z' represents the z-component of the power density when k_z is real and the medium is lossless. In general, however, S_z, as defined in (5.10c), is complex valued.

In a homogeneous region with constant $k_x \equiv (k^2 - k_z^2)^{1/2}$ and constant κ, (5.6) has the general solution

$$\begin{aligned} V &= V^+ \exp(ik_x x) + V^- \exp(-ik_x x) \\ Y^{-1}I &= V^+ \exp(ik_x x) - V^- \exp(-ik_x x) \end{aligned} \tag{5.11}$$

where $+$ and $-$ superscripts refer to waves propagating in the $+x$- and $-x$-directions, respectively.

For a homogeneous section of material, the relation between V, I at one end of the section and V, I at the other end can be written in matrix form

$$\begin{bmatrix} V(x) \\ I(x) \end{bmatrix} = \begin{bmatrix} \cos(k_x x) & iY^{-1}\sin(k_x x) \\ iY\sin(k_x x) & \cos(k_x x) \end{bmatrix} \begin{bmatrix} V(0) \\ I(0) \end{bmatrix} \quad (5.12)$$

If the medium is stepwise homogeneous, the field transformation is obtained by multiplying 2×2 matrices, similar to that in (5.12), relative to the successive sections. The formalism necessary to investigate stratified media of various degrees of complexity, with one, two, or more interfaces, has now been set up. The physical consequences of that formalism depend very much on whether the transverse wavenumber k_x is real or imaginary in the various regions and on the boundary conditions that are imposed at infinity, that is, as $|x| \to \infty$. For example, for the case of a single boundary, we may be interested in three different situations pictured in Fig. 5-2a–c. In the problem of Fresnel reflection, k_x is real in both media. If the wave is incident from the lower medium, we assume that there is no reflected wave in the upper medium, and we investigate the amplitude of the reflected wave in the first medium (Fig. 5-1c). The opposite case is

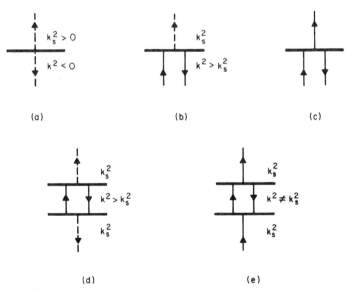

Fig. 5-2 Schematic representation of wave propagation with one interface (a–c) and two interfaces (d, e). (a) Surface polaritons. (b) Total reflection. (c) Fresnel reflection. (d) Slab waveguide. (e) Fabry–Perot etalon (or leaky waves).

when k_x is imaginary in both media with the wave decaying exponentially away from the boundary (Fig. 5-2a). Such trapped waves, called surface polaritons, can be found only when one of the two media has a negative k^2 or permittivity, a situation encountered with metals.[1] Still another case is when k_x is real in one medium and imaginary in the other (Fig. 5-2b). This case is encountered in the phenomenon of total reflection, which is investigated in the next section.

5.2 Total Reflection; The Goos–Hänchen Shift

Let the medium parameters be ϵ, μ for $x < 0$, and ϵ_0, μ_0 for $x > 0$ (perhaps free space), and let a plane wave be incident on the boundary between the two media from below (Fig. 5-3a). Depending on the angle of incidence and the polarization, the wave is either totally transmitted, partially transmitted and partially reflected, or totally reflected.

Let (5.11) be applicable to the region $x < 0$. For $x > 0$, we have, similarly,

$$V_0 = V_0^+ \exp(ik_{x0}x) + V_0^- \exp(-ik_{x0}x)$$

$$Y_0^{-1}I_0 = V_0^+ \exp(ik_{x0}x) - V_0^- \exp(-ik_{x0}x) \tag{5.13}$$

where

$$k_{x0}^2 \equiv k_0^2 - k_z^2, \qquad \mathrm{Re}(k_{x0}) > 0 \tag{5.14}$$

Note that k_z in (5.14) must be the same as k_z in (5.4). Because we have assumed that no wave is reflected from the upper boundary, we have $V_0^- = 0$. The continuity of V, I at $x = 0$ imposes that

$$V^+ + V^- = V_0^+, \qquad Y(V^+ - V^-) = Y_0 V_0^+ \tag{5.15}$$

Thus the reflection from the boundary is

$$\frac{V^-}{V^+} = \frac{Y/Y_0 - 1}{Y/Y_0 + 1} \tag{5.16}$$

From this expression, we conclude that the reflected wave has zero amplitude when $Y = Y_0$, that is, when

$$\frac{k_x}{\kappa} = \frac{k_{x0}}{\kappa_0} \tag{5.17a}$$

In particular, for E-waves in dielectrics ($\kappa = \epsilon$, $\kappa_0 = \epsilon_0$), this condition is

$$\frac{k_x}{k_{x0}} = \frac{\epsilon}{\epsilon_0} \tag{5.17b}$$

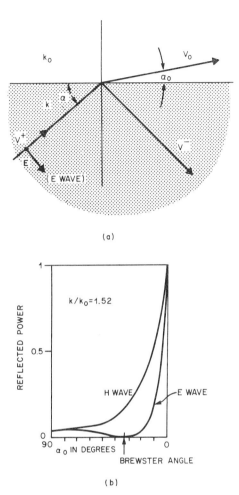

(a)

(b)

Fig. 5-3 (a) Transmission and reflection at an interface between two media. (b) Reflected power for *E*- and *H*-waves. At the Brewster angle, *E*-waves are completely transmitted.

Introducing the angles α, α_0 that the ray makes with the boundary $\tan(\alpha) = k_x / k_z$, $\tan(\alpha_0) = k_{x0} / k_z$, and using the relations $k_x^2 + k_z^2 = k^2$, $k_{x0}^2 + k_z^2 = k_0^2$, we obtain

$$\tan(\alpha_0) = \left(\frac{\epsilon}{\epsilon_0} \right)^{1/2} = \frac{1}{\tan(\alpha)} \tag{5.17c}$$

The angle α defined in (5.17c) is called the Brewster angle (see Fig. 5-3b). For that particular angle of incidence, the reflected ray, if it existed, would be parallel to the electric dipoles in the refractive medium. Because electric dipoles do not radiate along their own axes, there is no radiation from the dielectric that can supply power to a reflected wave.[2] A dielectric slab oriented at that particular angle is, in principle, free of reflection loss for E-waves. Thus gas lasers incorporating windows at the Brewster angle tend to oscillate with the electric field in the plane of incidence (E-waves) for which the loss is minimum. Optical waveguides incorporating lenses oriented at the Brewster angle have been proposed. The curvature of the lens is usually sufficiently small that the Brewster condition (5.17) be approximately satisfied over the lens area. This arrangement requires, of course, that the source be linearly polarized and that the waveguide path not be twisted.

In the foregoing, it was implicitly assumed that k_{x0} is real, that is, $k_0^2 > k_z^2$. If k_z is larger than k_0, the wave undergoes total reflection and decays in the less dense medium according to an exponential law

$$V_0^+ = \exp(-k_{x0i}x) \equiv \exp\left[-\left(k_z^2 - k_0^2\right)^{1/2}x\right]$$

where we have set $k_{x0} = ik_{x0i}$ (Fig. 5-4a). In that case, the ratio Y/Y_0 is imaginary, and the reflection V^-/V^+, given in (5.16), has modulus unity. The phase ϕ_0 of the reflection V^-/V^+ is

$$\phi_0 = \text{phase}\left(\frac{Y/Y_0 - 1}{Y/Y_0 + 1}\right) = -2\arctan\left(\frac{\kappa k_{x0i}}{\kappa_0 k_x}\right) \qquad (5.18)$$

Note that ϕ_0 is negative. ϕ_0 varies from 0, at the critical angle $k_{x0i} = 0$, to π under grazing incidence $k_x \approx 0$. The same π phase shift is suffered by H-waves reflected from electric walls and by E-waves reflected from magnetic walls.

A beam of light, perhaps a gaussian beam, can be represented by a spectrum of plane waves. When a beam is incident on an interface, each plane wave in the spectrum has a slightly different angle of incidence. Because the phase of a reflected wave depends on the angle of incidence, as (5.18) shows, the reflected beam is not exactly reconstituted after undergoing total reflection, although the total power is preserved. The beam is *shifted* in the forward direction with respect to the position that one would expect from geometrical optics, as shown in Fig. 5-4b. This shift is known as the Goos–Hänchen shift. To obtain an estimate of this effect, we can assume that the spectrum of the incident beam consists of just two plane waves with slightly different angles of incidence and evaluate the displacement of the point where the phases of the two waves coincide.[3] This approach is very similar to the one used in the time domain to define the group velocity of a pulse. The Goos–Hänchen axial displacement δz is

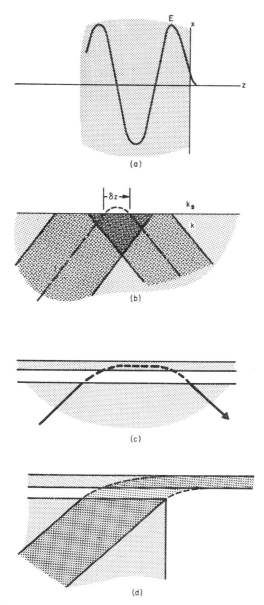

Fig. 5-4 Total reflection at an interface. (a) At grazing angles of incidence, the field at the boundary is very small compared with the field in the bulk. (b) A beam with finite cross section, e.g., a gaussian beam, experiences an axial shift δz (Goos–Hänchen) when undergoing total reflection. (c) The Goos–Hänchen effect is enhanced by a thin film. (d) If the first medium is interrupted, the power remains in the film. This is the principle of the prism coupler.

given by the stationarity of $k_z z + \phi_0$, that is (Artman[3]),

$$\delta z = -\frac{\partial \phi_0(k_z, \omega)}{\partial k_z} \tag{5.19}$$

Using for ϕ_0 the expression in (5.18), we obtain, for H-waves in dielectrics ($\kappa = \kappa_0 = \mu_0$),

$$\delta z = 2 \frac{\partial}{\partial k_z} \arctan \frac{\left(k_z^2 - k_0^2\right)^{1/2}}{\left(k^2 - k_z^2\right)^{1/2}} = \frac{2k_z}{k_{x0i} k_x} \tag{5.20a}$$

$$k_{x0i} \equiv \left(k_z^2 - k_0^2\right)^{1/2}$$

According to (5.20a), the shift δz is infinite at the critical angle, $k_{x0i} = 0$, $k_z = k_0$. The method, however, becomes questionable near that angle.

Similarly, the time delay δt is given by the stationarity of $-\omega t + \phi_0$. If we differentiate that quantity, we obtain

$$\delta t = 2\left(k_x k_{x0i}\right)^{-1}\left(k_0^{-2} - k^{-2}\right)^{-1}\left[\left(k_0 u_0\right)^{-1} - \left(ku\right)^{-1}\right]$$

$$+ 2 \frac{k_z^2}{k_x k_{x0i}}\left(k^2 - k_0^2\right)^{-1}\left(\frac{k}{u} - \frac{k_0}{u_0}\right) \tag{5.20b}$$

where $u \equiv \partial \omega / \partial k$, $u_0 \equiv \partial \omega / \partial k_0$ denote the group velocities of free waves in the two media. If material dispersion is neglected, we have $k_0 u_0 = ku = \omega$, and (5.20b) reduces to

$$\delta t = \frac{2k_z^2}{\omega k_{x0i} k_x} \tag{5.20c}$$

In that case, the simple relation (Agudin[3])

$$\frac{\delta z}{\delta t} = v_z \tag{5.20d}$$

where $v_z \equiv \omega / k_z$, holds. Equation (5.20d) holds also for the E-waves.

The Goos–Hänchen shift can be interpreted by picturing the incident beam as traveling some distance in the less dense region before reentering the denser region, as shown in Fig. 5-4b. In the case where the incident beam is circularly polarized, a small transverse shift along the y-axis has been predicted and observed (Imbert shift).[4] The Goos–Hänchen shift is strongly enhanced if a surface wave can be guided in the less dense medium by a reactive surface or a thin dielectric film, as shown in Fig.

5-4c. In fact, the shift δz may become as large as the beam itself. We can, in that case, picture the situation as the beam power being transferred to the thin film and transferred back to the denser medium after some distance. If the denser medium is interrupted, as shown in Fig. 5-4d, most of the power remains in the film. This is the principle of the prism coupler.[5]

Let us now go back to plane incident waves. The total field can be written

$$V(x) = -\sin[k_x x - \arctan(a)], \qquad x < 0$$

$$V(x) = a(1 + a^2)^{-1/2} \exp(-k_{x0i}x), \qquad x > 0$$

(5.21a)

where we have set

$$a \equiv \frac{\kappa_0 k_x}{\kappa k_{x0i}} \tag{5.21b}$$

We easily verify that V and $\kappa^{-1} \, dV/dx$ in (5.21) are continuous at $x = 0$.

It is instructive to consider the form assumed by $V(x)$ under grazing incidence, that is, for $k_z \approx k$, $k_x \approx 0$. To first order in k_x, expression (5.21) is

$$V(x) = \begin{cases} -\sin(k_x x - a), & x < 0 \\ a \exp\left[-(k^2 - k_0^2)^{1/2} x\right], & x > 0 \end{cases} \tag{5.22}$$

Thus the ratio of V at the interface, $x = 0$, to the maximum value of V, which is unity here, is of the order of $a \ll 1$. In conclusion, the field tends to vanish at the boundary as the angle α to the z-axis tends to zero, for both E- and H-waves. For a dielectric with $\mu = \mu_0$, we have, more precisely,

$$V(0) \to \begin{cases} \dfrac{k_x}{\left(k^2 - k_0^2\right)^{1/2}} \approx \dfrac{\alpha}{\delta}, & H\text{-waves} \\[4mm] \dfrac{(\epsilon_0/\epsilon) k_x}{\left(k^2 - k_0^2\right)^{1/2}} \approx \dfrac{(\epsilon_0/\epsilon)\alpha}{\delta}, & E\text{-waves} \\[4mm] \delta^2 \equiv 1 - \dfrac{k_0^2}{k^2} \approx 2\dfrac{\Delta n}{n} \end{cases} \tag{5.23}$$

The fact that the field tends to zero at the boundary compared to the value in the bulk considerably simplifies the theory of optical fibers whose cross section is large compared with the wavelength. The approximate expression (5.23) is applicable provided

$$\alpha \ll \alpha_{\text{critical}} = \delta \tag{5.24}$$

For example, if $\Delta n / n = 0.005$, $\delta = 0.1$, the present approximation is applicable, provided α is much smaller than $5°$. For H-waves in dielectrics, the ratio of H_x to $E_y \equiv E$ is, in that limit, of the order of $(\epsilon / \mu_0)^{1/2}$. The ratio of H_z to H_x is of the order of δ. Thus for weakly guiding fibers ($\delta \ll 1$), the axial (z) component of H is small compared with the transverse components under grazing incidence, even at the boundary.

E-waves can be guided near a single boundary, provided the permittivity of one of the two media is negative. Negative permittivities are found, for instance, in metals near the plasma frequency. Such surface waves (called surface polaritons)[1] have been observed. We shall assume in the rest of this section that the permittivities are positive. In that case, at least two boundaries are required to guide surface waves.

5.3 The Dielectric Slab

Let us consider a slab with thickness $2d$ and free wavenumber k, supported by a dielectric medium, called a substrate, with free wavenumber k_s. The upper medium, perhaps free space, has free wavenumber k_0 (Fig. 5-5). The phase shift through the slab is $2k_x d$. The phase shifts ϕ_s and ϕ_0 experienced by the wave totally reflected at the slab boundaries are given in (5.18). For consistency, we must have [5,6]

$$4k_x d + \phi_s + \phi_0 = 2m\pi \tag{5.25a}$$

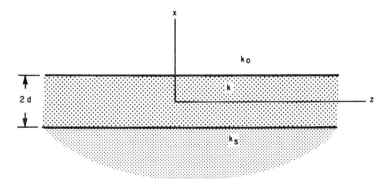

Fig. 5-5 Slab waveguide. The wavenumber of free waves in the slab material is denoted k. Free wavenumbers in the lower and upper media are k_s and k_0, respectively.

where m denotes an integer, and from (5.18),

$$\phi_s = -2 \arctan \frac{\kappa k_{xsi}}{\kappa_s k_x}, \qquad \phi_0 = -2 \arctan \frac{\kappa k_{x0i}}{\kappa_0 k_x} \qquad (5.25b)$$

$$k_{xsi}^2 = k_z^2 - k_s^2, \qquad k_{x0i}^2 = k_z^2 - k_0^2, \qquad k_x^2 = k^2 - k_z^2 \qquad (5.25c)$$

The condition (5.25a) is the same as that used for optical resonators if we replace the wavenumber squared k^2 by $k_x^2 \equiv k^2 - k_z^2$. The expression $2k_x d$ for the phase shift is exact because k does not depend on x until the boundary is reached. If k were slowly varying with x within the slab, we could use the first term of the WKB expansion in (5.25a) in place of $4k_x d$: $2\int_{-d}^{+d} k_x(x)\,dx$. Equation (5.25) defines the axial wavenumber k_z of the waves that are trapped by the dielectric slab.

Let the symmetrical slab with $k_0 = k_s$, $\kappa_0 = \kappa_s$ be considered first. The dispersion equation (5.25) becomes

$$2k_x d = m\pi + 2 \arctan \frac{\kappa k_{xsi}}{k_s k_x} \qquad (5.26a)$$

This dispersion relation can be written alternatively

$$\tan(k_x d) = \frac{\kappa k_{xsi}}{\kappa_s k_x}, \qquad m \text{ even}$$

$$\tan(k_x d) = \frac{-\kappa_s k_x}{\kappa k_{xsi}}, \qquad m \text{ odd} \qquad (5.26b)$$

In most cases, the form (5.26a) is more convenient than the forms in (5.26b). If we neglect the dispersion of the material, the frequency can be written as an explicit function of the axial phase velocity $v_z \equiv \omega/k_z$. To see that, let us introduce a normalized frequency

$$F \equiv \left(k^2 - k_s^2\right)^{1/2} d \qquad (5.27)$$

(F is sometimes denoted V) and a normalized phase velocity parameter b

$$b \equiv \frac{k_z^2 - k_s^2}{k^2 - k_s^2} = \frac{(v_s/v_z)^2 - 1}{(v_s/v_f)^2 - 1} \qquad (5.28)$$

where v_f and v_s denote the velocities of free waves in the slab (film) and the surrounding medium, respectively. With this notation (5.26a) becomes

$$F = (1 - b)^{-1/2}\left(\frac{m\pi}{2} + \arctan\left\{\frac{\kappa}{\kappa_s}\left[\frac{b}{(1 - b)}\right]^{1/2}\right\}\right) \qquad (5.29)$$

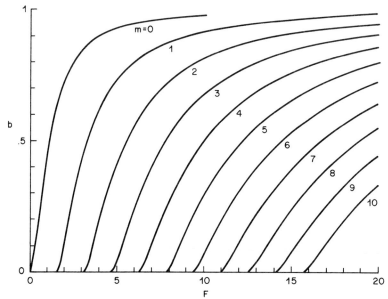

Fig. 5-6 Dispersion of the H-waves of symmetrical slabs. The parameters

$$b \equiv (k_z^2 - k_s^2)/(k^2 - k_s^2) \quad \text{and} \quad F \equiv (k^2 - k_s^2)^{1/2}d$$

are normalized phase velocities and frequencies, respectively.

where m even corresponds to even modes, and m odd to odd modes. Let us recall that $\kappa/\kappa_s = 1$ for H-waves, and $\kappa/\kappa_s = (k/k_s)^2$ for E-waves. The variation of b with F for H-waves, and $m = 0, 1, 2, \ldots$, is shown in Fig. 5-6. Let us make the following numerical example: The slab thickness $2d = 1.32$ μm, and the slab refractive index is equal to 1.45. The refractive index of the surrounding medium, or cladding, is equal to 1.4. We first evaluate $\delta = [1 - (1.4/1.45)^2]^{1/2} = 0.26$. If the free-space wavelength is $\lambda_0 = 1$ μm, the normalized frequency if $F = \delta kd = 0.26 \times 2\pi \times 1.45 \times 0.66 = \pi/2$. Figure 5-6 then shows that the normalized phase velocity for $m = 0$ is $b = 0.67$. Thus $\phi_i = (k_z^2 - k_s^2)^{1/2}d = b^{1/2}F = 1.28$. We also find $(k^2 - k_z^2)^{1/2}d = (1 - b)^{1/2}F = 0.902$. These numerical results will be used in subsequent examples.

The fundamental H- and E-modes have no low-frequency cutoff. Near the low-frequency cutoff, $b \ll 1$, the dispersion equation (5.29) takes the form

$$F = \frac{m\pi}{2} + \begin{cases} b^{1/2}, & H\text{-waves} \\ \left(\dfrac{k}{k_s}\right)^2 b^{1/2}, & E\text{-waves} \end{cases} \tag{5.30}$$

For the fundamental H-wave (H_0-mode), we have simply

$$F = b^{1/2} \tag{5.31}$$

or

$$\left(\frac{v_s}{v_z} \right)^2 = 1 + k_s^2 d^2 \left[\left(\frac{k}{k_s} \right)^2 - 1 \right]^2 \tag{5.32}$$

Note that the assumption $b \ll 1$ does not necessarily imply that v_z is close to v_s. An alternative is that k/k_s tends to infinity. In that limit, $kd \ll 1$, but $kd(k/k_s)$ comparable to unity, the slab behaves as a reactive surface with the boundary condition

$$\frac{dE}{dx} + sE = 0 \tag{5.33}$$

where $s \equiv k^2 d$, at $x = 0$.

Let us now go to the other extreme and consider modes that are far from cutoff. In that limit, the waves are incident on the boundary under grazing incidence and $k_z \approx k$. As we have seen before, the fields at the boundary become very small compared with their values in the bulk. Mathematically, this limit case is obtained by letting b tend to unity in (5.29). We obtain for the slab phase shift ϕ:

$$\phi \equiv F(1 - b)^{1/2} = k_x d \approx m \frac{\pi}{2} + \arctan\left[\frac{\kappa}{\kappa_s} (1 - b)^{-1/2} \right]$$

$$\approx (m + 1) \frac{\pi}{2} - \frac{\kappa_s}{\kappa} \frac{\phi}{F} \approx (m + 1) \frac{\pi}{2} \left[1 - \frac{\kappa_s}{\kappa} F^{-1} \right] \tag{5.34}$$

The zeroth-order term in (5.34)

$$\phi = (m + 1) \frac{\pi}{2} \tag{5.35}$$

could have been obtained as well by specifying that the field vanishes at the boundary. This is the boundary condition that is applicable to waveguides with electric walls (for H-waves) or magnetic walls (for E-waves).

If we now consider the first-order term in (5.34), we find a slight difference between the wavenumbers of H-waves and those of the corresponding E-waves. It is interesting to evaluate this difference. We have from (5.34), setting $n \equiv k/k_s$,

$$\phi_E^2 - \phi_H^2 \approx 2(m + 1)^2 \left(\frac{\pi}{2} \right)^2 (1 - n^{-2}) F^{-1} \tag{5.36}$$

Thus

$$(k_{zH} - k_{zE})d \approx \left[\frac{\pi(m + 1)}{2kd} \right]^2 (1 - n^{-2})^{1/2} \tag{5.37}$$

Similar expressions can be obtained for other types of dielectric waveguides, such as rods with elliptic cross section, d being replaced by a dimension of the order of the smallest dimension in the cross section. The result (5.37) helps define the range of application of the scalar approximation. Indeed, if a field is launched with some arbitrary polarization (e.g., circular), this polarization is preserved over a length of the order of $2\pi/(k_{zH} - k_{zE})$. This is the distance that it takes for the two near-degenerate electromagnetic modes to get out of phase. It is important to observe that $(k_{zH} - k_{zE})d$ is small compared with unity when either $k \approx k_s$ (weakly guiding fiber) or when kd is large compared to unity (overmoded fiber). These two conditions are distinct. In some cases, however, both may be applicable. Let us give a numerical example. For the fundamental mode $m = 0$, $k = 2\pi \ \mu m^{-1}$, $2d = 10 \ \mu m$, and $n = 1.01$, the beat wavelength previously defined is equal to 90 mm. This length is large compared to the size of the devices used in integrated optics but small compared to the length of fibers used in long-distance communication. Whether the scalar approximation is applicable or not therefore depends on the application and problem considered.

Let us now evaluate the group velocity $u_z = \partial\omega/\partial k_z$ of the slab modes. We differentiate the dispersion equation

$$2\phi \equiv 2k_x d = m\pi - \phi_s \tag{5.38}$$

given earlier, with respect to ω. In (5.38), ϕ and ϕ_s are known functions of k_z and ω. We use the relations $\partial k_x/\partial k_z = -k_z/k_x$ and $\partial k_x/\partial\omega = (k/k_x) \ \partial k/\partial\omega$, which follow from the definition $k_x = [k^2(\omega) - k_z^2]^{1/2}$, and obtain readily (Kogelnik and Weber[3])

$$u_z = \frac{\partial\omega}{\partial k_z} = \frac{2(k_z/k_x)d + \delta z}{2(k/k_x)d/u + \delta t} \tag{5.39}$$

where $u = \partial\omega/\partial k$ denotes the group velocity in the slab material. We have introduced the Goos–Hänchen shift

$$\delta z = \frac{-\partial\phi_s}{\partial k_z} \tag{5.40}$$

and the reflection time lag $\delta t = \partial\phi_s/\partial\omega$ from their definition. Their expressions are given in (5.20). The term $2(k_z/k_x)d$ in (5.39) represents the axial distance that a ray must travel to go from one boundary of the slab to the other. Similarly, the term $2(k/k_x)d/u$ represents the time of flight

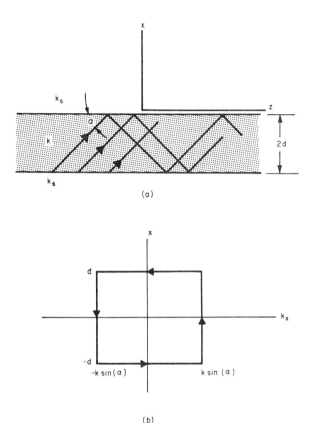

Fig. 5-7   WKB approximation of the slab waveguide. (a) Ray manifold. (b) Ray trajectory in phase space.

from one boundary to the other, because the distance to travel is equal to $2(k/k_x)d$ and the pulse velocity is u. Result (5.39) is easily generalized to asymmetrical slabs.

For oversized waveguides, δz and δt in (5.39) can be neglected. In that approximation, the axial group velocity can be obtained from the simple zigzag ray picture commonly used for metallic waveguides (see Fig. 5-7).

The ratio of axial phase to group velocity for H-waves is obtained from (5.39) and (5.20). If we neglect material dispersion,[7] we have (replacing the subscripts 0 by subscripts s),

$$\left(\frac{v_z}{u_z}\right)_{H\text{-waves}} = \frac{1 + (k/k_z)^2 k_{xsi}d}{1 + k_{xsi}d} \tag{5.41}$$

Alternatively, v_z/u_z can be expressed as a function of $\phi \equiv k_x d$ and

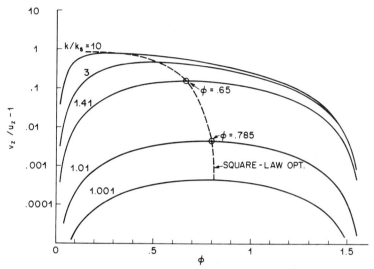

Fig. 5-8 Ratio of phase-to-group velocity of a slab. The slab refractive index is a parameter. Note that there is a region where v_z/u_z is stationary. The quantity $\phi \equiv (k^2 - k_z^2)^{1/2}d$ is monotonically increasing with d. [See Eq. (5.44).]

$n = k/k_s$, using (5.26b). We have first

$$k_{xsi}d = \phi \tan\left(\phi - \frac{m\pi}{2}\right) \qquad (5.42)$$

The expression for k_z/k_s is obtained by manipulating (5.42) and the definitions of k_x and k_{xsi}. We obtain

$$\frac{k_z^2}{k_s^2} \equiv \frac{n^2 k_z^2}{k^2} = 1 + (n^2 - 1)\sin^2\left(\phi - \frac{m\pi}{2}\right) \qquad (5.43)$$

Thus by substituting (5.43) and (5.42) in (5.41), v_z/u_z can be expressed as a function of ϕ and n only. We shall also need, in subsequent calculations, the relation

$$k_s d = \frac{(n^2 - 1)^{-1/2}\phi}{\cos(\phi - m\pi/2)} \qquad (5.44)$$

that follows from (5.42). The variation of v_z/u_z as a function of ϕ for various values of $n \equiv k/k_s$ and $m = 0$ is shown in Fig. 5-8. It is interesting that v_z/u_z reaches a maximum. For example, v_z/u_z is stationary near $\phi = 0.785$ for $n = 1.01$ and $m = 0$. Near that maximum, times of flight are proportional to phase shifts (or optical lengths). The usefulness of this observation will be clarified in Section 5.6, dealing with tapered slabs.

Let us now evaluate the axial component of the power in the slab and in the cladding. Because of symmetry, these powers are twice the integrals of S_z', defined in (5.10), from $x = 0$ to $x = d$, and from $x = d$ to $x = \infty$, respectively. We assume here that the origin of x is at the slab center. The result will be derived only for the *H*-waves. Let E denote the field at the slab boundary. The axial component of the power in the slab is

$$
P = \frac{k_z}{\omega\mu_0} E^2 \int_0^d \frac{\cos^2(k_x x - m\pi/2)}{\cos^2(k_x d - m\pi/2)} \, dx
$$

$$
= \frac{(k_z d/2\omega\mu_0) E^2 (1 + b/\phi_i)}{1 - b} \tag{5.45}
$$

where $\phi_i = b^{1/2}F$. F and b are defined in (5.27) and (5.28), respectively. The power in the cladding is

$$
P_s = \frac{k_z}{\omega\mu_0} E^2 \int_0^\infty \exp(-2k_{xsi}x) \, dx = \frac{k_z d}{2\omega\mu_0} E^2 \phi_i^{-1} \tag{5.46}
$$

Thus the total power is

$$
P_t = P + P_s = \frac{(k_z d/2\omega\mu_0) E^2 (1 + \phi_i^{-1})}{1 - b} \tag{5.47}
$$

The group velocity in the slab can be obtained alternatively from the Hellmann–Feynman theorem derived in Chapter 3 from Brown's identity on p. 218, and the above expressions for P and P_s. The detailed derivation is left as a problem.

Up to now, we have considered in detail only symmetrical slabs. In integrated optics, the upper medium, usually free space, has a wavenumber k_0 different from that, k_s, of the lower medium. We shall discuss that case for *H*-waves. Equation (5.25) can be written[8]

$$
F = (1 - b)^{-1/2}\left[\frac{m\pi}{2} + \arctan\left(\frac{b}{1 - b}\right)^{1/2} + \arctan\left(\frac{b + a}{1 - b}\right)^{1/2}\right]
$$

$$
\tag{5.48}
$$

In (5.48) we have defined, as before,

$$
F \equiv (k^2 - k_s^2)^{1/2} d \tag{5.49}
$$

$$
b \equiv \frac{k_z^2 - k_s^2}{k^2 - k_s^2} \tag{5.50}
$$

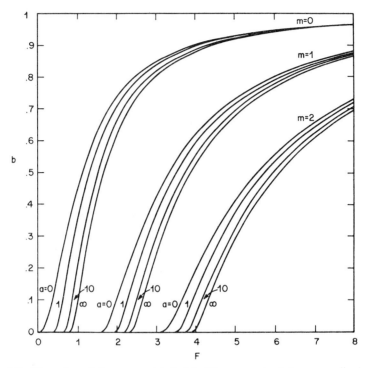

Fig. 5-9 Dispersion of the asymmetrical slab. The parameter b is a normalized phase velocity, and F is a normalized frequency. The parameter a expresses the degree of asymmetry. For symmetrical slabs, $a = 0$. (From Kogelnik and Ramaswamy,[8] by permission of the American Optical Society.)

and introduced an asymmetry parameter

$$a \equiv \frac{k_s^2 - k_0^2}{k^2 - k_s^2} \tag{5.51}$$

which is equal to zero for symmetrical slabs, and reaches infinity for strong asymmetries. The dependence of the normalized phase velocity b on the normalized frequency F for various degrees of asymmetry is shown in Figs. 5-9 and 5-10. Note that for asymmetrical waveguides $a \neq 0$, there is always a low-frequency cutoff.

A situation of practical importance in integrated optics is when k is only slightly larger than k_s, both k and k_s being much larger than k_0. For H-waves, the situation is almost the same as if the electric field were required to vanish at the upper boundary (usually the air–dielectric inter-

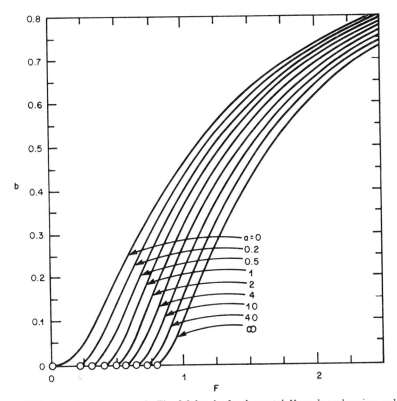

Fig. 5-10   Detail of the curves in Fig. 5-9 for the fundamental H-mode and various values of the asymmetry parameter a. (From Kogelnik and Ramaswamy,[8] by permission of the American Optical Society.)

face). The solutions are then the same as the antisymmetrical modes of the symmetrical slab in (5.29), with d changed to $2d$. The case of anisotropic slabs is discussed in Russo and Harris.[9]

5.4 Periodic Layers

Periodic layers of dielectrics with alternately high and low refractive indices provide high reflectivities to incident plane waves, up to 99.9% in the visible. The mechanism is similar to that of Bragg reflection. When $k_x x_0$—where $k_x \equiv (k^2 - k_z^2)^{1/2}$ denotes the component of the wave vector normal to the layers in the medium and x_0 the period—is of the order of π, reflections from the successive boundaries add up in phase, and a large total reflection is obtained even if the individual reflections are small. Such a periodic medium exhibits a narrow stopband. Equal ripple response in a narrow band, or other desirable response characteristics, can be obtained by properly selecting the thicknesses and indices of the first layers. Many excellent books discuss the optimum design of multilayer coatings.

It is not as well known that periodic layers can guide surface waves.[10] The propagation of surface waves along semi-infinite periodic dielectric layers is discussed in this section. This arrangement may find applications in integrated optics. In any event, it will serve the purpose of illustrating the general results given above. The dispersion equation of a semi-infinite sequence of periodic layers is derived on the basis of the transmission line representation discussed before. Simple approximate expressions are subsequently given, applicable to loosely bound waves.

Let us consider the two-dimensional configuration shown in Fig. 5-11, where an infinite number of periodic dielectric layers occupy the half-space $x < 0$. We are interested in waves propagating in the z-direction and decaying exponentially in the x-direction. The two eigenstates of polarization are the E- and H-waves discussed before, with the electric fields in the xz plane, or perpendicular to it, respectively. To obtain the axial wavenumber, we equate to zero the sum of the admittances observed at the plane $x = 0$, looking toward the periodic layers ($x < 0$) and looking toward free space ($x > 0$). The admittance of the periodic layers is obtained by evaluating the voltage–current transfer matrix for one section (i.e., two layers), multiplying two matrices such as those in (5.12) and

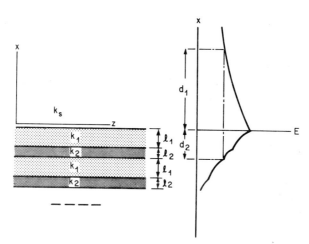

Fig. 5-11 Multilayer waveguide. Waves can be guided near the interface between a system of periodic layers and a homogeneous space with free wavenumber k_s. (From Arnaud and Saleh,[10] by permission of the American Optical Society.)

specifying that the admittance is the same at the input and output planes of that section. This admittance is

$$Y = B^{-1}\left\{ \tfrac{1}{2}(A - D) \pm \left[\tfrac{1}{4}(A + D)^2 - 1 \right]^{1/2} \right\} \qquad (5.52a)$$

where

$$A \equiv \cos(k_{x1}l_1)\cos(k_{x2}l_2) - Y_1^{-1}Y_2 \sin(k_{x1}l_1)\sin(k_{x2}l_2)$$

$$D \equiv \cos(k_{x1}l_1)\cos(k_{x2}l_2) - Y_2^{-1}Y_1 \sin(k_{x1}l_1)\sin(k_{x2}l_2) \qquad (5.52b)$$

$$B \equiv -i\left[Y_1^{-1}\sin(k_{x1}l_1)\cos(k_{x2}l_2) + Y_2^{-1}\sin(k_{x2}l_2)\cos(k_{x1}l_1) \right]$$

$$k_{x\alpha} = (k_\alpha^2 - k_z^2)^{1/2}, \qquad \alpha = 1, 2 \qquad (5.52c)$$

and

$$Y_\alpha = \frac{k_{x\alpha}}{n^2}, \qquad \alpha = 1, 2 \qquad (E\text{-waves}) \qquad (5.52d)$$

where $n_\alpha = k_\alpha/k_0$ denotes the refractive index of the layers, and

$$Y_\alpha = k_{x\alpha}, \qquad \alpha = 1, 2 \qquad (H\text{-waves}) \qquad (5.52e)$$

Unimportant constants have been omitted in the Y. In the above equations, k_1 and k_2 denote the free wavenumbers of the first and second layers, respectively, and l_1 and l_2 their thicknesses. Since we are interested in waves that decay as $x \to -\infty$, the sign of the square root in (5.52a) must be the same as that of the real quantity $A + D$.

The condition for transverse resonance is

$$k_{x0} + Y = 0 \qquad (5.53a)$$

where

$$k_{x0} = (k_0^2 - k_z^2)^{1/2} \equiv ik_{x0i} \qquad (5.53b)$$

Equations (5.53) and (5.52) define the axial wavenumber k_z.

A simple approximate solution of these equations can be obtained if we assume that the surface wave is loosely bound to the layers, i.e., that the decay per wavelength above the layers is small. In that case, k_{x0i}/k_0 is a small positive quantity, which tends to zero as k_0 tends to the cutoff propagation constant k_c. It follows from (5.53) that, as $k_0 \to k_c$, $Y \to i\infty$ for E-waves and $Y \to 0$ for H-waves. At cutoff, the dielectric layers act as an electric wall for E-waves and as a magnetic wall for H-waves.

It can be shown from (5.52) that, at cutoff ($k_{x0} = 0$), $k_{x1}l_1$ and $k_{x2}l_2$ are odd multiples of $\pi/2$. Thus, if λ_c is the cutoff wavelength, the smallest

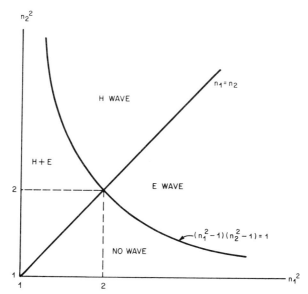

Fig. 5-12 Condition of propagation of *E*- and *H*-waves in a multilayer waveguide. We have defined $n_1 \equiv k_1/k_s$, $n_2 \equiv k_2/k_s$. (From Arnaud and Saleh,[10] by permission of the American Optical Society.)

values of l_1 and l_2 are

$$l_\alpha = \left(\tfrac{1}{4}\lambda_c\right)\left(n_\alpha^2 - 1\right)^{-1/2}, \qquad \alpha = 1, 2 \tag{5.54}$$

Equation (5.52) also shows that a solution exists only if $Y_1 < Y_2$ for both *E*- and *H*-waves. Using (5.52d) and (5.52e), we find that n_1 and n_2 must satisfy the condition

$$n_1^{-2}\left[n_1^2 - 1\right]^{1/2} < n_2^{-2}\left[n_2^2 - 1\right]^{1/2} \qquad (\text{*E*-wave}) \tag{5.55a}$$

or

$$n_1 < n_2 \qquad (\text{*H*-wave}) \tag{5.55b}$$

The regions in the n_1, n_2 plane that allow the existence of surface waves are shown in Fig. 5-12. The figure indicates that, depending on the refractive indices, periodic layers may support both *E*- and *H*-waves, *E*-waves alone, *H*-waves alone, or no surface wave at all.

To study the propagation near the cutoff wavelength λ_c, let us set

$$\lambda = \lambda_c - \Delta\lambda \tag{5.56}$$

The approximate solutions for k_{x0i} are

$$k_{x0i} \approx \lambda_c^{-1}\pi^2\left(Y_1^{-1} - Y_2^{-1}\right)^{-1}\frac{\Delta\lambda}{\lambda_c} \qquad (\text{*E*-wave}) \tag{5.57a}$$

and

$$k_{x0i} \approx \lambda_c^{-1} \pi^2 (Y_1 - Y_2)^{-1} \frac{\Delta\lambda}{\lambda_c} \qquad (H\text{-wave}) \qquad (5.57b)$$

The layers thus support surface waves for wavelengths shorter than the cutoff wavelength.

The distance d_1 above the layers, where the field amplitude is reduced by a factor $e = 2.718 \ldots$, is

$$d_1 = k_{x0i}^{-1} \qquad (5.58)$$

This distance is a sensitive function of the operating wavelength since it is proportional to $\lambda_c/\Delta\lambda$. The distance d_2 inside the dielectric layers, where the field amplitude is reduced by a factor $e = 2.718 \ldots$, can be calculated from the decay constant per pair of layers

$$\text{Re}\{\arg \cosh[\tfrac{1}{2}(A + D)]\} \qquad (5.59)$$

Neglecting terms of the order of $(\Delta\lambda/\lambda_c)^2$, we obtain from (5.59) and (5.52)

$$d_2 \approx (l_1 + l_2) \cosh^{-1}\left[\frac{1}{2}\left(\frac{Y_1}{Y_2} + \frac{Y_2}{Y_1} \right) \right]^{-1} \qquad (5.60)$$

where l_1 and l_2 are given in (5.54) and Y_1 and Y_2 in (5.57). Unlike d_1, the distance d_2 is almost independent of the frequency in the domain of interest.

Note that the above equations are based on the assumption of infinitely many layers. A rough estimate of the number of layers required in practice is obtained by assuming that the total thickness is equal to twice the distance d_2. From this rule, the approximate number of pairs of layers is found to be

$$N \approx \frac{2d_2}{l_1 + l_2} \qquad (5.61)$$

To illustrate these results, let the cutoff wavelength be 0.6328 μm (He–Ne line), $n_1 = 1.4$, and $n_2 = 1.6$. Figure 5-12 indicates that the layers can support H-waves. The above equations give $l_1 = 0.162$ μm, $l_2 = 0.127$ μm, $d_1 \approx (89/\Delta\lambda)$ μm, where $\Delta\lambda$ is the difference in angstroms between the cutoff and operating wavelengths, $d_2 \approx 1.2$ μm, and $N \approx 8$. If $\Delta\lambda = 5$ Å, $d_1 \approx 18$ μm. In that case, the losses are roughly ten times smaller (in decibels) than the bulk losses in the dielectric, because about 90% of the power is propagating in air and only 10% in the dielectric. If the order of the layers were reversed, i.e., if $n_1 = 1.6$ and $n_2 = 1.4$, the structure would support E-waves. The number of pairs of layers required in that case is prohibitively large, about 80. Note that the conditions that allow the propagation of surface waves would be significantly affected if the first layer had a thickness and refractive index different from that of the other odd-numbered layers.

The multilayer waveguide described in this section has the unique property of lying on a substrate and having cutoff behavior in air. Layers with alternately high and low refractive indices can be deposited on quartz capillaries, for example, by the vapor phase deposition technique. However, it is not known whether the accuracy required for low-loss propagation can be attained by that technique.

It is interesting that a mechanism of surface wave guidance between a periodic structure and free space similar to the one described in this section exists in solid state physics. The so-called Tamm surface states are electron surface waves guided near the boundary between a perfect crystal and free space. The electron wave function decays away from the boundary, exponentially in free space, and in a more complicated manner inside the crystal lattice. Similar effects could be observed on superlattices obtained by molecular beam epitaxy.[11]

5.5 Propagation along Contacting Dielectric Tubes

The dielectric slabs considered in the previous section are not true optical waveguides because the field is not confined in the transverse direction (y-axis). A mechanism for transverse confinement is now investigated. The confinement is provided by slow changes of the slab structure in the transverse direction.[12-14] There is confinement if the local wavenumber β exhibits a maximum along the z-axis. One possibility is that the slab thickness decreases away from the z-axis. (This will be discussed in Section 5.6.) An alternative solution considered here consists of using a pair of dielectric slabs whose thickness if fixed but whose spacing varies with the y-coordinate.[13]

Let us clarify our notation. In the theory of the uniform slab, we have selected arbitrarily a z-axis in the plane of the slab and evaluated the wavenumber k_z, the phase velocity $v_z \equiv \omega/k_z$, and the group velocity $u_z \equiv \partial\omega/\partial k_z$ with respect to that axis. Because the propagation is isotropic in the plane of the slab, we can, however, free the direction of the z-axis in that plane. We shall now denote by β the wavenumber in the plane of the slab, a quantity denoted earlier by k_z. The wave vector in the plane of the slab, with magnitude β, is denoted $\boldsymbol{\beta}$. We now select the z-axis along the direction of translational invariance of the (nonuniform) fiber. Thus the component β_z of $\boldsymbol{\beta}$ is a constant of motion. The local group velocity is

$\partial\omega/\partial\beta$. We shall evaluate the group velocity $v_g \equiv \partial\omega/\partial\beta_z$ of modes propagating along the z-axis.

Let us now go back to the specific problem of contacting-dielectric tubes of constant thicknesses $2d$ but variable spacing $2D(y)$. When the two walls are in contact ($D = 0$), the total thickness is $4d$. On the other hand, when the walls are far apart ($D \gg \lambda$), they are uncoupled, and the effective wall thickness is only $2d$. Because the wavenumber of a slab decreases with thickness, the local wavenumber β of the pair of slabs decreases as the spacing $2D$ increases. This reduction ensures confinement of the beams near the contact line, $y = 0$. The depth of the "potential well" $\beta^2(y)$ determines how many modes (with mode numbers $n = 0, 1, 2, \ldots$) can exist in the y-direction. If the well is sufficiently shallow, only one mode ($n = 0$) propagates.

The wavenumber β of the fundamental symmetrical H-mode guided by two parallel dielectric slabs with refractive index k/k_0, thickness $2d$, and spacing $2D$, is easily found from (5.12). It is defined by the equation

$$\tanh(k_{x0i}D) = \frac{(k_x/k_{x0i})\tan(2k_x d) - 1}{(k_{x0i}/k_x)\tan(2k_x d) + 1} \tag{5.62a}$$

where

$$k_x^2 = k^2 - \beta^2 \tag{5.62b}$$

$$k_{x0i}^2 = \beta^2 - k_0^2 \tag{5.62c}$$

Equation (5.62) gives D as an explicit function of the wavenumber β. The variation of the effective local permittivity $(\beta/k_0)^2$ obtained from (5.62) is plotted in Fig. 5-13 as a function of y for a relative dielectric permittivity $(k/k_0)^2 \equiv \epsilon/\epsilon_0 = 9.5$, a wall thickness $2d = 0.76$ mm, and a radius of curvature of the wall $\rho = 4.77$ mm. The spacing $2D$ is related to y by $2D = 2\rho\{1 - [1 - (y/\rho)^2]^{1/2}$.

The next step in the evaluation of the axial wavenumber β_z is to solve the eigenvalue equation

$$\frac{d^2 E}{dy^2} + \beta^2(y)E = \beta_z^2 E \tag{5.63}$$

where $\beta(y)$ is the local wavenumber just obtained from (5.62). An approximate solution of (5.63) is found if we note that for small $|y|$, $\beta^2(y)$ can be written approximately

$$\beta^2 \approx \beta_0^2(1 - \Omega^2 y^2) \tag{5.64}$$

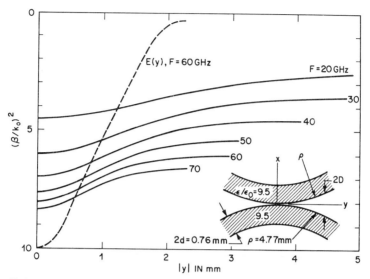

Fig. 5-13 Variation of the local slab wavenumber β with the distance $|y|$ from the contacting line between two dielectric tubes. The dashed line gives the variation of the electric field with y. (From Arnaud,[13] by permission of the Institution of Electrical Engineers.)

With this square-law approximation, the fundamental solution of (5.63) is

$$E = \exp\left(-\tfrac{1}{2}\beta_0\Omega y^2\right) \tag{5.65a}$$

as we verify by substituting (5.65a) in (5.63) with β^2 in (5.64), and we have

$$\beta_z^2 = \beta_0^2 - \beta_0\Omega \tag{5.65b}$$

At 60 GHz, for instance, we find from the curves in Fig. 5-13 that $\beta_0^2 = 12.68$ mm^{-2} and $\beta_0\Omega = 1.23$ mm^{-2}. Thus $k_z = 3.4$ mm^{-1}, and the beam half-width is $(\beta_0\Omega)^{-1/2} = 1.1$ mm. Once the variation of β_z with ω has been obtained, the group delay c/v_g follows by differentiation. The variation of c/v_g with frequency is shown in Fig. 5-14d. To check the accuracy of the above square-law approximation, (5.63) has been solved numerically with the initial conditions $E = 1$, $dE/dy = 0$ at $y = 0$. The parameter β_z is varied until $E \to 0$ as $y \to \infty$. At 60 GHz, the exact value of β_z is found by this numerical technique to be 3.395 mm^{-1} in close agreement with the one obtained from the square-law approximation $\beta_z = 3.4$ mm^{-1}. The variation of E with y, obtained by numerical integration, is shown in Fig. 5-13 as a dashed line.

These theoretical results are in good agreement with experiment. The experimental setup is shown in Fig. 5-14a. Two 254-mm-long alumina tubes (outer diameter = 9.55 mm, wall thickness = 0.75 mm, relative

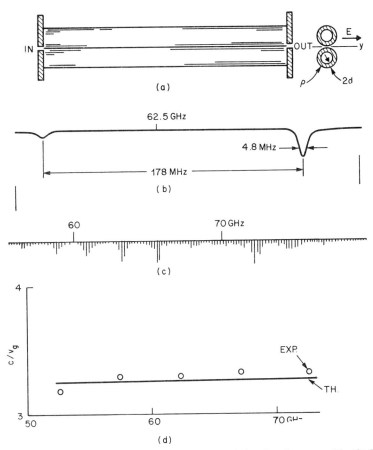

Fig. 5-14 (a) Experimental setup of the contacting dielectric tube waveguide. (b) Transmission as a function of frequency over a free spectral range. (c) Transmission from 55 to 75 GHz. (d) Group delay, theoretical (solid lines) and experimental (circles). (From Arnaud,[13] by permission of the Institution of Electric Engineers.)

permittivity = 9.5) are attached side by side. They are held between two brass plates having small (0.76 mm) coupling apertures for measurement purposes. The electric field E is along the y-axis. Figure 5-14b,c shows the response of the resonator when the generator frequency is swept from 50 to 80 GHz. The perfect regularity in spacing of the resonances demonstrates that only one mode propagates, at least up to 70 GHz. The fact that the tubes can be touched with the hand without significantly affecting the resonance shows that the field is concentrated near the contact line between the two cylinders. Figure 5-14b shows two adjacent resonances separated by a frequency $\Delta F = 178$ MHz. The group velocity is obtained at any frequency from the measured ΔF with the help of the formula $v_g = 2L\,\Delta F$, where $L = 254$ mm is the tube length. The variation of c/v_g as a function of frequency is shown in Fig. 5-14d. We observe that the propagation is almost dispersion free. Because the variation of c/v_g over a 10-GHz band does not exceed 0.05, for a monochromatic carrier pulse broadening would not exceed $[c^{-1}\Delta(c/v_g)/\Delta\omega]^{1/2} = 1.6$ nsec/km$^{1/2}$. The c/v_g curve measured for a 1.6 mm diameter rod made with the same material exhibits a much larger dispersion ($c/v_g = 5.3$ at 50 GHz, $c/v_g = 4.2$ at 70 GHz). This configuration can be made single mode and almost dispersion free ($d^2\beta/d\omega^2 \approx 0$). This example demonstrates the validity of the semiclassical diffraction theory.

5.6 Tapered Slabs

In the arrangement previously considered, the slab thickness is sufficiently small and the "potential well" $\beta^2(y)$ is sufficiently shallow that only one mode can propagate. For optical communication, it is desirable, however, that the cross-sectional dimensions be large compared to the wavelength for easy splicing. Furthermore, multimode operation is desirable when the source is spatially incoherent. These modes are required to have almost the same group velocity to prevent pulse broadening. We have shown in Chapter 4 that such a result can be achieved with multimoded fibers whose refractive index varies approximately as the square of the distance from axis. In contradistinction, fibers with a homogeneous core and a large cross section exhibit large pulse broadening. The fiber described in this section and shown in Fig. 5-15b has the property that it requires only one homogeneous material and nevertheless exhibits low pulse broadening. However, the dimensions are large only along y-axis. The fiber is essentially single moded in the x-direction. We shall assume for simplicity that only one H-mode in the x-direction can propagate and investigate how the spread in group velocity of the various y-modes can be minimized.[15] Because there is a large number of y-modes, a ray technique can be applied in the yz plane of the slab. The results in this section thus combine results in geometrical optics—derived in Chapter 4 and applicable to the yz plane—and results in wave optics applicable to the dielectric slab, which were given earlier in this chapter.

It is interesting for comparison to consider two other types of planar guides, shown in Fig. 5-15a,c. The slab shown in Fig. 5-15a has a uniform thickness, but its permittivity is assumed to vary with y. Times of flight are proportional to optical lengths in such a structure, provided that the dispersion of the material can be neglected, as is sometimes the case. For the tapered homogeneous slab discussed in the present section, we shall

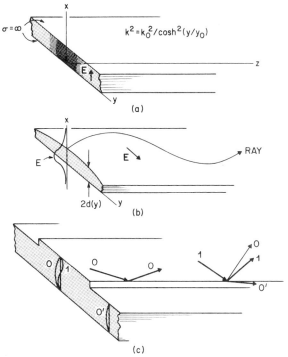

Fig. 5-15 (a) The slab thickness is uniform, but the permittivity varies with y. Times of flight are proportional to optical lengths if inhomogeneous dispersion can be neglected. (b) Tapered slab, with thickness a function of y. (c) Stepped slab. The discontinuity in thickness couples the various y-modes (e.g., 1 to 0′). (From Arnaud,[28] 1975, by permission of the American Optical Society.)

find that times of flight are *not* proportional to optical lengths, even in the absence of material dispersion ($dn/d\lambda_0 = 0$), because of the modal dispersion of the slab. The slab in Fig. 5-15c involves a step in thickness that introduces a significant coupling between the various x-modes (e.g., between the H_0 and H_2-modes). However, if the step is small, a theory that neglects this coupling may be adequate to evaluate the propagation constant of the fundamental mode. In the present section, x-mode couplings are assumed negligible.

We shall use the notation explained in the previous section. Let us write (5.41) to (5.44) with that notation for the H_0-modes. We have

$$\beta^2 = k_s^2[1 + (n^2 - 1)\sin^2 \phi], \qquad d = \frac{k_s^{-1}(n^2 - 1)^{-1/2}\phi}{\cos \phi}$$

$$\frac{1}{2}\frac{\partial \beta^2}{\partial \omega} = \frac{\omega^{-1}k_s^2(n^2\phi \tan \phi + n^2 \sin^2 \phi + \cos^2 \phi)}{\phi \tan \phi + 1} \qquad (5.66)$$

where k_s denotes, as before, the wavenumber of free waves in the medium surrounding the slab, $\phi \equiv k_x d$, and $n \equiv k/k_s$.

The equations describing the motion of pulses in isotropic media characterized by some wavenumber law $\beta(\omega, y)$, given in (4.62) and (4.63a), are (with the β-notation replacing the k-notation):

$$\frac{dy}{dz} = \frac{\beta_y}{\beta_z} \tag{5.67a}$$

$$\frac{d\beta_y}{dz} = \frac{1}{2} \frac{(\partial \beta^2 / \partial y)}{\beta_z} \tag{5.67b}$$

$$\frac{dt}{dz} = \frac{1}{2} \frac{(\partial \beta^2 / \partial \omega)}{\beta_z} \tag{5.67c}$$

and the time of flight of a pulse is, by integration of (5.67c),

$$t(z) = \frac{1}{2\beta_z} \int_0^z \left(\frac{\partial \beta^2}{\partial \omega} \right) dz \tag{5.68}$$

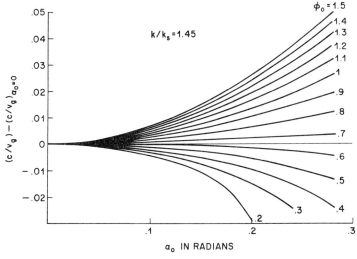

Fig. 5-16   Ratio of vacuum to axial group velocities (c/v_g) as a function of the ray angle α_0 at the origin, for a tapered dielectric slab with $n = 1.45$ and a quadratic variation of the characteristic angle $\phi = \phi_0 - 4 \times 10^{-5} y^2$, for various values of ϕ_0. Pulse broadening is $\Delta t = (10^4/3)(c/v_g)$ nsec/km. The slab phase shift on axis $\phi_0 = 0.65$ provides small variations of c/v_g over a large range of values of α_0. (From Arnaud,[15] by permission of the American Telegraph and Telephone Company.)

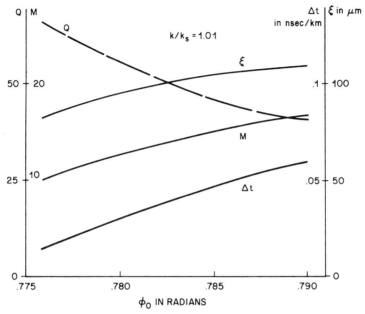

Fig. 5-17 Variation with the slab phase shift on axis ϕ_0 (or slab thickness on axis $2d_0$) of the quality factor Q, defined as the ratio of pulse broadening for an equivalent step-index fiber Δt_c to the actual pulse broadening Δt, for $n = 1.01$ and $\phi = \phi_0 - 10^{-5}y^2$. ξ denotes the maximum ray excursion, M the total number of modes, and Δt the pulse broadening. The ray period is 14 mm, and α_0 is equal to 2.6° for $\phi_0 = 0.785$, $\lambda_0 = 1$ μm. (From Arnaud.[15] Reprinted with permission from *The Bell System Technical Journal*, copyright 1974, The American Telegraph and Telephone Co.)

For square-law variations of $\beta^2(y)$, the optimum value ϕ_{opt} of ϕ on the axis of the tapered slab must be selected close to the maximum of the curve v_z/u_z versus ϕ shown in Fig. 5-8, because, near that maximum, times of flight are proportional to optical lengths. Thus we have a rule for the selection of the slab thickness on axis $2d_0 \equiv 2d(0)$. The optimum value of d_0 may be slightly different than the one given by the maximum of the curves in Fig. 5-8, because we want to minimize the variations of t in (5.68) over a finite range of $d(y)$.

For a detailed investigation, a numerical integration of the pulse equation (5.68) is necessary. Instead of specifying the slab profile $d(y)$, it is convenient to specify $\phi(y)$. Thus let us set in (5.66)

$$\phi = \phi_0 - Ky^2 \tag{5.69}$$

where K denotes a constant, and substitute the result in the ray equations (5.67a) to (5.68), where (5.66) is used.

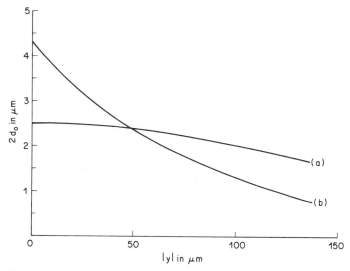

Fig. 5-18 Optimum slab profiles for $n = 1.01$. (a) Square-law profile $\phi = 0.785 - 10^{-5}y^2$. (b) Linear-law profile $\phi = 1 - 5 \times 10^{-3}|y|$. (From Arnaud.[15] Reprinted with permission from *The Bell System Technical Journal*, copyright 1974, The American Telegraph and Telephone Co.)

The variation of the average time of flight per unit length (v_g^{-1}) as a function of the angle α_0 that the ray makes with the z-axis at the origin is shown in Fig. 5-16 for ϕ_0 varying from 1.5 to 0.2 and $n = 1.45$, $\lambda_0 = 1$ μm. Large pulse broadening is observed when the slab is very thick on axis (e.g., $\phi_0 = 1.5$) or very thin (e.g., $\phi_0 = 0.2$). Optimum values are between 0.6 and 0.7. For $n = 1.01$ (refractive index of the slab 1% higher than that of the surrounding medium) and $\phi = \phi_0 - 10^{-5}y^2$, Fig. 5-17 shows that the tapered slab can be 50 times superior to the equivalent step fiber (factor Q). The profile of this fiber is shown in Fig. 5-18, the thickness on axis being equal to 2.5 μm. The optimum profile for a linear law $\phi = \phi_0 - 5 \times 10^{-3}|y|$ is shown in Fig. 5.18. For both quadratic and linear laws, a trade-off has to be made between the quality factor Q and the number M of y-modes. In conclusion, tapered homogeneous dielectric slabs can exhibit very low pulse broadening if they are properly dimensioned. If the slab material has a refractive index 1% higher than that of the surrounding medium, the thickness should be of the order of 2.5 $\pm$ 0.02 μm at a wavelength of 1 μm. In that case, the waveguide width would be of the order of 0.2 mm. Pulse broadening does not exceed 0.05 nsec/km for 15 modes. These optical waveguides can be stacked for multichannel operation. Their field structure matches fairly well that of injection lasers, which are usually multimoded in the plane of the junction. However, these guiding structures have not been experimented.

5.7 The Dielectric Rod; The Scalar Approximation

The circular dielectric rod is the simplest example of dielectric waveguide that has a finite cross section. The wavenumber of free waves in the rod material is assumed to be a constant k, and that of free waves in the surrounding medium to be another constant k_s, smaller than k. The surrounding medium is free space in the case of microwave dielectric antennas, the rod being sufficiently short and rigid to be supported from one end. In fiber optics, the dielectric rod (e.g., a fused silica fiber) is almost always surrounded by a material called the cladding. In most of this chapter, we assume that the cladding extends to infinity. We are considering modes whose field remains confined to the neighborhood of the core and ignore the cladding or radiation modes. In practical fibers, the cladding has a finite thickness and is covered with an absorbing material that attenuates the cladding modes.

The scalar approximation of wave optics has been investigated in Chapter 3. According to that approximation, each transverse component ψ of the electric field obeys approximately the scalar Helmholtz equation

$$[\nabla_t^2 + k^2(x, y)]\psi = k_z^2 \psi \tag{5.70}$$

in a uniform (z-invariant) medium. $\nabla_t^2 \equiv \partial^2/\partial x^2 + \partial^2/\partial y^2$ denotes the transverse Laplacian, and $k^2 \equiv \omega^2 \epsilon(x, y)\mu_0$. If ϵ has finite discontinuities, ψ and its first derivatives remain continuous.

For the step-index rod, the free wavenumber squared

$$k^2(r) = \begin{cases} k^2, & r < a \\ k_s^2, & r > a \end{cases} \tag{5.71}$$

is discontinuous. ψ and $\partial\psi/\partial r$, however, remain continuous at $r = a$. We shall need the Hellmann–Feyman theorem given in (3.202). For a step-index fiber free of material dispersion ($\partial k^2/\partial\omega^2 = k^2/\omega^2$), (3.202) can be written

$$\frac{dk_z^2}{d\omega^2} = \frac{\int\int \psi^2 [\partial k^2(x, y, \omega)/\partial\omega^2]\, dx\, dy}{\int\int \psi^2\, dx\, dy}$$

$$= \frac{k^2 P + k_s^2 P_s}{\omega^2 P_t} \tag{5.72a}$$

where P and P_s denote the integrals of ψ^2 over the core and cladding cross sections, respectively, and $P_t \equiv P + P_s$. It is usually easier to obtain the group velocity from (5.72a) than from a direct differentiation of k_z in (5.70). If $P_s = 0$, that is, if all the power is in the core, (5.72a) gives $u_z v_z = \bar{v}^2$, where $u_z \equiv d\omega/dk_z$, $v_z \equiv \omega/k_z$, $\bar{v} \equiv \omega/k$, a well-known relation applicable to waveguides. The relative time of flight τ, defined as the ratio of the time of flight of a pulse in the guide to that for free waves on axis, is in general, from (5.72a),

$$\tau \equiv \frac{\bar{v}}{u_z} = [1 + \delta^2(\eta - 1)]\frac{k}{k_z} \tag{5.72b}$$

where we have defined $\delta^2 \equiv 1 - k_s^2/k^2$, $\eta \equiv P/(P + P_s)$. Far above cutoff, $\eta = 1$ and $\tau = k/k_z$. If we further introduce $F^2 \equiv (k^2 - k_s^2)a^2 \equiv k^2 a^2 \delta^2$ and $u^2 \equiv (k^2 - k_z^2)a^2$, where a denotes, for the moment, an arbitrary length, we obtain for small δ (weakly guiding fibers) a normalized relative time of flight

$$\bar{\tau} \equiv \frac{\tau - 1}{\delta^2/2} = 2(\eta - 1) + \frac{u^2}{F^2} \tag{5.72c}$$

The maximum $\bar{\tau}$, within the geometrical optics limit, is obtained with $\eta \approx 1$, $u \approx F$, as $\bar{\tau} = 1$, in agreement with (4.210).

Let us now proceed to derive approximate expressions for the axial wavenumber and the normalized field. Because of the axial and circular symmetry of the fiber, solutions of (5.70) of the form

$$\psi(r, \varphi, z) = \psi(r)\exp[i(\mu\varphi + k_z z)] \tag{5.73}$$

where μ is an integer, can be found. Substituting (5.73) in the scalar Helmholtz equation (5.70), written in cylindrical coordinates, we obtain

$$\frac{d^2\psi}{dr^2} + r^{-1}\frac{d\psi}{dr} + \left(k^2 - k_z^2 - \frac{\mu^2}{r^2}\right)\psi = 0, \quad r < a$$

$$\frac{d^2\psi}{dr^2} + r^{-1}\frac{d\psi}{dr} + \left(k_s^2 - k_z^2 - \frac{\mu^2}{r^2}\right)\psi = 0, \quad r > a \tag{5.74}$$

Equations (5.74) are differential equations for Bessel functions of order μ. The bounded solutions of (5.74) are well known:

$$\psi(r) = J_\mu\left(\frac{ur}{a}\right), \quad r < a; \quad \psi(r) = AK_\mu\left(\frac{vr}{a}\right), \quad r > a \tag{5.75}$$

where we have defined

$$u^2 \equiv (k^2 - k_z^2)a^2 \tag{5.76a}$$

$$v^2 \equiv (k_z^2 - k_s^2)a^2 \tag{5.76b}$$

(v should not be confused with $\bar{v} \equiv \omega/k$) and we have

$$u^2 + v^2 = (k^2 - k_s^2)a^2 \equiv F^2 \tag{5.76c}$$

For trapped modes, u and v are real positive quantities: $k_s < k_z < k$. In (5.75), $J_\mu(\ )$ denotes the Bessel function of order μ and $K_\mu(\ )$ the modified Bessel function of the second kind of order μ. [Note the alternative notation $H_\mu^{(2)}(-ix) \equiv (2/\pi)i^{\mu+1}K_\mu(x)$.] Continuity of ψ and $d\psi/dr$ at the rod boundary $r = a$ requires that

$$\frac{uJ_\mu'(u)}{J_\mu(u)} = \frac{vK_\mu'(v)}{K_\mu(v)} \tag{5.77}$$

where primes denote differentiation with respect to the arguments. Using the identities

$$\mu J_\mu(x) \mp xJ_\mu'(x) = xJ_{\mu\pm1}(x), \qquad \mu K_\mu(x) \mp xK_\mu'(x) = \pm xK_{\mu\pm1}(x) \tag{5.78}$$

(5.74) can be written in the two following equivalent forms (upper or lower signs):

$$\frac{uJ_{\mu\pm1}(u)}{J_\mu(u)} = \frac{\pm vK_{\mu\pm1}(v)}{K_\mu(v)} \tag{5.79a}$$

$$u^2 + v^2 = F^2 \tag{5.79b}$$

Equations (5.79a) and (5.79b) define the axial wavenumber k_z for any given k, k_s, and a. Using the identities $J_{-\mu} = (-)^\mu J_\mu$ and $K_{-\mu} = K_\mu$, it can be shown that the solutions in (5.79) are the same for μ and for $-\mu$, as one expects from the rod symmetry, or from the fact that only the square of μ enters the wave equation (5.74). An alternative form is obtained by multiplying the expressions corresponding to the upper and lower signs in (5.79). We obtain a simple result first given by Biernson et al.[17] This result was recognized later to be a consequence of the usual scalar approximation.[16] We have

$$u^2\bar{J}_\mu(u) = -v^2\bar{K}_\mu(v) \tag{5.80}$$

where we have defined

$$\bar{J}_\mu(u) \equiv \frac{J_{\mu-1}(u)J_{\mu+1}(u)}{J_\mu^2(u)} \tag{5.81}$$

$$\bar{K}_\mu(v) \equiv \frac{K_{\mu-1}(v)K_{\mu+1}(v)}{K_\mu^2(v)} \tag{5.82}$$

Modal Power

Let us now evaluate the axial power in the core and in the cladding. Using a known identity for integrals of Bessel functions, the integration of the axial power density can be carried out explicitly. We obtain

$$P = \int_0^a \psi^2 2\pi r \, dr = 2\pi a^2 J_\mu^{-2}(u) \int_0^1 J_\mu^2(ux)x \, dx = \pi a^2 \left[1 - \bar{J}_\mu(u) \right] \tag{5.83}$$

if we assume that ψ is unity at the rod boundary $r = a$. Similarly, we have

$$P_s = \int_a^\infty \psi^2 2\pi r \, dr = 2\pi a^2 K_\mu^{-2}(v) \int_1^\infty K_\mu^2(vx)x \, dx = \pi a^2 \left[\bar{K}_\mu(v) - 1 \right] \tag{5.84}$$

The total axial power is obtained by adding the expressions in (5.83) and (5.84) and using (5.80):

$$P_t = \pi a^2 \frac{F^2}{u^2} \bar{K}_\mu(v) \tag{5.85}$$

The waveguide efficiency η defined earlier is, for the dielectric rod,

$$\eta = \frac{P}{P_t} = \left(1 - \frac{u^2}{F^2} \right) \left[1 - \frac{1}{\bar{J}_\mu(u)} \right] \tag{5.86}$$

Thus, once $u(F)$ has been obtained from (5.80), η is found readily from (5.86).

Finally, the normalized relative time of flight $\bar{\tau}$ is given in the absence of material dispersion by (5.72c) which we rewrite for convenience:

$$\bar{\tau} \equiv \frac{\bar{v}/u_z - 1}{\delta^2/2} = 2(\eta - 1) + \frac{u^2}{F^2} \tag{5.87}$$

The analytic results relative to P and P_s were first obtained by Snyder,[18] who also discussed the behavior of these quantities near cutoff. Equations (5.79), (5.83), (5.84), and (5.87) are our final general results.

Cutoff Condition

Let us now consider the cutoff condition $k_z = k_s$, $v = 0$. At cutoff, according (5.76c), we have $u = F$. The behavior of the $K_\mu(x)$ in the limit of small arguments ($x \to 0$) is as follows:

$$K_0(x) \approx -\log(x) \to \infty$$

$$K_{-\mu}(x) = K_\mu(x) \approx \frac{1}{2}(\mu - 1)! \left(\frac{2}{x} \right)^\mu, \qquad \mu > 0$$

$$\overline{K}_{-\mu}(x) = \overline{K}_\mu(x) = \begin{cases} (x \log x)^{-2} \to \infty, & \mu = 0 \\ -2\log(x) \to \infty, & \mu = 1 \\ \dfrac{\mu}{\mu - 1}, & \mu > 1 \end{cases} \qquad (5.88)$$

Thus, in the limit $v \to 0$, $\overline{J}_\mu(u)$ in (5.80) tends to zero for all values of μ, provided $u \ne 0$. The cutoff condition is therefore

$$J_{\mu-1}(u) = 0 \qquad (5.89)$$

with the reservation that the root $u = 0$ is acceptable only for $\mu = 0$.

The solution $\mu = 0$, $u = 0$ corresponds to the ground-state or fundamental mode ψ_{00}. This mode, in principle, can propagate at arbitrarily low frequencies. In fact, this mode is exceedingly sensitive to bending losses below $F \lesssim 0.5$.

The next mode in order of increasing cutoff frequencies is the ψ_{10} mode, whose cutoff frequency is given by the first root of $J_0(u)$, namely, $u = F = 2.405 \ldots$. Thus for single-mode operation, the normalized frequency F of a weakly guiding step-index fiber should be comprised between about 0.5 and 2.4. (Remember that only one state of polarization is considered.)

At cutoff, the axial powers P and P_s in the core and cladding are, respectively, from (5.83), (5.84), and (5.88),

$$P = \pi a^2 \qquad (5.90a)$$

$$P_s = \frac{\pi a^2}{\mu - 1}, \qquad \mu > 0 \qquad (5.90b)$$

Therefore

$$\frac{P}{P_s} = \mu - 1, \qquad \mu > 0 \qquad (5.90c)$$

$$\eta = \frac{P}{P + P_s} = 1 - \mu^{-1}, \qquad \mu > 0 \qquad (5.90d)$$

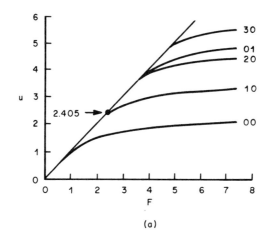

(a)

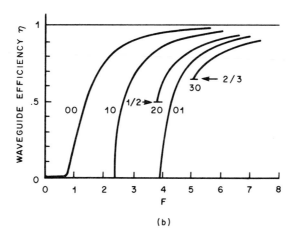

(b)

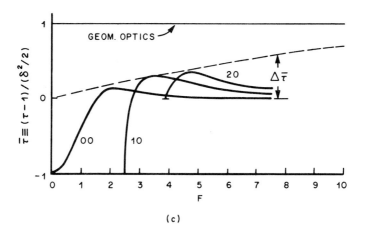

(c)

It is interesting that, for $\mu > 1$, the power in the rod at cutoff is not negligible compared with the power in the surrounding medium. As a result, the group velocity at cutoff is not equal to the velocity ω/k_s of free waves in the surrounding medium. To understand this important conclusion better, let us observe that the tangential wavenumber $\mathbf{k}_t$ at the surface of the rod has a z-component equal to k_z and the azimuthal component $k_\varphi = \mu/a$. Thus

$$k_t^2 \equiv k_z^2 + \left(\frac{\mu}{a}\right)^2 \tag{5.91}$$

At cutoff $k_z = k_s$, and the field decays approximately, just outside the rod, according to the law

$$\psi \approx \exp\left[-\left(k_t^2 - k_s^2\right)^{1/2} x\right] = \exp\left(-\frac{\mu x}{a}\right) \tag{5.92}$$

where we have used a local rectangular coordinate system with $x \equiv r - a$. If we integrate ψ^2 in (5.92) from $x = 0$ to $x = \infty$, we find that the power in the cladding is approximately $(a/2\mu)$ times the perimeter $2\pi a$ of the rod. Thus P_s is approximately equal to $\pi a^2/\mu$. This approximate result agrees with (5.90b) in the limit of large μ.

If $k_z < k_s$, the mode is below cutoff. It loses power continuously to the cladding. The attenuation may be extremely small, however, and for large normalized frequencies F, it may not be permissible to neglect the power carried by these leaky modes. We shall come back to that question later.

The normalized relative time of flight at cutoff is obtained by substituting (5.90d) and $u = F$ in (5.87). We obtain

$$\bar{\tau} \equiv \frac{\tau - 1}{\delta^2/2} = 1 - 2\mu^{-1}, \qquad \mu > 0 \tag{5.93}$$

In (5.93) we have assumed that

$$\delta^2 = 1 - \frac{k_s^2}{k^2} \ll 1 \tag{5.94}$$

Frequencies Well above Cutoff

At frequencies well above the cutoff frequency, we have $k_z \approx k_s$ and $v \approx F \to \infty$. In that limit, the modified Bessel function of the second kind

---

Fig. 5-19 Weakly guiding dielectric rod with radius a and free wavenumber k in a medium with free wavenumber k_s. We define the normalized frequency $F \equiv (k^2 - k_s^2)^{1/2}a$, $\delta^2 \equiv 1 - k_s^2/k^2 \equiv (F/ka)^2$. (a) Variation of the parameter $u \equiv (k^2 - k_z^2)^{1/2}a$. (b) Ratio $\eta \equiv P/P_t$ of the power in the core to the total power. (c) Relative times of flight $\tau \equiv$ time of flight in a mode divided by time of flight of free waves on axis. These curves follow from the curves in (a) and (b) by application of Brown's identity (see Section 3.17). (From Biernson and Kinsley,[17] by permission of the Institution of Electrical Engineers.)

Table 5.1   Range of the u Parameter^a

μ	α		
	0	1	2
0	0→2.4	3.83→5.52	7.01→ 8.65
1	2.4→3.83	5.52→7.01	8.65→10.17
2	3.83→5.13	7.01→8.41	10.17→11.61

^{a} The first number gives u at cutoff: $u_{\mu\alpha 0}$; the second gives u far from cutoff: $u_{\mu\alpha\infty}$.

is

$$K_\mu(v) \sim \left(\frac{\pi}{2v} \right)^{1/2} \exp(-v)\left(1 + \frac{4\mu^2 - 1}{8v} \right) \tag{5.95}$$

The last fractional term can be omitted in first approximation. From definition (5.82) we have therefore

$$\overline{K}_\mu(v) \approx 1 \tag{5.96}$$

Thus the rhs of (5.80) tends to infinity and

$$u^2 \overline{J}_\mu(u) \to \infty \tag{5.97}$$

From the definition (5.81) of $\overline{J}_\mu(u)$, if follows that

$$\frac{J_\mu(u)}{u} = 0 \tag{5.98}$$

In conclusion, u varies from a root of $J_{\mu-1}$, at cutoff, to the corresponding root of J_μ at high frequencies. Table 5.1 gives the range of variation of $u_{\mu\alpha}$ for each mode, for the first values of μ and α.

The variation of the parameter u as a function of the normalized frequency F is shown in Fig. 5-19a for various scalar modes. We give in Fig. 5-19b the ratio η of the power P in the core to the total power P_t at a function of F. The relative time of flight τ of various modes follows from the curves in (a) and (b) by application of (5.72c). τ is plotted as a function of F in Fig. 5-19c.

5.8 The Dielectric Rod; High-Order Modes

In this section we investigate the behavior of modes with large μ and α in oversized dielectric rods. For simplicity, the scalar approximation is maintained. In that limit, the wave optics results previously derived are related to the results in geometrical optics given in Chapter 4. As the mode numbers μ and α increase, the axial component k_z of the wave vector decreases and may even be less than k_s, the wavenumber in the cladding. We need not restrict ourselves to $k_z > k_s$, because, for large μ, the rays remain totally reflected, according to the laws of geometrical optics, even if $k_z < k_s$. We assume, however, that the angle that the rays make with the boundary remains much smaller than the critical angle defined by $k_z^2 + k_\varphi^2 = k_s^2$. In that limit, as we have seen for the case of a plane boundary, the field at the boundary tends to vanish in comparison with its value in the bulk. Thus a first approximation consists in setting the boundary condition $\psi = 0$ at $r = a$.

In the geometrical optics limit, a mode (μ, α) can be represented by a manifold of rays. These rays are straight lines in the present case because the core is homogeneous; they are repeatedly reflected at the cylindrical boundary and exhibit a helicoidal motion. For clarity, let us first consider rays in the cross section with $k_z = 0$ (no axial motion), as represented in Fig. 5-20. The system is an optical resonator with cylindrical boundary. The rays are tangent to an inner caustic line, a circle with radius a_0. The radius of the caustic is defined by the condition that the phase of the geometrical optics field is single valued. This involves two conditions. The first condition is that the phase shift along the caustic line be an integral number times 2π:

$$k2\pi a_0 = 2\pi\mu, \qquad \mu = 0,1,2 \qquad (5.99)$$

To set up the second condition, consider the closed path $ABOA$. The length of a ray from the caustic to the rod boundary AO or BO is $(a^2 - a_0^2)^{1/2}$. The arc length AB along the caustic is $2a_0$ arc $\cos(a_0/a)$. We must specify that the phase shift along the closed path $ABOA$ is an integral number times 2π. To be accurate, we must take into account also the $\pi/2$

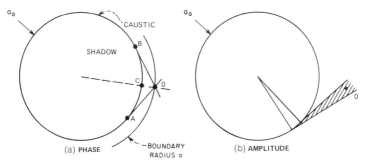

(a) PHASE   (b) AMPLITUDE

Fig. 5-20   Whispering-gallery modes guided by a cylindrical boundary between two dielectrics. In the geometrical optics (WKB) approximation, the modes can be represented by rays repeatedly reflected at the boundary and tangent to a circular caustic of radius a_0.

phase shift experienced at the caustic and the π phase shift experienced at the boundary. The origin of the $\pi/2$ phase shift was explained in Section 2.21. The π phase shift follows from the fact that ψ is required to vanish at $r = a$. We thus obtain the condition[20]

$$2k\left[(a^2 - a_0^2)^{1/2} - a_0 \text{ arc cos } \frac{a_0}{a}\right] = 2\pi(\alpha + \tfrac{3}{4}) \qquad (5.100)$$

where α denotes an integer. Introducing the value μ/k for a_0 obtained in (5.99), the eigenvalues for k (or for the frequency) follow from (5.100) and depend on the integers μ and α.

Let us now go back to our original problem. The frequency and the free wavenumber k is now assumed to be known. What we are looking for is the axial wavenumber k_z. To obtain k_z, we need only replace k^2 by $k^2 - k_z^2 \equiv (u/a)^2$ in (5.100). Thus, using the same notation as before, the equation defining k_z is, in the geometrical optics approximation,

$$(u^2 - \mu^2)^{1/2} - \mu \text{ arc cos } \frac{\mu}{u} = \pi(\alpha + \tfrac{3}{4}) \qquad (5.101)$$

To illustrate the accuracy of that approximation, the value of $u_{\mu\alpha}$ from (5.101) is compared in Table 5.2 to the corresponding root of $J_\mu(u)$. The latter is applicable as we have seen earlier, at frequencies much higher than the cutoff frequency. We shall omit the derivation of the geometrical optics field. The result is[20]

$$\psi(r) = \begin{cases} \left(\dfrac{2}{\pi}\right)^{1/2}\left[\left(\dfrac{ur}{a}\right)^2 - \mu^2\right]^{-1/4} \cos\left\{\left[\left(\dfrac{ur}{a}\right)^2 - \mu^2\right]^{1/2} \right. \\ \qquad \left. - \mu \text{ arc cos}\left(\dfrac{\mu a}{ur}\right) - \dfrac{\pi}{4}\right\}, \qquad a_0 < r < a \\[2em] \dfrac{1}{2}\left(\dfrac{2}{\pi}\right)^{1/2}\left[\mu^2 - \left(\dfrac{ur}{a}\right)^2\right]^{-1/4} \\ \qquad \times \exp\left\{-\mu \text{ arg cosh}\left(\dfrac{\mu a}{ur}\right)\right. \\ \qquad \left. + \left[\mu^2 - \left(\dfrac{ur}{a}\right)^2\right]^{1/2}\right\}, \qquad r < a_0 \end{cases} \qquad (5.102)$$

where the factor $\exp[i(\mu\varphi + k_z z)]$ has been omitted. Results (5.101) and (5.102) can also be obtained from the Debye expansion of $J_\mu(ur/a)$.

Table 5.2   **Comparison between** $u_{\mu\alpha}$ **from Geometrical Optics (5.101) (First Number) and the Roots of** $J_\mu(u)$ **(Second Number)**

		α	
μ	0	1	2
0	2.356	5.498	8.639
	2.405	5.520	8.654
1	3.795	6.997	10.161
	3.832	7.016	10.173
2	5.101	8.401	11.609
	5.136	8.417	11.620

Equation (5.101) can be solved explicitly when either $\mu \ll u$ or $\mu \approx u$. The first case corresponds to modes with small azimuthal wavenumber (including the fundamental mode) in oversized rods. We have, for that case,

$$u_{\mu\alpha} \approx \frac{\pi}{2}\,(2\alpha + \mu + \tfrac{3}{2}), \qquad \mu \ll u \tag{5.103}$$

In particular, for the ground state (fundamental mode) $\mu = \alpha = 0$, (5.103) gives the first number in Table 5.2: $u_{00} = 3\pi/4 \approx 2.356$.

The case $\mu = u$ is of much greater interest. It corresponds to the so-called whispering-gallery modes that cling to the boundary. The caustic radius a_0 is, in that case, very close to the boundary radius a. In the limit $\mu \approx u$, (5.101) becomes

$$u_{\mu\alpha} \approx \mu + b_\alpha \mu^{1/3} \tag{5.104a}$$

where

$$b_\alpha = \tfrac{1}{2}\big[3\pi(\alpha + \tfrac{3}{4})\big]^{2/3} \tag{5.104b}$$

Equation (5.104) is an approximate form for the zeros of $J_\mu(u)$ for $\mu \approx u$. The numerical value of b_α is given in Table 5.3 in Section 5.9, where it is compared to the roots of the Airy function.

The distance $a - a_0$ between the boundary and the caustic, which is a measure of the thickness of the mode in the radial direction, follows from (5.104a). We have

$$u_{\mu\alpha} = \big(k^2 - k_z^2\big)^{1/2} a = \mu + b_\alpha \mu^{1/3} \tag{5.105}$$

and $(k^2 - k_z^2)^{1/2} a_0 = \mu$, from (5.99) with k changed to $(k^2 - k_z^2)^{1/2}$. Thus the product of k and the mode thickness is

$$k(a - a_0) = b_\alpha \left(\frac{ka_0}{1 - k_z^2/k^2}\right)^{1/3} \equiv b_\alpha (k\rho)^{1/3} \tag{5.106}$$

The quantity

$$\rho \equiv \frac{a_0}{1 - k_z^2/k^2} \equiv \frac{a_0}{\sin^2 \theta} \qquad (5.107)$$

represents the radius of curvature of the average helical path followed by the rays. This path makes an angle $\theta = \arccos(k_z/k)$ with the z-axis. When $\theta = \pi/2$ (or $k_z = 0$), we have $\rho = a_0 \approx a$, but usually $\theta \approx 0$ and $\rho \gg a$. The mode thickness at any wavelength depends only on the helix radius of curvature ρ.

5.9 The Dielectric Rod; Whispering-Gallery Modes

We have seen in Chapter 1 that a mild curvature of the guide axis is equivalent to a constant transverse gradient of refractive index. The solution of the scalar parabolic wave equation for such a medium with constant transverse gradient is described by Airy functions. Thus the solutions of the wave equation applicable to waves clinging to curved boundaries can be represented by Airy functions. (See Wait[32] and Ref. 59.)

The Airy function solution can be derived from the exact Bessel function given earlier. When the argument of a Bessel function is close to its order μ, the following approximation can be used:

$$J_\mu(\mu + z\mu^{1/3}) \approx \left(\frac{2}{\mu} \right)^{1/3} \mathrm{Ai}(-2^{1/3}z) \qquad (5.108)$$

where $\mathrm{Ai}(\ )$ denotes the Airy function. Let the exact field in (5.75) be written with the help of (5.108) as

$$\psi(r) = J_\mu\left(\frac{ur}{a} \right) \approx \left(\frac{2}{\mu} \right)^{1/3} \mathrm{Ai}\left[-2^{1/3}\mu^{-1/3}\left(\frac{ur}{a} - \mu \right) \right] \qquad (5.109)$$

The requirement that $\psi(a) = 0$ is therefore

$$u = \mu - \frac{a_\alpha}{2^{1/3}}\mu^{1/3} \qquad (5.110)$$

where a_α denotes the αth zero of $\mathrm{Ai}(\ )$, a negative quantity. Thus the quantity b_α obtained in (5.104) from geometrical optics can be compared

Table 5.3   Values of the b-Parameter

α	b_α in (5.104b)	$-a_\alpha/2^{1/3}$, $\mathrm{Ai}(a_\alpha) = 0$
0	1.841	1.855
1	3.239	3.244
2	4.379	4.381

with the numbers $-a_\alpha/2^{1/3}$ obtained from wave optics. This comparison is made in Table 5.3.

Now let u from (5.110) be substituted in (5.109), and let us set

$$x \equiv r - a_0', \qquad a_0' = a + a\,\frac{a_\alpha}{2^{1/3}}\,\mu^{-2/3} \tag{5.111}$$

Table 5.3 shows that a_0' is almost equal to the caustic radius a_0. With that notation, the field in (5.109) is, if $\mu \gg 1$,

$$\psi(x) \approx (2/\mu)^{1/3}\,\mathrm{Ai}\left[-(2\mu^2)^{1/3}\,\frac{x}{a}\right] \tag{5.112}$$

When the argument of the Airy function is large compared with unity, the Airy function has the form

$$\mathrm{Ai}(z) \sim \begin{cases} \pi^{-1/2}(-z)^{-1/4}\sin\left[\tfrac{2}{3}(-z)^{3/2} + \tfrac{1}{4}\pi\right], & z \ll 0 \\[2mm] \tfrac{1}{2}\pi^{-1/2}z^{-1/4}\exp\left(-\tfrac{2}{3}z^{3/2}\right), & z \gg 0 \end{cases} \tag{5.113}$$

Thus, for large $|x|$, the field in (5.112) has the geometrical optics form

$$\psi(x) \sim \begin{cases} \left(\dfrac{a}{x}\right)^{1/4}\sin\left[\left(\dfrac{2^{3/2}}{3}\right)\mu\left(\dfrac{x}{a}\right)^{3/2} + \dfrac{\pi}{4}\right], & x \gg \dfrac{a}{\mu^{2/3}} \\[4mm] \dfrac{1}{2}\left(-\dfrac{a}{x}\right)^{1/4}\exp\left[-\left(\dfrac{2^{3/2}}{3}\right)\mu\left(-\dfrac{x}{a}\right)^{3/2}\right], & x \ll -\dfrac{a}{\mu^{2/3}} \end{cases} \tag{5.114}$$

This form exhibits the oscillatory behavior of the field at radii larger than the caustic radius and the exponential decay of the field inside the caustic.

Finally, let us give an expression that will be useful in a subsequent section. We have[21]

$$P_\alpha \equiv \int_{a_\alpha}^{\infty} \mathrm{Ai}^2(z)\,dz = \left[\mathrm{Ai}'(a_\alpha)\right]^2 \tag{5.115}$$

Table 5.4   Value of the Power Parameter P_α

α	$P_\alpha = [\text{Ai}'(a_\alpha)]^2$	$(-a_\alpha)^{1/2}/\pi$	$-a_\alpha$
0	0.491	0.487	2.338
1	0.645	0.643	4.087
2	0.748	0.748	5.520
3	0.828	0.829	6.786

where a_α denotes as before the αth zero of Ai( ), and Ai'( ) the derivative of Ai( ) with respect to the argument. Result (5.115) can be obtained from an integration by parts and use of the differential equation $A_i''(z) - z\,\text{Ai}(z) = 0$, which defines the Airy function. For large α we have approximately

$$P_\alpha \approx (-a_\alpha)^{1/2}/\pi \tag{5.116}$$

The exact value of P_α in (5.115) and its approximation in (5.116) are compared in Table 5.4.

5.10 The Dielectric Rod; Exact Solutions

Exact analytical solutions of the Maxwell equations can be obtained for the modes of propagation guided by homogeneous dielectric rods with circular cross section.[22-25] The equation that defines k_z, however, is more complicated than the one given in (5.79) which was based on the scalar approximation. To each scalar mode $\psi_{\mu\alpha}$ defined earlier, we need to associate two electromagnetic modes, corresponding to two different states of polarization. In many practical cases, the two electromagnetic modes are almost degenerate; that is, they have almost exactly the same k_z at a given frequency.

The ground state, or fundamental mode, of the rod, denoted earlier ψ_{00}, is similar to the HE_{11} dipole mode. For that mode, the two polarization states have exactly the same k_z. The dipole mode is shown in Fig. 5-21a. The field is almost uniform within the rod. It resembles the electric field in a dielectric rod located between two plane-parallel electrodes at different dc potentials, as shown in Fig. 5-21b. For dc fields, the electric field is exactly a constant inside the dielectric rod. This is only approximately true at the high frequencies considered. The notation HE is meant to suggest that the axial component of the H-field is larger than the axial component of the E-field. It should be noted, however, that the axial **E**- and **H**-field components are usually both very small compared with the transverse

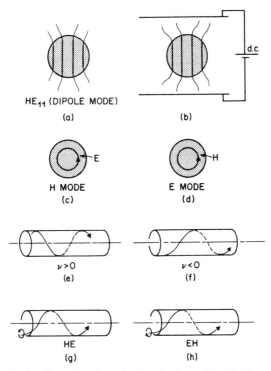

HE$_{11}$ (DIPOLE MODE)

(a)

(b)

H MODE
(c)

E MODE
(d)

$\nu > 0$
(e)

$\nu < 0$
(f)

HE
(g)

EH
(h)

Fig. 5-21   Modes in the dielectric rod. (a) Dipole mode. (b) Electrostatic field. (c) *H*-modes. (d) *E*-modes. (e) $\nu > 0$ modes. (f) $\nu < 0$ modes. (g) *HE*-modes. (h) *EH*-modes.

components. The first subscript 1 in HE_{11} refers to the azimuthal variation of the (say, radial) component of the field. For a rotation of π, this component clearly changes sign according to Fig. 5-21a; that is, the field points toward the rod center at $\varphi = \pi$ if it points away from the center at $\varphi = 0$. The absolute direction of the field, therefore, is the same at $\varphi = 0$ and $\varphi = \pi$. The second subscript 1 in HE_{11} refers to the radial variation of the field., In electromagnetism, it is customary to begin numbering the radial waves at 1, rather than 0 as in scalar optics (or quantum mechanics). Although this is somewhat unfortunate, we shall follow the tradition in that respect.

A second group of modes of particular interest is the $H_{0\alpha}$-, $E_{0\alpha}$-mode group shown in Fig. 5-21c, d, respectively. These modes exhibit no azimuthal variation of the field. In the limit of large rod radii, or short wavelengths, the $H_{0\alpha}$-modes resemble the corresponding modes of the circular metallic pipe (the low-loss mode used in microwave communications) with the azimuthal electric field vanishing at the boundary.

To understand the basic features of modes with azimuthal number ν, $|\nu| > 1$, it is useful to consider first the ray picture in Fig. 5-21e, f, which shows helical paths that describe the so-called whispering-gallery modes with either positive ($\nu > 0$) or negative ($\nu < 0$) helicities, which we discussed in the previous section. By symmetry, it is clear that modes with opposite helicity have the same axial wavenumber. This ray picture is applicable to electromagnetic (EM) waves as well as to scalar waves. For the electromagnetic field, however, the rays can have two states of polarization. If the boundary is metallic, the polarization rotates along the helical path, with respect to the normal to the boundary. The rate of rotation is equal to the torsion of the helical path, as we have seen in Section 4.23. Because the state of polarization must be invariant, the state of polarization can only be circular, clockwise, or counterclockwise. This result is applicable to helical whispering-gallery modes guided by metallic boundaries. It approximately applies also to large dielectric rods. For *HE*-waves, the sense of rotation of the polarization is the same as that of the wave, as shown in Fig. 5-21g. For EH-waves, the sense of rotation of the polarization is opposite to that of the wave, as shown in Fig. 5-21h. This observation provides a more important distinction between *HE*- and *EH*-waves than the one given earlier, based on the strength of the axial *E*- and *H*-components. As indicated earlier, the first subscript of the *HE*- and *EH*-waves refers to azimuthal variations, and the second subscript to radial variations. A detailed picture of the field configuration for the lowest-order modes can be found in Snitzer.[23]

Let us now discuss the mathematical formalism. An interesting approach is that based on circularly polarized field components.[24] This approach, however, would require additional introductory material. We shall therefore follow a more conventional method.

Because of the circular symmetry of the rod, solutions of the form

$$\begin{bmatrix} E_z \\ H_z \end{bmatrix} = \begin{bmatrix} E_z(r) \\ H_z(r) \end{bmatrix} \exp[i(k_z z + \nu\varphi)] \qquad (5.117)$$

can be found. The z-components of **E** and **H** obey the scalar Helmholtz equation. In cylindrical coordinates, this wave equation is, for fields of the form (5.117), as in (5.74),

$$\left(\frac{d^2}{dr^2} + r^{-1} \frac{d}{dr} + k^2 - k_z^2 - \frac{\nu^2}{r^2} \right) \begin{bmatrix} E_z(r) \\ H_z(r) \end{bmatrix} = 0 \qquad (5.118)$$

Equation (5.118) is the equation that defines the Bessel functions. The solutions are therefore

$$\begin{bmatrix} E_z(r) \\ H_z(r) \end{bmatrix} = \begin{bmatrix} E \\ H \end{bmatrix} \times \begin{cases} J_\nu\left(\dfrac{ur}{a}\right), & r \leqslant a \\[2mm] K_\nu\left(\dfrac{vr}{a}\right), & r \geqslant a \end{cases} \tag{5.119}$$

where we have defined, as before,

$$u^2 \equiv (k^2 - k_z^2)a^2 \tag{5.120a}$$

$$v^2 \equiv (k_z^2 - k_s^2)a^2 \tag{5.120b}$$

$$u^2 + v^2 = (k^2 - k_s^2)a^2 \equiv F^2 \tag{5.120c}$$

a denotes the rod radius, and K_ν the modified Bessel function of the second kind. [An alternative notation is $H_\nu^{(2)}(-ix) \equiv (2/\pi)i^{\nu+1}K_\nu(x)$.]

The transverse form of the Maxwell equations in (3.171) gives, for $r < a$,

$$i\omega\epsilon \frac{\partial E_z}{\partial r} = k_z\nu r^{-1}H_z + \frac{u^2}{a^2}H_\varphi$$

$$i\omega\mu \frac{\partial H_z}{\partial r} = -k_z\nu r^{-1}E_z - \frac{u^2}{a^2}E_\varphi \tag{5.121}$$

or, equivalently,

$$\begin{bmatrix} -E_\varphi \\ H_\varphi \end{bmatrix} = \left(\frac{a}{u}\right)^2 \begin{bmatrix} k_z\nu r^{-1} & i\omega\mu\,\partial/\partial r \\ i\omega\epsilon\,\partial/\partial r & -k_z\nu r^{-1} \end{bmatrix} \begin{bmatrix} E_z \\ H_z \end{bmatrix} \tag{5.122a}$$

Similarly, for $r > a$, we have

$$\begin{bmatrix} -E_\varphi \\ H_\varphi \end{bmatrix} = -\left(\frac{a}{v}\right)^2 \begin{bmatrix} k_z\nu r^{-1} & i\omega\mu_s\,\partial/\partial r \\ i\omega\epsilon_s\,\partial/\partial r & -k_z\nu r^{-1} \end{bmatrix} \begin{bmatrix} E_z \\ H_z \end{bmatrix} \tag{5.122b}$$

Let us substitute (5.119) in (5.122) and set $r = a$. We obtain

$$\begin{bmatrix} -E_\varphi \\ H_\varphi \end{bmatrix}_a = a J_\nu(u) \begin{bmatrix} k_z\nu u^{-2} & i\omega\mu J \\ i\omega\epsilon J & -k_z\nu u^{-2} \end{bmatrix} \begin{bmatrix} E_z \\ H_z \end{bmatrix}_a \tag{5.123a}$$

$$\begin{bmatrix} -E_\varphi \\ H_\varphi \end{bmatrix}_a = -a K_\nu(v) \begin{bmatrix} k_z\nu v^{-2} & i\omega\mu_s K \\ i\omega\epsilon_s K & -k_z\nu v^{-2} \end{bmatrix} \begin{bmatrix} E_z \\ H_z \end{bmatrix}_a \tag{5.123b}$$

where we have defined

$$J \equiv \frac{J_\nu'(u)}{u J_\nu(u)} \tag{5.124a}$$

$$K \equiv \frac{K_\nu'(v)}{v K_\nu(v)} \tag{5.124b}$$

and primes denote derivatives with respect to the argument. [An alternative notation is $K = -H$, with H similarly defined from $H_\nu^{(2)}(-iv)$.]

Because E_z, H_z, E_φ, H_φ are continuous at the core–cladding interface, the system of equations (5.123) has nontrivial solutions only if the determinant of the sum of the two matrices is equal to zero. Thus

$$\det \begin{bmatrix} k_z \nu (u^{-2} + v^{-2}) & i\omega(\mu J + \mu_s K) \\ i\omega(\epsilon J + \epsilon_s K) & -k_z \nu (u^{-2} + v^{-2}) \end{bmatrix} = 0 \tag{5.125}$$

or, expanding the determinant,

$$k_z^2 \nu^2 (u^{-2} + v^{-2})^2 = \omega^2 (\epsilon J + \epsilon_s K)(\mu J + \mu_s K) \tag{5.126a}$$

$$u^2 + v^2 = F^2 \tag{5.126b}$$

Equations (5.126) define, for any ν, two sets of modes. These two sets correspond to two different states of polarization, the HE_ν- and the EH_ν-modes. A few transformations of (5.126) are useful in applications. They are now given.

Using the definition of u and v given before, we have

$$k_z^2 \nu u^{-2} = k^2 \nu u^{-2} - \nu a^{-2} \tag{5.127a}$$

$$k_z^2 \nu v^{-2} = k_s^2 \nu v^{-2} + \nu a^{-2} \tag{5.127b}$$

Thus, by addition,

$$k_z^2 \nu (u^{-2} + v^{-2}) = \omega^2 (\epsilon \mu \nu u^{-2} + \epsilon_s \mu_s \nu v^{-2}) \tag{5.127c}$$

The dispersion equations (5.126) can therefore be written in the form

$$\frac{\nu u^{-2} + \nu v^{-2}}{\mu J + \mu_s K} = \frac{\epsilon J + \epsilon_s K}{\epsilon \mu \nu u^{-2} + \epsilon_s \mu_s \nu v^{-2}} \equiv \chi \tag{5.128}$$

If we use the identities

$$\nu J_\nu(x) \mp x J_\nu'(x) = x J_{\nu \pm 1}(x) \tag{5.129a}$$

$$\nu K_\nu(x) \mp x K_\nu'(x) = \mp x K_{\nu \pm 1}(x) \tag{5.129b}$$

given earlier in (5.78), we find that

$$\nu u^{-2} = \tfrac{1}{2}(J_\nu^- + J_\nu^+) \tag{5.130a}$$

$$J = \tfrac{1}{2}(J_\nu^- - J_\nu^+) \tag{5.130b}$$

$$\nu v^{-2} = -\tfrac{1}{2}(K_\nu^- - K_\nu^+) \tag{5.130c}$$

$$K = -\tfrac{1}{2}(K_\nu^- + K_\nu^+) \tag{5.130d}$$

where we have defined

$$J_\nu^\pm \equiv \frac{J_{\nu\pm1}(u)}{u\,J_\nu(u)} \tag{5.131a}$$

$$K_\nu^\pm \equiv \frac{K_{\nu\pm1}(v)}{v\,K_\nu(v)} \tag{5.131b}$$

Note that $K_{-\nu}^- = K_\nu^+$. [An alternative notation is $H^+ \equiv -K^+$, $H^- = K^-$ with $H^\pm$ similarly defined from $H_\nu^{(2)}(-iv)$.]

If we substitute (5.130) in (5.128) and assume for simplicity that $\mu = \mu_s$, we obtain

$$\frac{\mu_s - \chi}{\mu_s + \chi} = -\frac{J_\nu^+ + K_\nu^+}{J_\nu^- - K_\nu^-} = \frac{\epsilon J_\nu^+ + \epsilon_s K_\nu^+}{\epsilon J_\nu^- - \epsilon_s K_\nu^-} \equiv \gamma \tag{5.132}$$

The parameter γ, incidentally, characterizes the state of polarization of the wave. This is the ratio of the two circularly polarized components of the field at the radius where the intensity is maximum.[24] For numerical values of the waveguide efficiency see Biernson and Kinsley.[17]

Cutoff Condition

At cutoff, $k_z = k_s$ and $v = 0$. Thus $u = F$. For small arguments $x \to 0$, we have, as in (5.88),

$$K_0(x) \approx -\log(x) \to \infty, \quad K_\nu(x) \approx \frac{1}{2}(\nu - 1)!\left(\frac{2}{x}\right)^\nu, \quad \nu > 0 \tag{5.133}$$

Thus, in the limit where $v \to 0$, we have

$$K_{-\nu}^+ = K_\nu^- \to \begin{cases} \infty, & \nu = 0 \\ [2(\nu - 1)]^{-1}, & \nu > 0 \end{cases} \tag{5.134}$$

and $K \to \infty$. Substituting (5.134) in (5.132), we obtain the cutoff conditions. We shall omit the details and give only the results. Because of symmetry, the solutions are the same for ν and $-\nu$. We have two sets of solutions. The first set is

$$J_\nu(u) = 0 \tag{5.135}$$

The solution $u = 0$ is acceptable only for $\nu = 1$. This set of solutions is identical to that obtained in (5.89) for scalar waves, with the relabeling $\mu = \nu + 1$. The second set of solutions can be written

$$\frac{(\nu - 1)^{-1} u J_\nu(u)}{J_{\nu-1}(u)} - 1 = \frac{\epsilon}{\epsilon_s}, \qquad \nu > 1 \tag{5.136a}$$

or equivalently, using (5.78),

$$\frac{J_{\nu-2}(u)}{J_\nu(u)} = \frac{\epsilon_s - \epsilon}{\epsilon_s + \epsilon} \tag{5.136b}$$

For weakly guiding fibers, the approximation $\epsilon/\epsilon_s \approx 1$ can be made, in which case we have $J_{\nu-2}(u) = 0$. This solution is the same as that given for scalar waves with the relabeling $\mu = \nu - 1$.

Frequencies Well above Cutoff

In the limit $a \to \infty$, for a given mode number, v tends to infinity. The wave propagates almost along the z-axis in the rod and is incident on the boundary at grazing angles. The second terms in the parentheses on the rhs of (5.126a) vanish because $K \to 0$ when $v \to \infty$. On the lhs of (5.126a), k_z can be replaced by $k = \omega(\epsilon\mu)^{1/2}$. Thus it is apparent that (5.126a) becomes, in the limit $a \to \infty$, using the definition (5.124a),

$$\frac{J_\nu'(u)}{J_\nu(u)} = \mp \frac{\nu}{u} \tag{5.137}$$

Equivalently, using the identity (5.129a), (5.137) is

$$J_{\nu \mp 1}(u) = 0 \tag{5.138}$$

For symmetry reasons, modes with opposite values of ν have the same axial wavenumbers. For $\nu = 2$, for example, the wavenumbers of the two sets of modes are given from (5.138) by the roots of J_1 and J_3. For $\nu = -2$, they are given by the roots of J_{-3} and J_{-1}. However, these are the same because $J_{-\nu} = (-)^\nu J_\nu$. For the fundamental HE_{11}-mode ($\nu = 1$)

and the H_{01}-modes ($\nu = 0$), the solutions are the roots of

$$J_0(u) = 0, \qquad u_{00} = 2.4 \ldots \qquad (5.139a)$$

and

$$J_1(u) = 0, \qquad u_{10} = 3.8 \ldots \qquad (5.139b)$$

respectively.

The modes H_{01}, E_{01}, and HE_{21} have almost the same wavenumber in that limit: $a \to \infty$. It is interesting to evaluate the actual splitting. The exact wavenumbers of the H_{01}- and E_{01}-modes are, setting $\nu = 0$ in (5.126a), $\mu_s = \mu$, and $n^2 = \epsilon/\epsilon_s$,

$$\frac{J_1(u)}{uJ_0(u)} + \frac{K_1(v)}{vK_0(v)} = 0, \qquad (H_0) \qquad (5.140a)$$

and

$$\frac{J_1(u)}{uJ_0(u)} + n^2 \frac{K_1(v)}{vK_0(v)} = 0, \qquad (E_0) \qquad (5.140b)$$

respectively. Setting

$$u = u_0 + \delta, \qquad \delta \ll 1 \qquad (5.141)$$

where u_0 is the root of (5.139) in (5.140a) and (5.140b), and $v = (k^2 - k_s^2)^{1/2}a$ on the rhs, the difference in wavenumber is found to be

$$\Delta k_z \equiv k_{zH} - k_{zE} = a^{-1}\left(\frac{3.8}{ka}\right)^2\left(1 - \frac{1}{n^2}\right)^{1/2} \qquad (5.142)$$

This expression has the same form as for a slab with $a \approx d$. If $a = 10\ \mu\text{m}$, $n = 1.41$, and $\lambda = 1\ \mu\text{m}$, the beat wavelength $2\pi/\Delta k_z$ is equal to 50 mm, a surprisingly large length.

The correspondence between electromagnetic modes ν and scalar modes μ is shown in Table 5.5. For instance, the ψ_0 scalar modes correspond to

**Table 5.5 Comparison between Scalar Modes μ
and Electromagnetic Modes ν at High Frequencies**[a]

		$\nu = -2$	$\nu = -1$	$\nu = 0$	$\nu = 1$	$\nu = 2$		
$\nu < 0$	EH	$J_{-3} = 0$	$J_{-2} = 0$	$J_{-1} = 0$	$J_0 = 0$	$J_1 = 0$	HE	$\nu > 0$
		$\mu = -2$	$\mu = -1$	$\mu = 0$	$\mu = 1$	$\mu = 2$		
	HE	$J_{-1} = 0$	$J_0 = 0$	$J_1 = 0$	$J_2 = 0$	$J_3 = 0$	EH	

[a] Note $J_{-\nu} = 0 \Leftrightarrow J_\nu = 0$.

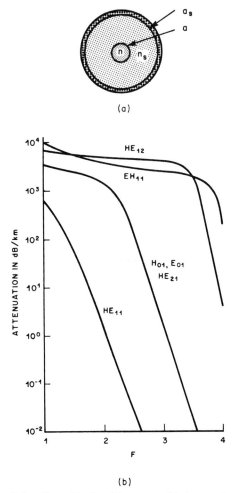

(a)

(b)

Fig. 5-22 (a) Step-index fiber with absorbing cover. (b) Attenuation for the fundamental HE_{11}-mode and higher-order modes. (From Clarricoats and Chan[25] by permission of the Institution of Electrical Engineers.)

the HE_1 electromagnetic modes. The physical significance of the scalar approximation is that, if, for instance, a circularly polarized field is launched into the fiber, this field configuration is approximately maintained over a certain length. Eventually, however, the polarization is transformed because the two electromagnetic modes have slightly different axial wavenumbers.

In the previous discussion, the cladding was assumed to extend to infinity. Figure 5-22 shows the effect of a finite cladding covered with an absorbing material, on the loss of the fundamental mode and higher-order modes.[25]

This completes the analyses of the two most important piecewise homogeneous dielectric guides, the slab and the rod. Dielectric guides with rectangular, elliptic, and tubular cross sections, and uniaxial circular rods, are treated by Kapany and Burke.[24] A review of shielded dielectric guides is made by Schlosser and Unger.[23] We shall now consider in some detail various combinations of slabs and rods, and the effects of coupling and bending.

5.11 Coupling between Trapped Modes; Cross Talk

Transverse coupling between optical waveguides is a useful effect in integrated optics. The fabrication of optical directional couplers is based on the transverse electromagnetic coupling. On the contrary, it is desirable to minimize the transverse coupling in a bundle of fibers because this coupling causes cross talk. The coupling mechanism is the same in both cases. However, the numerical values of the parameters are quite different. They require separate investigations. Although most fibers have circular symmetry, we shall first study slabs because of the greater mathematical simplicity.

For H-waves guided by symmetrical stratified identical guides, the coupling formulas (3.80)–(3.82) reduce to the simple form[26]

$$c = \frac{(k_{xsi}/k_z)E^2}{\displaystyle\int_{-\infty}^{+\infty} E^2 \, dx} \tag{5.143a}$$

where E denotes the electric field of the isolated guides, evaluated anywhere between the two guides, k_z denotes the axial wavenumber of the isolated guides, and $k_{xsi}^2 = k_z^2 - k_s^2$, where k_s denotes the wavenumber of free waves in the medium separating the guides (cladding) at the location where E is evaluated. An equivalent expression for c is

$$c = \frac{k_{xsi}}{2\omega\mu_0} E_n^2 \tag{5.143b}$$

where E_n is the normalized field $E_n \equiv E/P^{1/2}$

$$P = \frac{k_z}{2\omega\mu_0} \int_{-\infty}^{+\infty} E^2 \, dx \tag{5.143c}$$

When the fibers are identical, the axial wavenumbers of the normal modes of the coupled fibers (symmetrical and antisymmetrical fields) are simply $k_z + c$ and $k_z - c$, respectively.

Reactive Surface

Let us consider first two single-sided reactive surfaces with normalized susceptance s and spacing D. The boundary condition at a reactive surface is

$$\frac{dE}{dx} + sE = 0 \qquad (5.144)$$

The condition (5.33) given earlier for a thin slab with material free wavenumber k and thickness $2d$ is a special case of (5.144), with $s \equiv k^2 d$. In the absence of coupling ($D = \infty$), the field decays above the structure according to

$$E = \exp(-sx) \qquad (5.145a)$$

If we substitute this result in (5.143), we obtain, after a straightforward integration of $E^2(x)$ from $x = 0$ to $x = \infty$ and noting that $k_{xsi} = s$,

$$c = \frac{2s^2}{k_z} \exp(-2sD) \qquad (5.145b)$$

It is interesting to compare this result to the exact result. Using the boundary condition (5.144) at the two reactive surfaces, the axial wavenumbers of the symmetrical (S) and antisymmetrical (AS) modes can be obtained. The exact dispersion equations are found to be, respectively,

$$\tanh(k_{xsi}D) = \frac{s}{k_{xsi}} \qquad \text{(symmetrical)} \qquad (5.146a)$$

$$\tanh(k_{xsi}D) = \frac{k_{xsi}}{s} \qquad \text{(antisymmetrical)} \qquad (5.146b)$$

Because $\tanh(x) \approx 1 - 2\exp(-2x)$ for large x, (5.146a) and (5.146b) are, for large $k_{xsi}D$,

$$k_{zS}^2 - k_s^2 = s^2[1 + 4\exp(-2sD)] \qquad (5.147a)$$

$$k_{zAS}^2 - k_s^2 = s^2[1 - 4\exp(-2sD)] \qquad (5.147b)$$

respectively. Thus

$$c \equiv \tfrac{1}{2}(k_{zS} - k_{zAS}) \approx \frac{2s^2}{k_z} \exp(-2sD) \qquad (5.148)$$

in agreement with (5.145b). We have used the approximation $k_{zS} \approx k_{zAS} \equiv k_z$.

Slabs

Let us now consider the case of two identical dielectric slabs, not necessarily very thin. The slab material free wavenumber is denoted k, and the slab thickness $2d$. The square of the normalized field at the slab boundary follows from (5.47):

$$E_n^2 = \frac{2\omega\mu_0}{k_z d}(1 - b)\left(1 + \frac{1}{\phi_i}\right)^{-1} \tag{5.149}$$

Thus, if we substitute (5.149) in (5.143b), the coupling is found to be[26,28]

$$c = \frac{1}{k_z d^2}\phi_i^2(1 - b)(1 + \phi_i)^{-1}\exp\left(-\frac{2\phi_i D}{d}\right) \tag{5.150}$$

We have defined as before $\phi_i = k_{xsi}d = b^{1/2}F$. The normalized phase velocity b is a function of the normalized frequency F plotted in Fig. 5-6. If we select the largest slab thickness that allows the propagation of a single H-mode, we have $F = \pi/2$, and from Fig. 5-6, $b = 0.67$. Thus $\phi_i = 1.28$. We make the approximation $k_z \approx k$ in (5.150). The coupling is

$$c = \frac{0.24}{kd^2}\exp\left(-\frac{2.5 D}{d}\right) \tag{5.151}$$

Thus, for a constant relative spacing D/d, the coupling between two single-mode slabs with constant normalized frequency varies as the inverse of the square of the thickness.

Let us make the following numerical example:

$$2d = 1.32 \ \mu\text{m}, \qquad k = 2\pi \times 1.45 \ \mu\text{m}^{-1}, \qquad k_s = 2\pi \times 1.4 \ \mu\text{m}^{-1} \tag{5.152}$$

That is, the refractive index of the slab and that of the cladding are $n = 1.45$ and $n_s = 1.4$, respectively, if $\lambda_0 = 1 \ \mu\text{m}$. Substituting these values in (5.151), we find

$$c \ \text{m}^{-1} = 6 \times 10^4 \exp(-3.88 D_{\mu\text{m}}) \tag{5.153}$$

It follows from the results in Section 3.16 that the cross talk between two 1-km-long identical fibers is equal to -20 dB when $c = 10^{-4} \ \text{m}^{-1}$. According to (5.153) this coupling corresponds to a spacing $2D = 11 \ \mu\text{m}$ between the two slabs. If the fibers have slow sinusoidal irregularities with $\Delta k_z/k_z = 10^{-4}$, the -20-dB cross talk is obtained for a much larger coupling $c = 0.25 \ \text{m}^{-1}$. According to (5.153), this larger coupling corresponds to a smaller spacing $2D = 6.2 \ \mu\text{m}$. Other results concerning the coupling between fibers can be found in Marcatili.[27]

5.12 Reduction of Cross Talk between Dielectric Slabs

We investigate here whether the coupling between two dielectric slabs can be reduced by introducing a layer of metal, such as silver, without introducing excessive losses at the same time.[28] The cross-talk power, as we have seen in (3.184), is proportional to the square of the coupling c. The coupling c, in turn, is proportional to the square of the field of one slab at a point located halfway between the slabs. Thus the cross-talk power is proportional to the fourth power of that field.

We shall evaluate first the reduction in field introduced by a metallic layer under evanescent wave excitation (axial wavenumber k_z, transverse wavenumber ik_{xsi}).

Let the metallic layer free wavenumber be denoted $k' \equiv k'_r + ik'_i$, and its thickness be denoted d' (see Fig. 5-23). By writing the boundary conditions, we obtain the transmission of the layer, defined as the field in the presence of the metallic layer divided by the field in the absence of the metallic layer. The result is

$$t = 2[(1 + \bar{\kappa})e^+ + (1 - \bar{\kappa})e^-]^{-1} \exp(\kappa d') \qquad (5.154)$$

where we have defined

$$\kappa \equiv k_{xsi} = \left(k_z^2 - k_s^2\right)^{1/2} \qquad (5.155a)$$

$$\kappa' \equiv k'_{xsi} = \left(k_z^2 - k'^2\right)^{1/2} \qquad (5.155b)$$

$$\bar{\kappa} \equiv \frac{1}{2}\left(\frac{\kappa'}{\kappa} + \frac{\kappa}{\kappa'}\right) \qquad (5.156)$$

$$e^\pm \equiv \exp(\pm\kappa' d') \qquad (5.157)$$

Let us assume that the metallic layer is thick, that is, $\mathrm{Re}(\kappa' d') \gg 1$. The expression in (5.154) simplifies to

$$t = \frac{4\kappa\kappa'}{(\kappa + \kappa')^2} \exp[-(\kappa' - \kappa)d'] \qquad (5.158)$$

We have, of course, $t = 1$ if $\kappa' = \kappa$, that is, in the absence of the metallic layer.

Let us make the following numerical application. At a wavelength $\lambda_0 = 1$ μm, the free wavenumber of silver is[2]

$$k' \equiv k'_r + ik'_i = 0.2k_0 + i5k_0 \qquad (5.159)$$

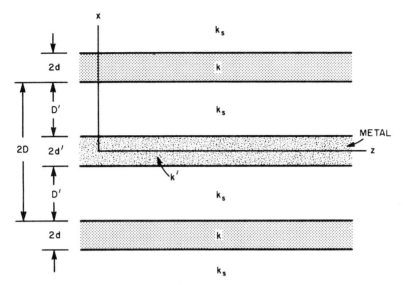

Fig. 5-23 Coupling between dielectric slabs. The coupling between dielectric slabs is drastically reduced by a thin layer of metal, such as silver.

Note the large imaginary part of k'. The free wavenumber of a typical glass is

$$k_s \equiv k_0 n_s = 1.4 k_0 \tag{5.160}$$

Because k' has a large imaginary part, we have approximately $\kappa' = k'_i$. With the above numerical values, the metallic layer transmission is

$$t \approx 0.25 \exp(-30 d'_{\mu m}) \tag{5.161}$$

We verify that, provided $d' \gg 0.1$ μm, the condition that the metallic layer be thick is well satisfied. Because the cross-talk power is proportional to the fourth power of t, the reduction of cross talk, in decibels, is

$$10 \log_{10} t^4 = 520 d'_{\mu m} \tag{5.162}$$

For example, if the silver layer thickness is $2d' = 0.5$ μm, the cross talk is reduced by 130 dB. This reduction of cross talk is independent of the initial value of the cross talk, within the approximations made. It is also independent of the location of the layer between the two slabs.

   The loss introduced by the metallic layer on the fiber propagation depends critically on the distance $D' = D - d'$ between the metallic layer and the dielectric slab. In the previous example, where the cross talk between identical slabs is assumed to be -20 dB in the absence of the

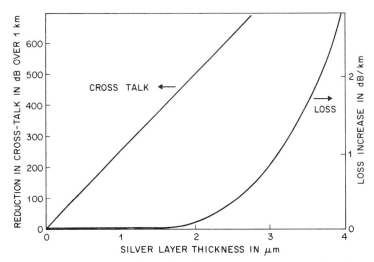

Fig. 5-24 Reduction of cross talk in decibels over a length of 1 km and increase of the loss, as a function of the thickness $2d'$ of a layer of silver located symmetrically between two dielectric slabs. The spacing $2D$ between the slabs is kept a constant as $2d'$ varies. Thus the distance between the layer and the slab varies. (From Arnaud.[28] Reprinted with permission from *The Bell System Technical Journal,* copyright 1975, The American Telegraph and Telephone Co.)

metallic layer, we found $2D = 11$ μm. A 130-dB reduction of cross talk requires $2d' = 0.5$ μm. Thus for that case, $D' = D - d' = 5.25$ μm. Note that the loss introduced by the metallic layer is due to the real part rather than the imaginary part of k'. The loss can be obtained from the expression of the complex reflection of the metallic layer

$$r \approx \frac{\kappa - \kappa'}{\kappa + \kappa'} \tag{5.163}$$

We find the imaginary part of r from (5.163) and (5.155b):

$$r_i \approx \frac{2\kappa k_r'}{k_i'^2} \tag{5.164}$$

with the same approximations as before. For silver, $r_i = 0.005$. The loss introduced on the dielectric slab propagation by this reflection can be obtained with the help of a perturbation formula. The details will be omitted. The imaginary part k_{zi} of k_z is related to r_i by

$$k_{zi} = r_i\kappa \frac{E_n^2}{2\omega\mu_0} \exp(-2\kappa D') \tag{5.165}$$

where E_n denotes as before the normalized field at the slab boundary given in (5.149). The slab loss, expressed in decibels per kilometer is, for the above numerical values,

$$\mathcal{L} \text{ dB/km} = 2.6 \times 10^6 \times \exp(-3.88 D'_{\mu m}) \qquad (5.166)$$

If $D' = 5.25 \ \mu m$, as assumed before, the loss introduced by the silver layer is

$$\mathcal{L} = 0.004 \text{ dB}/km \qquad (5.167)$$

It is negligible. However, this loss would drastically increase if D' were reduced below 4 μm. The variation of the reduction of the cross talk given in (5.162) and the loss increase given in (5.166) are plotted in Fig. 5-24 as functions of the silver thickness, the distance between the slabs being kept a constant $2D = 11 \ \mu m$.

The cross talk can be reduced further if the slabs are made systematically dissimilar. Using the above results, one finds that for a difference in thickness of 10%, the cross talk is less than -20 dB over a 1-km length if the ratio of slab spacing to slab thickness exceeds 1.5 for $F = \pi/2$ (e.g., $2D > 15 \ \mu m$ if $2d = 10 \ \mu m$).

5.13 Coupling between Round Fibers

The coupling between two slabs, whose field is independent of the y-coordinate, was obtained in Section 5.11. We now consider the more complicated case where the fields depend on y. We shall use the scalar approximation and a slightly different normalization. If $\psi_u(x, y)$ denotes the unnormalized field of a waveguide, the normalized field is defined as

$$\psi(y) = \psi_u(0, y) \left[k \int \int \psi_u(x, y) \, dx \, dy \right]^{-1/2} \qquad (5.168)$$

Because the variations of the free wavenumber in the cross section are small, the precise definition of k in (5.168) is unimportant. The Fourier components of the isolated guides that have the same dependence on y couple together. These Fourier components are

$$\hat{\psi}(k_y) = (2\pi)^{-1/2} \int_{-\infty}^{+\infty} \psi(y) \exp(-ik_y y) \, dy \qquad (5.169)$$

If we substitute fields $\psi_a(y)$ and $\psi_b(y)$ of the form given in (5.168) and

(5.169) in the general expression (3.83) of the coupling, we obtain

$$c = \int_{-\infty}^{+\infty} s(k_y)\hat{\psi}_b^*(k_y)\hat{\psi}_a(k_y)\, dk_y \qquad (5.170a)$$

where

$$s(k_y) = \left(k_z^2 + k_y^2 - k_s^2\right)^{1/2} \qquad (5.170b)$$

is the rate of decay of the spectral component $\hat{\psi}_a(k_y)$ of the field of guide a in the x-direction. This is also the rate of growth of the field of guide b because only near synchronous components couple together. To obtain (5.170) we have used the fact that the integral of $\exp(iky)$ from $y = -\infty$ to $y = +\infty$ is equal to $2\pi\delta(k)$, where $\delta(k)$ denotes the Dirac symbolic function.

If the dependence of s on k_y in (5.170) can be neglected $s(k_y) \approx s(0) \equiv s$, the expression for the coupling simplifies to

$$c = s\int_{-\infty}^{+\infty} \psi^2(y)\, dy \qquad (5.171)$$

Let us apply these general results to fundamental $\psi_{00} \approx HE_{11}$ modes, propagating along two identical oversized dielectric rods, with radii a, spaced a distance $2D$ apart. The distance between axes is $2D + 2a$ (see

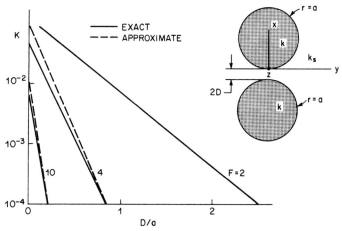

Fig. 5-25 Normalized coupling $K \equiv ca/\delta$ between two identical oversized dielectric rods with radii a as a function of spacing. The parameter is the normalized frequency $F = (k^2 - k_s^2)^{1/2}a$, and we have defined $\delta^2 \equiv 1 - (k_s/k)^2$. The length required for complete transfer of power from one fiber to the other is $\pi/2c$. Solid lines are from Snyder[27] and dashed lines from the approximate theory in the text.

Fig. 5-25). As before, k and k_s denote the free wavenumbers in the rod material and in the surrounding medium, respectively. Because the fibers are highly oversized, we can set $v = F \equiv (k^2 - k_s^2)^{1/2}a$, and $u = u_0 \approx 2.4 \ldots$ is the first root of J_0 in the expression (5.85) of the power. Thus the normalized field at the rod boundary is

$$\psi(r = a) = u_0(\pi^{1/2}k^{1/2}aF)^{-1} \qquad (5.172)$$

The decay of ψ away from the rod boundary (distance x) is approximately $\exp(-Fx/a)$. Because this distance varies approximately as $y^2/2a$, we obtain the variation of ψ with y

$$\psi(y) \approx u_0(\pi^{1/2}k^{1/2}aF)^{-1} \exp\left(-\frac{Fy^2}{2a^2}\right) \qquad (5.173)$$

The y-axis is assumed to be tangent to the rod at $x = y = 0$.

Let us now Fourier-transform $\psi(y)$. If we substitute (5.173) in (5.169), we obtain

$$\hat{\psi}(k_y) = \pi^{-1/2}u_0k^{-1/2}F^{-3/2} \exp\left(-\frac{k_y^2a^2}{2F}\right) \qquad (5.174)$$

Because the spectral component $\hat{\psi}(k_y)$ varies as $\exp(-sx)$ as a function of x, the coupling from (5.174) and (5.170) is

$$c = \frac{u_0^2}{\pi k F^3} \int_{-\infty}^{+\infty} s(k_y) \exp[-2s(k_y)D] \exp\left(-\frac{k_y^2a^2}{F}\right) dk_y \quad (5.175)$$

With the approximation $F \gg 1$, we can replace $s(k_y)$ by $s_0 \equiv (k_z^2 - k_s^2)^{1/2} \approx F/a$ and obtain a closed-form expression of the coupling between two oversized dielectric rods:

$$c = \frac{u_0^2}{\pi^{1/2}} k^{-1}F^{-3/2}a^{-2} \exp\left(-\frac{2FD}{a}\right) \qquad (5.176a)$$

or

$$K \equiv \frac{ca}{\delta} = \frac{u_0^2}{\pi^{1/2}} F^{-5/2} \exp\left(-\frac{2FD}{a}\right) \qquad (5.176b)$$

where

$$\delta^2 \equiv \left(\frac{F}{ka}\right)^2 \equiv 1 - \left(\frac{k_s}{k}\right)^2 \qquad (5.176c)$$

The variation of K with F and D/a is shown in Fig. 5-25. It is interesting to compare this result to a more accurate expression obtained by Snyder[27]:

$$K = \frac{u^2}{v^3} \frac{K_0[2v(D+a)/a]}{K_1^2(v)}$$

$$\approx \frac{u_0^2}{\pi^{1/2}} \left(1 + \frac{D}{a}\right)^{-1/2} F^{-5/2}\left(1 - \frac{1}{F}\right)^2 \exp\left(-\frac{2FD}{a}\right) \quad (5.177)$$

Equation (5.177) clearly goes to (5.176) for large F and $D \ll a$. The comparison between the approximate expression for K in (5.176b) and the exact result is shown in Fig. 5-25.

5.14 Coupling to Mode Sinks

When two guides with finite cross section carrying trapped modes approach each other, modes of the isolated guides that have almost the same axial wavenumber interact strongly, as we have seen in the previous sections. We now investigate the coupling of a guide with finite cross section and a guide with infinite cross section.[16] The latter will be called a substrate. The substrate may be a semi-infinite block of dielectric, a dielectric slab, or simply free space. A substrate does not guide modes with a definite axial wavenumber because the waves are free to propagate in any direction in a plane (for the case of a dielectric slab) or in a volume (for the case of free space). The propagation in the substrate is characterized by a dispersion surface that we denote $H(k_y, k_z) = 0$. If, for some k_y, a k_z can be found for the substrate waves that is equal to the axial wavenumber of the mode trapped by the guide, the guided mode loses power continuously into the substrate. This effect is easily understood from conventional antenna theory. The field of the isolated guide can be replaced by a surface distribution of electric and magnetic currents, having an $\exp(ik_z z)$ dependence on z. The radiation of these currents adds in phase in a direction defined by the angle $\alpha_0 = \cos^{-1}(k_z/k_s)$ if the surrounding medium is isotropic with wavenumber k_s. A similar mechanism is responsible for the Čerenkov radiation observed when a fast electron traverses a piece of dielectric. For that reason we shall call α_0 the Čerenkov angle.

The most straightforward approach to the evaluation of the radiation loss is to use the conventional antenna theory. The radiation field is expressed as an integral over the equivalent currents at the antenna (or fiber) boundary. We shall use a different approach based on the evaluation of the coupling between the guided modes and the (continuous) substrate modes. This approach is most useful when the substrate, instead of being simply free space, has a rather complicated structure and is weakly coupled to the guide. This approach is therefore more general and powerful than the conventional antenna theory, and yet it remains simple and provides closed-form expressions for many configurations of interest.

The significance of the coupled mode approach will be better appreciated if we proceed by steps, considering first an oversized lossy guide in place of an infinite substrate. Let us consider a lossless guide that carries a trapped mode with real axial wavenumber k_{z0}. Let this guide be coupled to a lossy guide carrying a trapped mode with complex axial wavenumber k_{zs}. The axial wavenumbers k_z of the normal modes are given, as we have seen, by the second-degree equation

$$(k_z - k_{z0})(k_z - k_{zs}) = c^2 \qquad (5.178)$$

where c denotes the coupling. If we solve (5.178) for k_z, we obtain

$$k_z = k_{z0} + \tfrac{1}{2}(k_{zs} - k_{z0}) - \left[\tfrac{1}{4}(k_{zs} - k_{z0})^2 + c^2\right]^{1/2} \qquad (5.179)$$

The minus sign before the square root was selected in order that, in the limit $c \to 0$, $k_z \to k_{z0}$.

To first order in c^2, k_z is

$$k_z = k_{z0} - \frac{c^2}{k_{zs} - k_{z0}} \qquad (5.180)$$

Thus, setting $k_{zs} \equiv k_{zsr} + i k_{zsi}$, the imaginary part k_{zi} of k_z is

$$k_{zi} = c^2 k_{zsi}\left[(k_{zsr} - k_{z0})^2 + k_{zsi}^2\right]^{-1} \qquad (5.181)$$

This result is valid, provided

$$c \ll k_{zsi} \qquad (5.182)$$

The imaginary part of c can be neglected.

Let us now assume that the lossy waveguide is highly overmoded. The axial wavenumbers of many of the modes that it carries are very close to k_{z0}. Thus many modes need to be considered in evaluating k_{zi}. We need to sum the rhs of (5.181) over these modes, which are labeled with the subscript α. We thus replace (5.181) by

$$k_{zi} = \sum_\alpha c_\alpha^2 k_i \left[(k_\alpha - k_{z0})^2 + k_i^2\right]^{-1} \qquad (5.183)$$

where, for brevity, we have set $k_\alpha \equiv k_{zsr\alpha}$. We have also assumed that $k_i \equiv k_{zsi\alpha}$ is independent of α.

In the limit where the cross section S of the lossy guide tends to infinity, the wavenumbers k_α become denser and denser, and the summation in (5.183) can be replaced by an integral

$$k_{zi} = \lim_{S \to \infty} \sum_\alpha c_\alpha^2 k_i \left[(k_\alpha - k_{z0})^2 + k_i^2 \right]^{-1}$$

$$= k_i \int C(k_z) \left[(k_\alpha - k_{z0})^2 + k_i^2 \right]^{-1} dk_z \qquad (5.184)$$

where we have defined a coupling density C by

$$C(k_z) \, dk_z = \lim_{S \to \infty} (c_\alpha^2) \qquad (5.185)$$

the range of α being defined by the condition

$$k_z < k_\alpha < k_z + dk_z \qquad (5.186)$$

The interval dk_z is small. There are nevertheless many modes in that range because S is very large. The density C exists because, as $S \to \infty$, the coupling c decreases at least as fast as S^{-1}. Indeed, the axial power in the lossy waveguide (which we shall now call a substrate) is proportional to S if the energy density is kept a constant.

In (5.185) it was implicitly assumed that, as the range dk_z of k_z narrows down, all the couplings c_α tend to be the same. In fact, α labels different types of substrate modes, for example, those corresponding to the H-waves and the E-waves of a dielectric slab. These two types of modes usually do not have the same coupling with the guided mode for a given k_z. Thus the coupling density C should be considered a sum over the different types of substrate modes that have the same k_z, for example, $C = C_H + C_E$, where the subscripts H and E refer to the H-waves and E-waves, respectively.

In the limit of small dk_z in (5.186), we have $k_\alpha = k_z$. If k_i is small, the integrand in (5.184) is sharply peaked near $k_\alpha = k_z \approx k_{z0}$ and behaves as a Dirac symbolic function. A straightforward integration gives

$$k_{zi} = \pi C(k_{z0}) \qquad (5.187)$$

This result is independent of k_i. Thus we have obtained the loss suffered by a trapped mode of the guide. This loss does not result from the dissipation losses of the substrate, which we may now assume to vanish, but from radiation into the substrate. The result (5.187) is exceedingly simple. The rather lengthy derivation that we gave bypasses mathematical difficulties and clarifies the transition from oversized slightly lossy waveguides to lossless substrates having an infinite cross section.

Radiation Loss of a Reactive Surface[16]

To illustrate the previous general results, let us evaluate the radiation of a surface wave, guided by a reactive surface, with normalized susceptance s, into a semi-infinite piece of dielectric as shown in Fig. 5-26. We consider H-modes and assume that the field is independent of the y-coordinate. In the absence of the dielectric, the reactive surface guides surface waves with axial wavenumber $k_z = k_{z0} = (k_0^2 + s^2)^{1/2}$. Radiation takes place if $k_{z0} < k'$, where k' denotes the wavenumber in the dielectric, because, as we have discussed earlier, radiation from various points of the reactive surface adds up in phase in the (real) Čerenkov direction defined by $\alpha_0 = \cos^{-1}(k_z/k')$. The condition $k_{z0} < k'$ rests on the assumption that the coupling is weak.

The configuration presently considered is so simple that an exact solution can be obtained by writing the boundary conditions. Let this exact result be derived first. The boundary condition at the reactive surface is

$$\frac{dE}{dx} + sE = 0 \tag{5.188}$$

The field in the dielectric has the form

$$E(x) = \exp(ik'_x x) \tag{5.189a}$$

$$\frac{dE}{dx} - ik'_x E = 0 \tag{5.189b}$$

$$k'^2_x = k'^2 - k^2_z \tag{5.190}$$

We have omitted the solution $\exp(-ik'_x x)$ because the wave is assumed to radiate away from the reactive surface. Note that k_z is expected to have a small positive imaginary part because of the radiation loss. Assuming that k' is real, that is, that the dielectric is free of dissipation losses, (5.190) shows that k'_x has a small negative imaginary part. Thus the wave amplitude grows exponentially in the dielectric as the distance from the boundary increases. Solutions of the Maxwell equations of that type are called leaky waves. It is not difficult to show that the curves of constant irradiance in the dielectric are straight lines making with the z-axis the Čerenkov angle $\alpha_0 = \cos^{-1}(k_z/k')$. In that expression, the imaginary part of k_z is neglected. Between the reactive surface and the dielectric, the medium (e.g., free space) has wavenumber k_0. Thus the solution has the form

$$E = E^+ \exp(ik_x x) + E^- \exp(-ik_x x) \tag{5.191}$$

$$k^2_x = k^2_0 - k^2_z \tag{5.192}$$

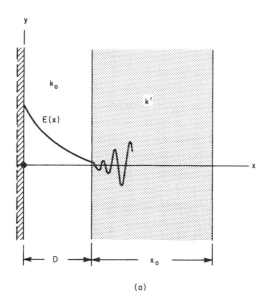

(a)

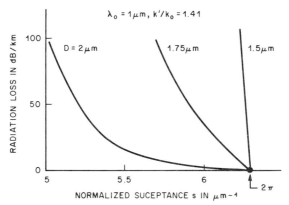

Fig. 5-26 (a) Radiative coupling between a reactive surface with normalized susceptance s and a semi-infinite dielectric with free wavenumber $k' > (k_0^2 + s^2)^{1/2}$, where k_0 denotes the free wavenumber in the intermediate medium. For the calculation of the radiation loss, the dielectric is first assumed finite with thickness x_0. (b) Variation of the radiation loss as a function of the normalized susceptance of the surface for various surface–dielectric spacings (D). (From Arnaud.[16] Reprinted with permission from *The Bell System Technical Journal*, copyright 1974, The American Telegraph and Telephone Co.)

Note that the real part of k_x is expected to be small compared with the imaginary part and that the field essentially decays from the reactive surface until the dielectric boundary is reached.

If we specify that E and dE/dx are continuous, we obtain the equation defining k_z:

$$(k_x - is)(k_x + k_x') = (k_x + is)(k_x - k_x') \exp(2ik_x D) \quad (5.193)$$

where D denotes the spacing between the reactive surface and the dielectric. Because this equation cannot be solved explicitly for k_z, we shall consider the limit where $sD \to \infty$. Let us expand k_x and k_x' in power series of the small parameter

$$\delta \equiv \exp(-2sD) \quad (5.194)$$

We write

$$k_x = is + k_{x1}\delta + \cdots \quad (5.195)$$

$$k_x' = k_{x0}' + k_{x1}'\delta + \cdots \quad (5.196)$$

where

$$k_{x0}'^2 \equiv k'^2 - k_0^2 - s^2 \quad (5.197)$$

For negligible couplings ($\delta = 0$), we have $k_x = is$, and $k_x'^2 = k'^2 - k_{z0}^2 \equiv k_{x0}'^2$, where $k_{z0}^2 = k_0^2 + s^2$ was given earlier.

If we substitute (5.195)–(5.197) in (5.193) and collect terms of first order in δ, we obtain

$$k_{x1} = \frac{2is(is - k_{x0}')}{is + k_{x0}'} \quad (5.198)$$

On the other hand, from the definition $k_x \equiv (k_0^2 - k_z^2)^{1/2}$ and (5.195), we have

$$\text{Im}(k_z) = -\frac{s\delta}{k_{z0}} \text{Re}(k_{x1}) \quad (5.199)$$

Thus, if we substitute (5.198) in (5.199), we obtain

$$k_{zi} = \frac{4s^3 k_{x0}'}{(k'^2 - k_0^2)k_{z0}} \exp\{-2sD) \quad (5.200)$$

More explicitly, (5.200) is

$$k_{zi} = 4s^3 [k_0^2(n^2 - 1)]^{-1}(k_0^2 + s^2)^{-1/2}[k_0^2(n^2 - 1) - s^2]^{1/2} \exp(-2sD) \quad (5.201)$$

where $n \equiv k'/k_0$. Equation (5.201) gives the radiation loss in nepers per unit length, explicitly as a function of $k_0 \equiv 2\pi/\lambda_0$, n, s, and D.

This result can be obtained much more readily with the method outlined at the beginning of this section. The general expression for the coupling between H-waves is

$$c^2 = \left(\frac{s}{k_{z0}}\right)^2 \frac{E^2}{\int E^2 \, dx} \frac{E_s^2}{\int E_s^2 \, dx} \tag{5.202}$$

where E denotes the field of the mode guided by the isolated reactive surface, and E_s the field of a mode of the isolated dielectric. The integrals are over the full cross section, and E and E_s are evaluated anywhere between the reactive surface and the dielectric. We select the location $x = 0$ of the dielectric boundary. The field of the mode guided by the isolated reactive surface is $\exp(-sx)$. Thus

$$\frac{E^2}{\int E^2 \, dx} = 2s \exp(-2sD) \tag{5.203}$$

Let us consider now the dielectric alone and first assume that its thickness (in the x-direction) is large but finite and equal to x_0. Because x_0 is large, we can use the theory of propagation of H-waves in a dielectric slab at frequencies well above cutoff given in (5.35). In that limit

$$\frac{E_s^2}{\int_{-\infty}^{+\infty} E_s^2 \, dx} \approx \frac{E_s^2}{\int_0^{x_0} E_s^2 \, dx} = \frac{2k_{x0}'^2}{(k'^2 - k_0^2)x_0} \tag{5.204}$$

Let us now evaluate the number $N \, dk_z$ of modes in the dielectric whose axial wavenumber lies between k_z and $k_z + dk_z$. We shall use again the approximation that the frequency is much larger than the cutoff frequency

$$k_x' = \frac{(m+1)\pi}{x_0}, \qquad m = 0, 1, 2, \ldots \tag{5.205}$$

and the definition $k_x'^2 = k'^2 - k_z^2$. By differentiation, we obtain

$$N = \frac{k_{z0}x_0}{\pi k_{x0}'} \tag{5.206}$$

In the final expression, the substitution $k_x' \to k_{x0}'$, $k_z \to k_{z0}$ can be made.

The general expression (5.197), $k_{zi} = \pi C$ for the radiation loss, can now be used, where $C = c^2 N$, N is given in (5.206), and c^2 is obtained by substituting (5.203) and (5.204) in (5.202). The result is

$$k_{zi} = \pi C = \pi c^2 N = \frac{4s^3 k_{x0}'}{(k'^2 - k_0^2)k_{z0}} \exp(-2sD) \tag{5.207}$$

in agreement with expression (5.200) obtained by letting $D \to \infty$ in the exact solution (5.193). The loss in decibels per kilometer is obtained by multiplying k_{zi} in (5.201) by 8.7×10^9. It is shown in Fig. 5-26 as a function of the normalized susceptance s of the surface, for $\lambda_0 = 1$ μm, $n = 1.41$, and spacings $D = 1.5$, 1.75, and 2 μm. The loss when the dielectric is replaced by a lossy foam that attenuates the wave because of dissipation, rather than because of radiation, in contradistinction, varies smoothly.

The reactive surface considered can support only one mode. However, similar results can be obtained if the reactive surface is replaced by a moderately thick dielectric slab supporting a few H-modes. In that case, we may be able to choose the refractive index of the dielectric in such a way that all the slab modes, except the fundamental one, are attenuated by radiation into the substrate. There is a clear advantage in using a mode sink (the substrate) for mode selection rather than a dissipative material.

The radiation loss is proportional to c^2 and therefore to the square of the normalized field of the guide, if we assume that the substrate remains the same. The radiation loss of a slab is smaller than that of a reactive surface having the same k_z, because the thicker the slab, the larger the axial power for the same field at the boundary.

The approach used in this section is most effective to analyze the radiation of a glass fiber into a slab substrate (illustrated in Fig. 5-33). We only need to know the field of the isolated fiber at its boundary $r = a$ and the field of the isolated slab. These results were given in the previous section. We shall give without derivation the expression of the radiation loss when the slab and fiber material have the same refractive index n, the radius a of the fiber is much larger than λ_0, and the slab thickness $2d \approx a$ is so chosen that only the fundamental HE_{11}-mode propagates without radiation loss. The loss suffered by the first spurious mode H_{01} is[16]

$$\mathcal{L} \text{ dB/km} = 340 n^{-1} (n^2 - 1)^{-3/2} (k_0^4 a^5)^{-1} \exp\left[-2(n^2 - 1)^{1/2} k_0 D \right] \quad (5.208)$$

where D denotes the spacing between the fiber and the slab. Because of the very fast dependence of $\mathcal{L}$ on a ($\mathcal{L} \propto a^{-5}$), the radiation loss of the spurious modes becomes negligible, even over a 1-km length, when the rod radius exceeds about 15 wavelengths if $n \approx 1.45$. Larger rod radii can be used if $n \approx 1.01$, for example, but the bending losses of the fundamental mode become large, as we shall see in Section 5.17.

The configurations shown in Fig. 5-27, where small perturbations are introduced on surface waves, need to be analyzed by different methods. A straightforward application of perturbation formulas does not give any meaningful result because the unperturbed wave (plane wave) carries infinite power. We can, however, proceed the other way around and start from the perturbed state, assumed to be known, and remove the perturbation. It can be shown that the results obtained for the structures in Fig.

5-27b, d coincide with those obtained by matching the k_z of the three regions. Figure 5-28 illustrates in more detail the mode configuration for axially bounded whispering-gallery modes. This configuration explains the mechanism of guidance and mode selection of the helical fiber that we shall describe later. The reverse perturbation method sketched above is useful also to obtain the wavenumber of waves guided by dielectric rods of arbitrary cross section in the limit of small F. The rod is considered, from that point of view, a perturbation of free space.

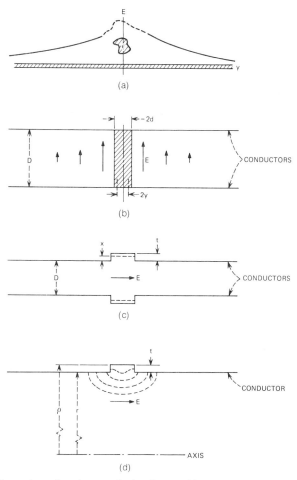

Fig. 5-27 Illustration of various methods of perturbing a guiding surface by a one-dimensional guide. (a) Dielectric rod coupled to a reactive surface. (b) Slab coupled to a TEM wave. (c) The groove guide. (d) Confinement of whispering-gallery modes. (From Arnaud.[7] Reprinted with permission from *The Bell System Technical Journal,* copyright 1974, The American Telegraph and Telephone Co.)

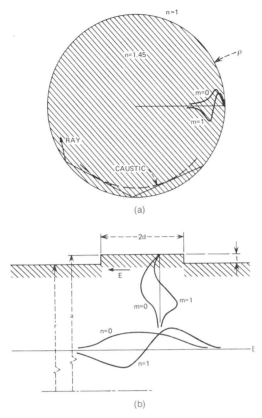

(a)

(b)

Fig. 5-28   Detailed view of Fig. 5-27d. Whispering-gallery modes propagating along a cylindrical boundary are described by an Airy function ($m = 0, 1, \ldots$). Axial confinement is provided by a reduction of the cylindrical boundary on both sides of the central region (step t of the radius). The axial-mode number is $n = 0, 1, \ldots$. This configuration has application to the helical single-material fiber. (From Arnaud.[7] Reprinted with permission from *The Bell System Technical Journal*, copyright 1974, The American Telegraph and Telephone Co.)

5.15 Bending Loss of a Reactive Surface

Open waveguides support modes whose phase velocity is smaller than the velocity of plane waves in the surrounding medium. Thus no radiation takes place under normal conditions. If the fiber is bent, however, the phase velocity increases in proportion to the distance from the curvature center. At some radius, it exceeds the velocity of plane waves in the medium, and a radiation loss is suffered. This effect is of great practical importance in fiber communication because it sets a limit on how sharp bends can be made without resulting in untolerable loss. For most single-mode optical glass fibers, a radius of curvature of the order of 1 m can be tolerated. For gas lenses and weakly guiding millimeter wave systems, the minimum radius is sometimes as large as 100 m. The relative insensitivity of glass fibers to bends results from the rather large change in refractive index in the cross section that they provide. It constitutes their main advantage compared to other guiding systems such as gas lenses.

Different methods have been used to evaluate the radiation losses of curved waveguides. For simple geometries, it is possible to solve the boundary-value problem using a cylindrical coordinate system, the loss being given by the imaginary part of the propagation constant.[29,30] Another method consists in evaluating the power radiated at the radius where the phase velocity of the guided mode becomes equal to the velocity in the surrounding medium.[31] Ray pictures can also be used to describe the propagation in curved dielectric fibers[32] or antenna theory.[33]

We shall use the following approach.[34] For curved geometries, the guided mode of the open waveguide is coupled to whispering-gallery type radiation modes. We first assume that the surrounding medium (called a substrate) is bound at some radius b and has finite dissipation losses. The loss suffered by the waveguide mode as a result of its coupling to the substrate is evaluated. The radius b is subsequently allowed to reach infinity and the dissipation loss to vanish. We are then left with the bending losses. Whispering-gallery modes are characterized by a circular caustic with radius a_0. The behavior of the field is oscillatory outside the caustic and exponential inside the caustic. At the caustic, the phase velocity is just equal to the phase velocity of free waves. It is therefore not difficult to find for what values of a_0 synchronism with the waveguide mode is achieved. Although a_0 exceeds only slightly the rod curvature radius ρ, the distance $y_0 \equiv a_0 - \rho$ is the most critical parameter that influences the loss. Once the synchronism conditions have been obtained, the coupling coefficients and the mode number densities can be evaluated. We shall use for the waveguide field the expressions applicable to scalar fields, which are slightly simpler than the expressions applicable to the Maxwell field.

In most practical situations, the fiber curvature is not a constant, but varies along the fiber axis z. This causes additional radiation losses suffered at the transition between the straight and the curved sections of the fiber. Because the radiation from the bend itself has the form $\exp(-\rho/\rho_0)$, where ρ denotes the radius of curvature and ρ_0 a constant, and the radiation at the junctions has the form $1/\rho^2$, the latter becomes significant for large ρ, that is, for very low radiation losses. The mechanical strength of the fiber usually prevents the curvature of the fiber from changing abruptly. The transition radiation is nevertheless important for single-mode communication fibers (see Kuhn[57] for detailed results).

AXIS

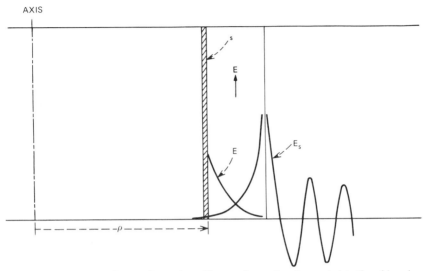

Fig. 5-29 Bending of a reactive surface. The reactive surface is coupled to the whispering-gallery modes guided by a (fictitious) boundary.

When a highly multimoded fiber is bent in some random manner, the ray slopes associated with the guided waves tend to increase in proportion to the square root of the length of the fiber. Eventually, the rays cease to be totally reflected. This is easily understood if we observe that randomly bent fibers are analogous to mechanical oscillators (e.g., harmonic oscillators for the case of square-law fibers) driven by random forces $f(t)$ proportional to the curvature of the fiber $C(z)$. The equivalent mechanical oscillator gains energy as time goes on, that is, the amplitude and momentum increase until the limit is reached. Note that even in nominally single-mode fibers, higher-order modes can be excited and propagate over a certain length and cause significant pulse broadening. On the other hand, the fact that the ray period is not a constant when irregularities are present along the fiber may have a favorable consequence when many modes are excited. The difference in path length that normally exists between axial rays and rays making large angles with the axis tends to be smoothed out, and the pulse does not broaden with distance as fast as it would in the absence of irregularities. It is often observed that beyond a certain length, which depends on the number of modes being excited, pulse broadening in multimode fibers increase as the square root of the path length, rather than in direct proportion to the path length.

In this section, we shall concentrate on the problem of radiation loss of a single mode when the curvature is a constant and consider first the radiation loss suffered by a curved (single-sided) reactive surface with normalized susceptance s. Let us first assume that the reactive surface is straight. Because the field decays as $\exp(-sx)$, where x denotes the distance from the surface, we have

$$R(0) = E(0)^2 \left(\int_0^\infty E^2 \, dx \right)^{-1} = 2s \qquad (5.209)$$

The axial wavenumber is $k_z = (k_s^2 + s^2)^{1/2}$, where k_s denotes the free wavenumber in the medium. When the reactive surface is bent, with a large radius of curvature ρ, it can be assumed in first approximation that the azimuthal wavenumber k_φ at the radius of the reactive surface is equal to k_z. Thus k_z is now denoted k_φ. The azimuthal mode number is $\mu \equiv k_\varphi \rho$. The radius a_0 where the phase velocity of the guided wave equals the free-space velocity is given by

$$\mu = k_\varphi \rho = k_s a_0 \qquad (5.210)$$

and the difference between the radius a_0 and that of the reactive surface is

$$D = a_0 - \rho \approx \frac{s^2 \rho}{2k_s^2} \qquad (5.211)$$

where we have used the relation $k_\varphi^2 = k_s^2 + s^2$ and the approximation $D \ll \rho$.

Let us now turn our attention to the radiation modes. Let us assume that the space surrounding the reactive surface is restricted to the interior of a cylindrical boundary of radius b, with $b - \rho \ll \rho$.

The boundary guides whispering-gallery modes that have a caustic radius a_0. At that radius, the azimuthal wavenumber equals the free wavenumber k_s. Now let x denote radial distances from the caustic (positive x being defined outward). The field in (5.114) has the form

$$E(x) \sim \begin{cases} x^{-1/4} \sin\left[\dfrac{2^{3/2}}{3} k_s a_0^{-1/2} x^{3/2} + \dfrac{\pi}{4} \right], & x \gg 0 \\[3mm] \frac{1}{2}(-x)^{-1/4} \exp\left[-\left(\dfrac{2^{3/2}}{3} \right) k_s a_0^{-1/2}(-x)^{3/2} \right], & x \ll 0 \end{cases}$$

$$(5.212)$$

if we replace μ by $k_s a_0$.

The field vanishes at the boundary $\rho = b$, or $x = b - a_0 \equiv L$. Thus

$$\frac{2^{3/2}}{3} k_s a_0^{-1/2}(b - a_0)^{3/2} + \frac{\pi}{4} = (m + 1)\pi \qquad (5.213)$$

where m denotes an integer. We now wish to know how many modes have an azimuthal wavenumber between k_φ and $k_\varphi + dk_\varphi$ at the reactive surface, with radius ρ. Because the boundary radius b and k_s are constants, the caustic radius a_0 defined in (5.213) varies when m varies. The variation of the caustic radius a_0 causes a variation of k_φ that follows from the equality $k_\varphi \rho = k_s a_0$. Thus the mode number density N, defined as the number of modes whose azimuthal wavenumber is between k_φ and $k_\varphi + dk_\varphi$, divided by dk_φ, is

$$N = \frac{dm}{dk_\varphi} = \frac{\rho}{k_s} \frac{dm}{da_0} = \frac{\rho}{k_s} \sqrt{2}\, \pi^{-1} k_s a_0^{-1/2} L^{1/2} = \sqrt{2}\, \pi^{-1} \rho^{1/2} L^{1/2} \quad (5.214)$$

Because $b - a_0 \ll a_0$, we have neglected the variation of the term $a_0^{-1/2}$ compared with the variation of $(b - a_0)^{3/2}$. This approximation may at first seem inconsistent with the fact that we eventually let b tend to infinity. However, we should remember that the radiation modes become extremely dense for values of $L \equiv b - a_0$ small compared with the curvature radius. This condition is required by the paraxial approximation used for the field. In the final expression in (5.214), we have made the approximation $a_0 \approx \rho$.

Let us now evaluate the power carried by a radiation mode. Because there are many oscillations of the sine function in (5.212) between $\rho = a_0$ and $\rho = b$, we can ignore the contribution of the exponentially decaying field ($x < 0$) to the power, replace the $\sin^2$ term in the expression of E^2 by its averaged value $\frac{1}{2}$. Thus

$$\int_{-\infty}^{L} E^2(x)\, dx \approx \frac{1}{2} \int_0^L x^{-1/2}\, dx = L^{1/2} \quad (5.215)$$

The normalized squared field inside the caustic

$$S(x) \equiv NE^2(x) \Big/ \int_{-\infty}^{L} E^2\, dx \quad (5.216)$$

is therefore, from (5.212), (5.214), and (5.215),

$$S(x) = \frac{\sqrt{2}}{4\pi} \rho^{1/2}(-x)^{-1/2} \exp\left[-2(2^{3/2}/3) k_s a_0^{-1/2}(-x)^{3/2} \right] \quad (5.217)$$

Note the cancellation of L.

The radiation loss is now obtained by the general coupling formula

$$k_{zi} = \frac{\pi s^2}{k_s^2} R(x) S(x) \quad (5.218)$$

Let R and S be evaluated at the reactive surface. R has the value in (5.209), and S has the value in (5.217) with $-x = D$ the distance between the reactive surface and the caustic of the radiation modes in (5.211).

Thus the radiation loss is

$$k_{zi} = \frac{s^2}{k_s^2} \, 2s \, \frac{\sqrt{2}}{4} \, \rho^{1/2} \left(\frac{s^2 \rho}{2k_s^2} \right)^{1/2} \exp\left[-2\left(\frac{2^{3/2}}{3} \right) k_s a_0^{-1/2} \left(\frac{s^2 \rho}{2k_s^2} \right)^{3/2} \right]$$

(5.219)

This expression can be written with $a_0 \approx \rho$

$$k_{zi} = \left(\frac{s^2}{k_s} \right) \exp\left(-\frac{\rho}{\rho_0} \right)$$

(5.220a)

$$\rho_0 \equiv \frac{3}{2} \, \frac{k_s^2}{s^3}$$

(5.220b)

This very simple and important result was first obtained by Miller and Talanov[29] by solving the exact boundary-value problem. The approach used here is approximate but much more convenient for weakly guiding fibers, particularly if the fiber is coupled to a substrate, such as a dielectric slab, as we shall see later.

5.16 Bending Loss of a Dielectric Slab

The method described in the previous section is applicable, in principle, to a dielectric slab. We shall assume that the field of the bent slab is almost that of the straight slab (ignoring for a moment the fact that the field of the bent slab is leaking). Because a radius of curvature ρ is equivalent to a gradient of refractive index $1/\rho$, the curvature introduces a change d/ρ of effective refractive index from the slab center to the slab boundary. On the other hand, for the fundamental mode of the slab, far from cutoff, we have $k_x d \approx \pi/2$, or $1 - k_z/k \approx (\lambda/6d)^2$. Thus the field in or near the slab can be considered essentially unperturbed by the bending when $(\lambda/6d)^2 \gg d/\rho$, or $\rho \gg 36d^3/\lambda^2$. We assume in the following that this condition is met. In that approximation, the azimuthal wavenumber of the bent slab, at the center of the slab, is approximately equal to k_z, the axial wavenumber of the straight slab. We need to know the azimuthal wavenumber k_φ not at the slab center, but at the outer boundary of the slab, with radius ρ. Because of the invariance of the angular phase velocity, these quantities are related by $k_\varphi \rho = k_z(\rho - d)$. As it turns out, the difference between k_φ and k_z cannot be neglected, even for very small curvatures. Substituting the above expression for k_φ in the exponential factor $\exp(-\rho/\rho_0)$ in (5.220), where $\rho_0 = \frac{3}{2}(k_s^2/s^3)$, $s^2 \equiv k^2 - k_s^2 \equiv k_{xsi}^2$, we obtain $\exp(-\rho/\rho_0) \approx \exp(2sd) \exp(-\rho/\rho_0')$, where ρ_0' is defined with respect to

k_z: $\rho_0' \equiv (\frac{3}{2})(k_s^2/s'^3)$, $s'^2 \equiv k_z^2 - k_s^2$, and $sd \approx F$. The correction factor $\exp(2sd) \approx \exp(2F)$ is independent of the radius of curvature ρ. It therefore cannot be neglected, even for large ρ.

We have seen that the bending loss is proportional to the square of the normalized field of the optical guide. Thus, for a dielectric slab,

$$k_{zi} = \frac{s}{2k} \exp\left(-\frac{\rho}{\rho_0'}\right) \exp(2sd)R \qquad (5.221)$$

where the expression of ρ_0' is given above and, from (5.47),

$$R = \frac{E^2}{\int_{-\infty}^{+\infty} E^2\, dx} = \frac{d^{-1}(1-b)}{1 + \phi_i^{-1}} \qquad (5.222)$$

The slab parameter $\phi_i = b^{1/2}F$, where b is given as a function of the normalized frequency F in Fig. 5-6. Note, incidentally, that in the limit $d \to 0$ but s finite, $R \to s$. This is half the value given in (5.209) for a (single-sided) reactive surface. The factor 2 results from the fact that, in the limit $d \to 0$, a dielectric slab can be considered a double-sided, rather than single-sided, reactive surface. That is, for the same field, the power carried by the dielectric slab is twice as large as that carried by a reactive surface.

As an example, let us assume that the slab is single moded for H-modes; that is, the normalized frequency $F \equiv (k^2 - k_s^2)^{1/2}d$ is equal to $\pi/2$. Figure 5-6 shows that the normalized phase velocity $b \equiv (k_z^2 - k_s^2)/(k^2 - k_s^2) = 0.67$. We have further that $\phi_i = b^{1/2}F = 1.28$. Assuming that $\lambda_s = 1$ μm, $k_s \equiv 2\pi/\lambda_s = 2\pi$ μm^{-1}, the bending loss of the slab is, from (5.221) and (5.222),

$$\mathcal{L} \text{ dB/km} = 40 \times 10^8 d^{-2} \exp\left[-\frac{\rho}{(29d^3)}\right] \qquad (5.223)$$

where ρ and d are in micrometers. The slab refractive index is assumed to vary in such a way that $F = \pi/2$, according to the law

$$\frac{n}{n_s} = \left[1 + \left(\frac{F}{k_s d}\right)^2\right]^{1/2} = \left[1 + \left(\frac{1}{4d}\right)^2\right]^{1/2} \qquad (5.224)$$

Note that as d increases, the prefactor in (5.223) decreases because of the reduction of the normalized field of the slab. This effect is, however, almost negligible compared with the opposite effect due to the exponential term.

The preceding expressions are based on the assumption that the field of the bent slab is essentially the same as that of the straight slab (ρ

$\gg 36d^3/\lambda^2$). Let us now go to the other extreme and assume that the bending is strong. In that limit, the wave clings to the outer boundary of the slab and becomes a whispering-gallery mode. The inner boundary of the slab plays essentially no role in the propagation. The expression of the coupling resistance R of whispering-gallery modes guided by a circular dielectric boundary follows from (5.115) and the continuity of E and its first derivative

$$R \equiv \frac{E^2}{\int E^2 \, dx} = \frac{2k^2}{s^2 \rho} \tag{5.225}$$

It is interesting that Airy functions do not appear in this expression. We also obtain the azimuthal wavenumber k_φ at the outer radius ρ of the slab,

$$\frac{k_\varphi}{k} = 1 - 1.85(k\rho)^{-2/3} \tag{5.226}$$

If we substitute these expressions in (5.221) with $d = 0$, we obtain the loss of whispering-gallery modes guided by a circular dielectric boundary

$$k_{zi} = \frac{k}{s\rho} \exp\left(- \frac{\rho}{\rho_0}\right) \tag{5.227}$$

where

$$\rho_0 \equiv \frac{3}{2} \frac{k_s^2}{s^3} , \qquad s^2 = k_\varphi^2 - k_s^2, \qquad k_\varphi = k\left[1 - 1.85(k\rho)^{-2/3}\right] \tag{5.228}$$

Thus the loss of whispering-gallery modes guided by a dielectric rod can be written as an explicit function of k, k_s, and ρ.

5.17 Bending Loss of the Round Fiber

Let us now evaluate the bending loss of a guide whose field is not independent of y.[30,34] The azimuthal wavenumber at the outer radius ρ of the guide is denoted k_φ as before. The Fourier transform $\hat{\psi}(k_y)$ of the field is defined as in Section 5.13. Here again, we consider independently the radiation loss for each k_y and integrate over k_y. As k_y increases, the caustic of the whispering-gallery modes guided by the surrounding medium (radiation modes) recedes from the optical guide, and the radiation loss becomes negligible for some small value of k_y. Summing the contributions from each k_y-component, we obtain the total bending loss

$$k_{\varphi i} = \int_{-\infty}^{+\infty} \tfrac{1}{2} s(k_y) \exp\left[- \frac{\rho}{\rho_0(k_y)}\right] \hat{\psi}^2(k_y) \, dk_y \tag{5.229}$$

where

$$s^2(k_y) = k_\varphi^2 + k_y^2 - k_s^2, \qquad \rho_0(k_y) = \frac{3}{2}\frac{k_s^2}{s^3(k_y)} \qquad (5.230)$$

In many cases, $\hat{\psi}^2(k_y)$ has a negligible amplitude beyond some small value of k_y, and we can use the approximation

$$\frac{\rho}{\rho_0(k_y)} \approx \frac{\rho}{\rho_0(0)} + \frac{\rho s k_y^2}{k_s^2}, \qquad s(k_y) \approx s(0) \equiv s \qquad (5.231)$$

Thus we finally obtain

$$k_{\varphi i} = \tfrac{1}{2} s \, \exp\!\left(-\frac{\rho}{\rho_0}\right) \int_{-\infty}^{+\infty} \hat{\psi}^2(k_y) \exp\!\left(-\frac{\rho s k_y^2}{k_s^2}\right) dk_y \qquad (5.232)$$

The bending loss is expressed in (5.232) as the Gauss transform of the straight waveguide spatial spectral density (in the direction of the bending axis).

Let us apply this result to the dielectric rod[§] whose normalized field is given in (5.173). Substituting this expression in (5.232) and integrating we obtain, setting $s_0 \approx F/a$,

$$k_{\varphi i} = \frac{u_0^2}{2\pi^{1/2}} \frac{1}{F^2 a} \left(\frac{F\rho}{a} + \frac{k_s^2 a^2}{F}\right)^{-1/2} \exp\!\left(-\frac{\rho}{\rho_0}\right) \qquad (5.233)$$

where

$$\rho_0 = \frac{3}{2}\frac{k_s^2}{s^3}, \qquad u_0 = 2.4 \ldots \qquad (5.234)$$

If we introduce the critical radius $\rho_0' = (\tfrac{3}{2})(k_s^2/s'^3)$, where $s'^2 = k_z^2 - k_s^2$ and k_z is the axial wavenumber of the straight waveguide, (5.233) becomes

$$k_{\varphi i} = \frac{u_0^2}{2\pi^{1/2}} \frac{1}{F^2 a} \left(\frac{F\rho}{a}\right)^{-1/2} \exp(2F) \exp\!\left(-\frac{\rho}{\rho_0'}\right) \qquad (5.235)$$

We have neglected the second term inside parentheses preceding the exponential in (5.223) because ρ is large. An interesting experimental study

[§] The reader is cautioned that bending (and torsion) of a round fiber breaks the degeneracy between $\pm \mu$ modes. The bending loss can be defined only for the modes of the deformed fiber.

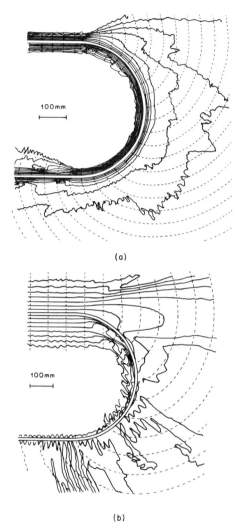

(a)

(b)

Fig. 5-30 Experimentally measured amplitude and phase of the field of an *H* plane bent dielectric rod with cross section 9.5 mm × 12 mm, $\epsilon/\epsilon_0 = 2.3$; radius of curvature, 248 mm; lines of constant magnitude, —; lines of constant phase, ---. (a) 12.7 GHz, (b) 8.17 GHz. (From Neumann and Rudolph,[35] by permission of the IEEE.)

of single-mode fibers is given in Neumann and Rudolph.[35] The measured radiation of a bent rectangular fiber at two frequencies is shown in Fig. 5-30. The experimental bending loss is about half the theoretical prediction (see Fig. 5-31). The reason for this discrepancy has not yet been established. The theory may be slightly defective in that it does not take into

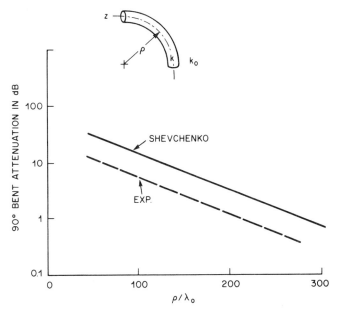

Fig. 5-31 Attenuation of a 90° bent as a function of bending radius. Shevchenko's theory,[31] —; experimental results, ---. (From Neumann and Rudolph,[35] by permission of the IEEE.)

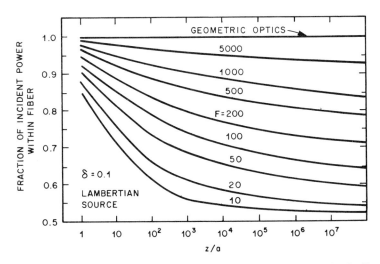

Fig. 5-32 Fraction power incident in a round dielectric fiber that remains in the fiber, as a function of normalized axial coordinate. The slow decay is associated with the leak of modes having large azimuthal numbers (slightly leaky rays). $F \equiv (k^2 - k_s^2)^{1/2}a$ denotes the normalized frequency.

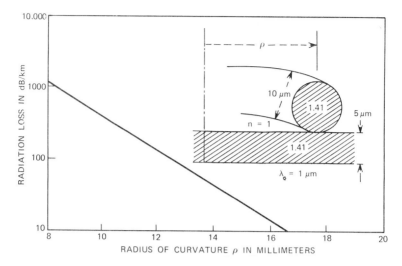

Fig. 5-33 Bending loss of a slab-coupled rod as a function of bend radius ρ. (From Arnaud.[34] Reprinted with permission from *The Bell System Technical Journal*, copyright 1974, American Telegraph and Telephone Co.)

account the distortion of the fiber field due to bending nor the elastooptic coefficient.

The radiation loss of slightly leaky helicoidal modes ($\mu \neq 0$) of dielectric rods with straight axis could be obtained along the same lines. The result obtained by Snyder[27] from a direct perturbation of the equation defining k_z is shown in Fig. 5-32. The power left in the fiber is shown as a function of the axial distance from the source. A uniform excitation of the modes is assumed.

The mode-selection mechanism afforded by the coupling of an oversized fiber to a dielectric slab that allows the fiber to be oversized and yet single moded has been discussed in Section 5.14. Slab-loaded fibers are more sensitive to bends than conventional fibers if the bend is in the plane of the slab. Closed-form expressions for the bending loss of slab-loaded fibers can be found in Arnaud.[34] The dependence of the loss on the radius of curvature is shown in Fig. 5-33. The dependence of the bending loss on the axial wavenumber of the various modes of the isolated fiber is shown in Figs. 5-34 and 5-35. The arrows give the axial wavenumbers of the fundamental mode and of the higher-order modes of the fiber. The "sound barrier" effect pointed out in Section 5.14 that takes place when the axial wavenumber of the isolated fiber is just equal to that of waves in the slab is less pronounced for the bent fiber than for the straight fiber, but is still very significant. (Note that the vertical scales in Figs. 5-34 and 5-35 are logarithmic in decibels.)

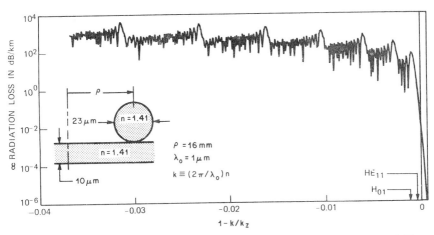

Fig. 5-34   Bending loss of the slab-coupled rod for various modes (HE_{11}, H_{01}, ...). Note the "sound barrier" jumps in loss that occur when the rod axial wavenumber equals the slab free wavenumber. (From Arnaud.[34] Reprinted with permission from *The Bell System Technical Journal,* copyright 1974, The American Telegraph and Telephone Co.)

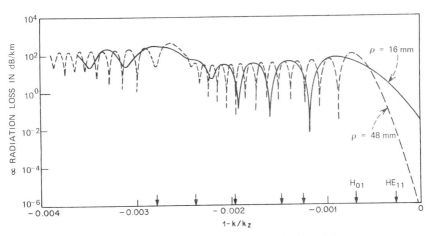

Fig. 5-35   Detail of Fig. 5–34 near the critical condition.

5.18 Radiation Losses due to a Wall Perturbation

In the preceding sections, it was assumed that the fiber is perfectly uniform along the z-axis. There are, however, unavoidable fluctuations in the diameter of fibers along the axis. These fluctuations have two effects: They couple the modes guided by the fiber, and they cause the guided modes to radiate (in the language of coupled mode theory, the fluctuations couple trapped modes to radiation modes.)[38]

Two trapped modes are coupled significantly whenever the wavenumber of the irregularities (2π divided by the period), assumed sinusoidal, equals their difference in wavenumber. If the irregularity is not sinusoidal, it can be expressed as a Fourier integral over a finite length of the fiber. Each component can be considered independently, provided the amplitude of the irregularity is sufficiently small. If the wavenumber Ω_0 of the geometrical irregularities is precisely equal to $k_{z1} - k_{z2}$, where k_{z1} and k_{z2} are the axial wavenumbers of two guided modes, there is complete transfer of power from one mode to the other after a length that depends on the amplitude of the sinusoidal irregularity. The situation is therefore similar to that of two uniform and identical coupled guides. A similar problem is encountered in quantum mechanics in the theory of stimulated emission from one level to another. The perturbation in that case is caused by the external electric field, whose angular frequency is close to $\omega_1 - \omega_2$, the difference in energy level between the two atomic states. A trapped mode with wavenumber k_z radiates into free space if the wavenumber Ω_0 of the irregularity exceeds $k_z - k_s$, that is, if $k_z - \Omega_0$ reaches the wavenumbers of radiation modes. A rather rigorous approach is based on the coupling between guided modes and radiation modes. The radiation loss is found to be enhanced when $k_z - \Omega_0$ corresponds to the axial wavenumber of a leaky mode.

We shall give here only an order of magnitude of the radiation loss. Let us assume that the slab is thick and consider a guided mode near cutoff, with k_z only slightly larger than k_s. Thus a relatively small wavenumber Ω_0 of the irregularity is sufficient to have $k_z - \Omega_0$ coincide with the wavenumber of a radiation mode.

We have seen that the field of the unperturbed slab can be represented by two plane waves going upward and downward between the slab boundaries. To within a constant factor $(\epsilon/\mu_0)^{1/2}$ that we omit for brevity, the power in the slab is approximately dE^2, where $2d$ is the slab thickness and E the field of the plane waves. Consider the plane wave incident on the slab boundary. If this boundary is displaced by a distance x, the field just outside the boundary is multiplied by

$$\exp[i(k_x - k_{xs})x] \tag{5.236}$$

as we have seen in Chapter 4. Near cutoff, k_{xs} (which is imaginary) is much smaller than k_x. For a small sinusoidal deformation $x(z) = x_0 \sin(\Omega_0 z)$, the field just outside the slab boundary is therefore

$$E \exp(ik_z z) \exp[ik_x x(z)] \approx E[1 + ik_x x(z)] \exp(ik_z z)$$

$$\approx E \exp(ik_z z) + \tfrac{1}{2} k_x x_0 E[\exp(i\Omega_0 z) - \exp(-i\Omega_0 z)] \exp(ik_z z) \tag{5.237}$$

We are interested only in the component that has an $\exp[i(k_z - \Omega_0)z]$ $\equiv \exp(i\beta z)$ dependence on z. Its squared normalized amplitude is

$$E_p^2 = \frac{k_x^2 x_0^2}{4d} \tag{5.238}$$

The radiation of this perturbed field is obtained from conventional antenna theory. The perturbed field radiates in a direction defined by the vector $\mathbf{k}_s$, whose length is k_s and whose axial component is $\beta \equiv k_z - \Omega_0$. The power radiated, omitting a factor $(\epsilon_s/\mu_0)^{1/2} \approx (\epsilon/\mu_0)^{1/2}$, is $L(k_{xs}'/k_s)E_p^2$, where $L \gg d$ denotes the length of the sinusoidal irregularity, and $k_{xs}'^2 \equiv k_s^2 - \beta^2$. The attenuation k_{zi} of the guided mode caused by the sinusoidal irregularity is half the ratio of the power lost per unit length divided by the power carried by the mode. We thus obtain

$$k_{zi} \text{ nepers/unit length} = \frac{k_{xs}'k_x^2x_0^2}{8k_sd} \tag{5.239}$$

where

$$k_{xs}'^2 \equiv k_s^2 - \beta^2, \qquad \beta \equiv k_z - \Omega_0, \qquad k_x^2 = k^2 - k_z^2 \tag{5.240}$$

The radiation loss is zero when the radiation condition $\beta < k_s$ is just met ($k_{xs}' = 0$) and increases as the square root of $k_s - \beta$. (See Fig. 5-36.)

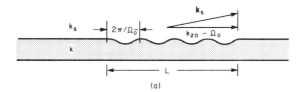

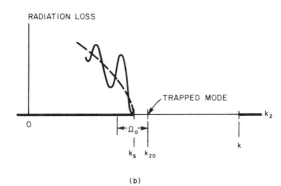

Fig. 5-36 Radiation loss of a slab with sinusoidal variation of one boundary (period $2\pi/\Omega_0$). (a) Slab geometry. (b) Schematic description of the loss. The calculation in the text ignores the excitation of the leaky modes and gives the dotted line, free of oscillations.

In deriving (5.239) we have omitted the fact that the field of the wave reflected (rather than transmitted) at the sinusoidal boundary is perturbed too and multiplied by $\exp(2ik_x x)$. This reflected wave eventually emerges from the slab and contributes to the radiation loss. When $\beta \equiv k_z - \Omega_0$ is close to the axial wavenumber of a leaky mode, the contributions of the two waves add up in phase, and radiation loss exceeds that given in (5.239). Our result (5.239) gives an average value of the radiation loss, but ignores the fast fluctuations associated with the excitation of leaky modes.

More advanced theories of nonuniform fibers are listed in Refs. 38 and 39. When the period of the irregularity is of the order of half a wavelength, forward waves are coupled to backward waves. This is the principle of the "distributed feedback laser" proposed by Kogelnik and Shank.[39] This important subject cannot be dealt with adequately here. We therefore refer the reader to the papers cited in Yariv and Gover.[39] For smooth transitions between two open waveguides, the reader should consult Shevchenko's work (note the warning on pp. 11, 12 of that book).[31]

5.19 Optical Fibers for Communication

Problems, methods, and results most relevant to communication by optical fibers[40] are summarized in this section. Some practical aspects of fiber optics are placed in perspective. We briefly outline the advantages and drawbacks of the various types of optical fibers that have been considered earlier, characterize the materials available, and describe simple numerical methods for the evaluation of pulse broadening. Finally, a few essential results concerning communication systems using glass fibers are presented. Recent reviews are listed in Ref. 41.

Types of Fibers

Optical fibers of potential value for communication are sketched in Fig. 5-37. Because present fabrication techniques strongly favor circularly symmetric configurations, the most important fibers are those shown in the first line of the figure and in (m) and (k).

The best-known fiber is the multimode step-index fiber shown in (a). The coupling of this fiber to light-emitting diodes is good if $\Delta n/n$ is large, but optical pulses broaden very rapidly. This type of fiber seems to be of main interest for short distances, of the order of 100 m.

The multimode near-square law fiber shown in (b) is the most important at the present time, because this type of fiber exhibits small pulse broadening (though not as small as theory predicts for the optimum profile).

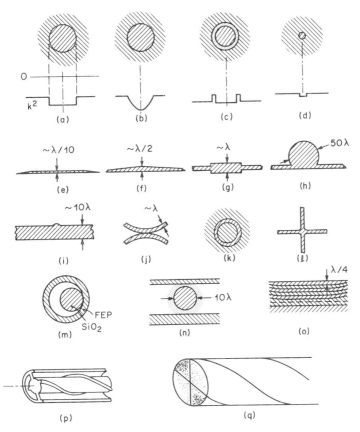

Fig. 5-37   Schematic representation of various optical fibers. Two asterisks indicate that the fiber was experimented in optics, one asterisk that it was experimented at microwave frequencies. (a) **Multimode step-index fiber. (b) **Multimode near square-law fiber. (c) **"W" fiber. (d) **Single-mode fiber. (e) **Thin film fiber. (f) Multimode tapered fiber with optimum profile. (g) Ridge fiber. (h) **Single-material fiber. (i) Perturbed slab. (j) *Contacting dielectric tubes. (k) Annular fiber. (l) "X" fiber. (m) **FEP clad fiber. (n) Slab-coupled fiber. (o) Periodic layers with slow transverse taper. (p) Helicoidal fiber. (q) Helical fiber.

Numerical techniques that are particularly useful for that type of fiber will be given in this section. Experimental results were reported in Section 4.17.

The "W" fiber in (c) is based on a new concept.[36,37] It carries fewer modes for a given core radius than the previously considered fibers because high-order modes tunnel to the cladding, through the ring-shaped refractive index barrier. We are not aware of conclusive experimental results.

The fiber in (d) is dimensioned in such a way that it carries only one mode, or only a few modes. The details of the refractive index profile are rather unimportant. Single-mode fibers can be used only with laser sources. This type of fiber has been found capable of low loss and large transmission capacity in laboratory conditions but cabling remains difficult. By letting the fiber carry a few modes, rather than just one, the sensitivity to bends is reduced.

Optical fields can be trapped in annular regions of higher index,[58] as shown in (k). Concentric rings of high refractive index materials can provide multichannel operation, with the circular symmetry being preserved.

The fiber shown in (m) is similar in principle to that in (a), but the cladding is a high-loss plastic material rather than glass. In order to reduce the influence of the cladding loss on the fiber loss, the plastic envelope is loosely fitted to the core. This type of fiber, described in more detail in Section 4.17, is rugged and has low loss, but pulse broadening is large. It is mostly attractive for short-distance applications. Newly found silicon rubbers, with losses of the order of 1000 dB/km, allow the fitting of the cladding to the core to be tight, and thereby provide a good protection of the active core.

The helical fiber shown in (q) has a saddle-shaped graded-index profile that rotates spatially along the fiber axis (see Section 2.17). This helical fiber belongs to the class of near square-law fibers. Thus, pulse broadening is in principle small. Whether the helical fiber presents any advantage over more conventional graded-index fibers has not been established.

We shall now consider fibers that employ (or could, in principle, employ) only one material, for instance, pure silica. Single-material fibers were first proposed at a time when the lowest measured losses of solids were of the order of 1000 dB/km. It was then proposed to trap the optical field in a tapered film having a thickness much smaller than the wavelength as shown in (e), in order that most of the field propagate in air (Nishizawa and Otsuka[12]). Preliminary experimental results are reported in Ref. 12. After it was discovered that pure silica has low optical loss,[40] slab thicknesses comparable to the wavelength or larger were considered.[14,15] The single-material fibers in (f)–(j) are closely related to one another. Yet, the expected modes of operation are somewhat different.

Let us first consider the fibers in (h) and (i). Experimental results obtained with this type of fiber in 1973 by Kaiser et al.[42] are remarkable because of the length of fiber drawn (of the order of 1 km) and the low loss achieved. The profile of a single-material fiber for multimode operation is shown in (h) and for single-mode operation in (i). The envelope, which can be made with the same material as the guiding region, is not shown on the figure. Rules have been given for evaluating the number of modes that can be trapped in the structure.[14] This number depends on the geometric dimensions of the fiber, essentially the ratio of the diameter of the core to the thickness of the supporting slab; it is almost independent of wavelength. Deatiled theories of propagation of single-material fibers such as the one shown in (g) are given by Marcatili.[14]

We have shown in Section 5.6 that by properly selecting the profile, the modes of a tapered slab [such as the one shown in (f)] can be made to have almost equal group velocities. In tapered slabs, the transverse variation of thickness is so slow that there is no significant coupling between modes. Thus, it is essential to ensure that these modes have almost the same group velocity. The tapered slab is intended to be used in conjunction with a source that is multimoded in only one dimension, for instance, an injection laser having a wide junction. Because this structure lacks circular symmetry, its fabrication may be difficult.

Contacting tubes, shown in (j), and crosses, shown in (l), also can guide optical waves. Experimental results have been reported in the microwave[13] and optical[42] ranges.

The helicoidal single-mode single-material fiber[7] shown in (p) presents, over the fiber shown in (i), the advantage of not requiring a thin membrane. Its operation has not been demonstrated. It has been demonstrated, however, that a silica preform can be spun at the required rate during the fiber pulling process (J. A. Arnaud, unpublished). There are two general objections to single-material fibers: the difficulty of fabricating and splicing them, and the rather large amount of pure silica employed. However, fibers that employ only pure silica in the active region are much less sensitive to radiations than doped silica fibers. This is important in some applications.

The fiber shown in (n) is akin to single-material fibers. Indeed, the rod and the slab (which acts as a mode sink) can be made of the same material and be in physical contact with one another. The high-order modes of the rod are eliminated because they couple to the slab modes. This mechanism of mode selection is more efficient than that found in the W fiber [in (c)] because of the "sound barrier" effect discussed in Section 5.14. The slab-coupled rod should be less sensitive to bends than conventional fibers.

The multilayer guide in (o) discussed in Section 5.4 has the unique feature of guiding most of the optical power outside the dielectric material, and yet to possess a thick substrate [unlike the guide in (e) for example]. It can be tapered for transverse confinement. Transverse coupling between two such multilayer guides would probably be easier to achieve than with conventional guides. No experiment concerning the multilayer fiber has been reported, however, and it is not known how accurate the layer thickness needs to be in order that low-loss operation be achieved.

Bundles of closely spaced clad fibers are useful because of the large total cross section. Splicing is easy, but the splicing loss is high (e.g., ~ 3 dB) because, in that technique, no attempt is made to align individual cores.

Finally, let us mention the possibility of introducing an effective anisotropy in the fiber material with the help of fast spatial variations of the refractive index (e.g., with a period of about 0.5 μm in the radial direction). This fast oscillation can be superimposed on a slow variation. The resulting anisotropy provides an additional parameter that may help minimize pulse broadening.

Material Characteristics

The refractive index of fused silica (SiO_2) can be raised by mixing it, for example, with germania (GeO_2) and phosphor oxide (P_2O_5), and lowered by mixing it, for example, with boron oxide (B_2O_3). These compounds can be formed inside a capillary tube by a vapor-phase deposition technique. The tube is subsequently collapsed into a preform. The preform is heated (at about 2000°K) with a torch or a carbon dioxide laser and pulled into a fiber. The material characteristics discussed below are applicable to bulk samples. One must be cautioned that the optical properties of the fiber material may differ somewhat from those of bulk samples having the same composition.

The variation of the square of the refractive index n as a function of the free-space wavelength λ_0 is well described, for most transparent materials, by a three-term Sellmeier law

$$n^2 = 1 + \sum_{\gamma=1}^{3} A_\gamma (1 - \pi_\gamma)^{-1} \qquad (5.241\mathrm{a})$$

where

$$\pi_\gamma \equiv \left(\frac{l_\gamma}{\lambda_0} \right)^2 \qquad (5.241\mathrm{b})$$

This expression involves six coefficients, namely, A_1, A_2, A_3, l_1, l_2, l_3, that can be obtained by measuring n at six different wavelengths, or by best-fitting the experimental curve. In applications, the two following dispersion parameters are needed: The parameter

$$D \equiv \frac{\omega}{k} \frac{dk}{d\omega} \qquad (5.242)$$

and the material dispersion parameter

$$M \equiv \frac{\omega^2}{k} \frac{d^2k}{d\omega^2} = \frac{\lambda_0^2}{n} \frac{d^2n}{d\lambda_0^2} \tag{5.243}$$

where $k \equiv (\omega/c)n$, $\omega/c \equiv 2\pi/\lambda_0$. The radial variation of D is involved in the evaluation of ray broadening, and M is involved in the evaluation of the effect of the nonzero spectral width of the source. After a few transformations, D and M can be written, respectively,

$$D - 1 = n^{-2} \sum_{\gamma=1}^{3} A_\gamma \pi_\gamma (1 - \pi_\gamma)^{-2} \tag{5.244}$$

$$M = -(D-1)^2 + n^{-2} \sum_{\gamma=1}^{3} A_\gamma \pi_\gamma (1 - \pi_\gamma)^{-2} \frac{3 + \pi_\gamma}{1 - \pi_\gamma} \tag{5.245a}$$

The coefficients A_γ and l_γ of the Sellmeier law (5.241) have been measured recently by Fleming[43] for the ternary compound SiO_2–GeO_2–B_2O_3. The samples were prepared by rf plasma fusion of glass particulates and cooled slowly. The refractive index measurements were made with an Abbe refractometer with an accuracy of about 10^{-4}. The coefficients A_γ, l_γ, $\gamma = 1, 2, 3$, are given in Table 5.6 for various concentrations of germania (in %) and boron oxide (in %). The square of the refractive index is approximately proportional to dopant concentration for germania. In the presence of boron oxide, quadratic terms must be included.

Table 5.6 Computed Parameter for Sellmeier Fit[a]

Sample GeO₂(%)	B₂O₃(%)	A_1	l_1	A_2	l_2	A_3	l_3
0	0	0.6961663	0.0684043	0.4079426	0.1162414	0.8974794	9.896161
4.1	0	0.68671749	0.07267519	0.43481505	0.11514351	0.89656582	10.002398
7	0	0.6869829	0.078087582	0.44479505	0.11551840	0.79073512	10.436628
13.5	0	0.73454395	0.086976930	0.42710828	0.11195191	0.82103399	10.846540
9.1	7.7	0.72393884	0.085826532	0.41129541	0.10705260	0.79292034	9.3772959
0	25	0.70724622	0.080478054	0.39412616	0.10925792	0.63301929	7.8908063
0	25[b]	0.67626834	0.076053015	0.42213113	0.11329618	0.58339770	7.8486094
4	9.7	0.70420420	0.067974973	0.41289413	0.12147738	0.95238253	9.6436219
0	7	0.69200948	0.065669958	0.40329419	0.11628286	0.94625695	8.7506878

[a] Table taken from Fleming,[43] by permission.
[b] Quenched.

Table 5.7 Refractive Index and Dispersion Parameters at Three Wavelengths for Germania and Boron Oxide Doped Silica

λ_0 (μm)	GeO$_2$ (%)	B$_2$O$_3$ (%)	n	$10^3 \times (D - 1)$	$10^3 \times M$
0.9			1.45175	8.85	12.0
1.06	0	0	1.44967	8.79	6.1
1.2			1.44805	9.43	1.9
0.9			1.47295	9.00	16.6
1.06	13.5	0	1.47089	8.26	9.9
1.2			1.46939	8.30	5.7
0.9			1.46301	9.40	14.0
1.06	9.1	7.7	1.46082	9.13	7.7
1.2			1.45913	9.60	3.2
0.9			1.45073	9.60	12.0
1.06	0	25	1.44847	9.65	6.0
1.2		(quenched)	1.44667	10.50	1.3
0.9			1.44992	9.40	13.0
1.06	0	25	1.44773	9.30	6.5
1.2			1.44601	10.00	1.9
0.9			1.45617	9.57	12.9
1.06	4	9.7	1.45390	9.60	6.2
1.2			1.45212	10.30	1.7
0.9			1.44766	9.98	10.4
1.06	0	7	1.44526	10.50	3.7
1.2			1.44328	11.70	−1.0

The values of n, D, and M at three wavelengths $\lambda_0 = 0.9$, 1.06, and 1.2 μm obtained from Table 5.6 and (5.241), (5.244), and (5.245) are shown in Table 5.7 for various dopant concentrations.

For power-law media, there is a simple relation between the inhomogeneous dispersion parameter D_κ introduced in Chapter 4 and the radial variation of D in Table 5.7:

$$D_\kappa - 1 = \frac{[n_s^2 / D(0)][D(0) - D(a)]}{n_0^2 - n_s^2} \qquad (5.245b)$$

where n_0 and n_s denote the axial and cladding indices, and a the core radius. In typical multimode fibers, $D_\kappa - 1$ is about 50 times the radial variation of D.

The refractive index of the actual fiber can be estimated by a simple scaling from measurements made on the preform if the dopant has low diffusivity. This is the case, in particular, for germania. The diffusion of boron oxide, however, is high and, for that dopant, the scaling law may not be applicable. The refractive index of the actual fiber can be measured with limited accuracy with an interference microscope (essentially a Mach-Zehnder interferometer). A simple alternative technique consists of measuring the Fresnel reflection. An accuracy of 2×10^{-3} and a resolution of 1 μm have been obtained.[44] The losses of the material are mainly due to Rayleigh scattering and absortion by impurities (see Section 4.17). They are not known with precision as a function of radius.

Among other material characteristics that are important for the design of optical fibers, we mention the following.

(a) The onset of nonlinear effects, such as Raman and Brillouin stimulated scattering. For broadband sources such as light-emitting diodes these effects appear only for powers larger than 1 W and can usually be ignored.[45]

(b) Sensitivity to radiation. For a typical germania-doped silica fiber a 100 rad γ-ray irradiation (from a ^{60}Co source) increases the loss by 1 dB/km. Pure silica is much less sensitive to radiation than doped silica by about two orders of magnitudes. Glasses are much more sensitive to radiation than doped silica by about two orders of magnitudes.[46] Note that 20 rad is roughly the radiation level expected from background lifetime.

Problems of resistance of the fiber material to stress, corrosion, etc., are obviously of major importance[47] but they cannot be dealt with adequately here. Let us only note that large changes of refractive index in the fiber cross section usually imply large changes in coefficients of thermal expansion and high stresses during the cooling process.

Numerical Evaluation of Pulse Broadening in Optical Fibers

Closed-form expressions were given in Chapter 4 for the impulse response of fibers whose index squared profile is a constant plus a power of radius, or when the deviations of the profile from a square law are small. These analytic results were based on the scalar ray optics approximation.

The ray optics numerical method described below is to be used when the departure of the profile from square law is large. We shall subsequently give a numerical technique based on wave optics, which should be used when the fiber carries few modes, or when the profile is not well behaved, for example, when it varies by steps. Because the principle of these numerical techniques is straightforward, they can be implemented on programmable desk computers. The computing time, however, can be large. (Evaluation of the axial wavenumber, time of flight, and field pattern of the 1000 modes carried by a typical multimode fiber requires 5 min on an IBM 370 computer, if one uses the method described below.)

When an optical fiber is highly multimoded, and when the refractive index profile is well behaved, ray optics methods provide an efficient way of estimating the broadening of optical pulses. Ray optics (ør WKB) methods also provide useful guidance toward obtaining wave optics solutions.

We shall assume that the dopant concentrations d_1, d_2 have been measured as functions of the transverse coordinates x, y of the fiber, for instance with an electron microscope and an X-ray analyzer. The fiber is assumed to be uniform and the fiber material isotropic.

The refractive index n, the dispersion parameter D, and the material dispersion M are then obtained from (5.241), (5.244), and (5.245), respectively, as functions of x and y, at some wavelength λ_0.

Measurements can be made on the optical source (e.g., an LED) to obtain the radiation pattern of a small area $dx\,dy$ of the source centered at x, y, selected by a pinhole. The radiation is measured in the direction defined by the transverse components k_x, k_y of the wave vector **k**, to within dk_x, dk_y. Sometimes the fiber is sealed to an LED source. In that case, the measurement is made from the tip of the fiber section. Let us assume that the radiation from the small area $dx\,dy$ is displayed on a screen perpendicular to the fiber axis at a distance d from the source. The coordinates on the screen are denoted x_s, y_s. Let us also assume, for simplicity, that there is an index-matching fluid between the fiber and the observing screen, and that refraction at the fiber tip can be neglected. If the irradiance on the screen, proportional to the detected power, is $P(x_s, y_s)$, the power density $f(k_x, k_y)$ in the k_x, k_y space is obtained by evaluating the jacobian of the transformation $x_s/k_x = y_s/k_y = d/k_z$. We obtain

$$f(k_x, k_y) = \frac{k^2 d^2}{k_z^4}\, P\left(\frac{k_x d}{k_z}\, ,\, \frac{k_y d}{k_z} \right) \tag{5.246}$$

where $k_z^2 = k^2 - k_x^2 - k_y^2$. The arguments x, y of P and f have been omitted for brevity. The frequency spectrum of each ray pencil that originates from the source can be measured with a monochromator. It is generally a function of x, y, k_x, and k_y.

From this basic information concerning the fiber and the source, the impulse response can be obtained with the numerical technique described below, based on ray optics. For simplicity, the fiber is assumed to have circular symmetry. The basic equations of ray optics were derived in Chapter 4. They are reproduced below for the reader's convenience.

The time of flight $t(z)$ of a pulse along a ray defined by the initial conditions $x(0) \equiv x_0$, $y(0) \equiv y_0$, $k_x(0) = k_{x0}$, $k_y(0) = k_{y0}$ is obtained by integrating the pair of equations

$$\frac{dr(z)}{dz} = \frac{[k^2(r) - k_z^2 - \mu^2/r^2]^{1/2}}{k_z} \tag{5.247a}$$

$$\frac{dt(z)}{dz} = \frac{k^2(r)}{\omega k_z} D(r) \tag{5.247b}$$

where $k(r) \equiv (\omega/c)n(r)$, and $n(r)$ and $D(r)$ are obtained from (5.241) and (5.244), respectively. $k_z^2 = k^2(r_0) - k_{x0}^2 - k_{y0}^2$ and $\mu = x_0 k_{y0} - y_0 k_{x0}$ are obtained from the initial conditions of the ray. Both k_z and μ remain constant along any given ray. It is sufficient to carry the integration of (5.247) from the minimum root r_{min} of the square root in (5.247a) to the maximum root r_{max}. [Using the pair of equations in (5.247), t can be written as an integral over r from r_{min} to r_{max}. The numerical evaluation of that integral is more difficult than that of (5.247), however, because the integrand is singular at the limits of integration. We shall therefore not discuss this alternative formulation.]

The relative time of flight τ is defined as the ratio of the time of flight of a pulse on a ray to the corresponding time $t_0 \equiv [k_0 D(0)/\omega]z$ on axis. To minimize rounding errors, it is advisable to integrate the difference $t - t_0$, rather than t and t_0 separately.

If we set for convenience in (5.247)

$$k^2(r) \equiv k_0^2[1 - \delta^2(r)] \tag{5.248a}$$

with $\delta(0) = 0$, and

$$k_z^2 \equiv k_0^2(1 - B), \qquad k_0 r \equiv \bar{r}, \qquad k_0 z \equiv \bar{z}, \qquad \omega t \equiv \bar{t} \tag{5.248b}$$

the ray equations become

$$\frac{d\bar{r}}{d\bar{z}} = \left[\frac{B - \delta^2(\bar{r}) - \mu^2/\bar{r}^2}{1 - B} \right]^{1/2} \tag{5.248c}$$

$$\frac{d\bar{t}}{d\bar{z}} = [1 - \delta^2(\bar{r})](1 - B)^{-1/2} D(\bar{r}) \tag{5.248d}$$

which are easily integrated numerically for given initial conditions μ, B obtained by sampling the k_x, k_y, x, y space with the weight f given in (5.246).

The above equations provide the transit time of the pulse center. Because the pulse has finite duration, it has a nonzero spectral width, and some frequency components may arrive ahead of others. This source of pulse broadening is usually negligible in fiber optics. However, if the source has a broad frequency spectrum, as is the case for LEDs, much larger than the reciprocal of the pulse width, the pulse may broaden significantly even for plane waves. The broadening is proportional to the spectral width of the source and to the second derivative of the refractive index with respect to wavelength $\partial^2 n / \partial \lambda_0^2$. In first approximation, it is sufficient to consider this effect for axial rays. With the notation in (5.243), we find that pulse broadening due to material dispersion alone is

$$\Delta t \approx 5000 \, \frac{\Delta \lambda_0}{\lambda_0} \, M(0) \text{ nsec/km} \tag{5.249}$$

where $\Delta \lambda_0 / \lambda_0$ is the relative spectral width of the source. From the numerical values in Table 5.7 and a source spectral width of 0.04 μm, for example, we find from (5.249) that pulse broadening in pure silica is 2.7 nsec/km at $\lambda_0 = 0.9$ μm and 0.3 nsec/km at $\lambda_0 = 1.2$ μm. These values do not include ray broadening.

As we have seen, the parameter M vanishes for most materials at wavelengths in the 1.15–1.4 μm range. If a fiber is designed to operate near that range of wavelengths, the fact the M varies along a ray cannot be ignored. An integration needs to be performed in order to obtain a sufficiently accurate value of pulse broadening. The simplest method in that case is to evaluate the quasi-monochromatic impulse response at various frequencies, and convolve it with the source spectrum.

When the number of modes carried by the fiber is small, or when the refractive index profile is not well behaved, ray optics techniques are insufficiently accurate and one must resort to wave optics techniques. In most practical fibers, the variation of the refractive index in the cross section is small, typically less than 2%, and the scalar Helmholtz equation, applied to the transverse components of the field, is sufficiently accurate. (For a numerical technique based on the Maxwell equations, see Vassell[48] and Clarricoats and Chan.[25])

We shall assume again circular symmetry and factor out a term $\exp[i(\mu \varphi + k_z z - \omega t)]$. The basic formulas were given earlier in this chapter and are repeated here for the reader's convenience. The field $E(r)$ is a solution of

$$\frac{r^{-1} d(r \, dE/dr)}{dr} + [k^2(r) - k_z^2 - \mu^2/r^2] E = 0 \tag{5.250}$$

where $k(r) \equiv (\omega/c)n(r)$. If $k(r)$ has finite discontinuities, E and its first derivatives are continuous. With the notation in (5.248a) and (5.248b) the scalar Helmholtz equation (5.250) can be written as a pair of first-order equations[49]

$$\frac{dE}{d\bar{r}} = \frac{K}{\bar{r}}$$

$$\frac{dK}{d\bar{r}} = \bar{r}\left[\delta^2(\bar{r}) - B + \frac{\mu^2}{\bar{r}^2} \right] E \qquad (5.251)$$

where $K(\bar{r})$ and $E(\bar{r})$ are continuous functions of $\bar{r}$. The mode field has an oscillatory behavior when

$$A(\bar{r}) \equiv \delta^2(\bar{r}) - B + \frac{\mu^2}{\bar{r}^2} \qquad (5.252)$$

is positive. The zeros of $A(\bar{r})$ correspond to the turning points of rays in the WKB approximation. Let the first and last zeros of $A(\bar{r})$ be denoted $\bar{r}_{\min}$ and $\bar{r}_{\max}$, respectively. From $\bar{r} = 0$ to $\bar{r} = \bar{r}_{\min}$, it is convenient to integrate, instead of (5.251), the differential (Riccati) equation for the ratio $P(\bar{r}) \equiv K(\bar{r})/E(\bar{r})$. We have from (5.251)

$$\frac{dP}{d\bar{r}} = \bar{r}[\delta^2(\bar{r}) - B] + \frac{\mu^2 - P^2}{\bar{r}} \qquad (5.253)$$

with the initial condition $P(0) = \mu$. Note that the rhs of (5.253) is well behaved at $\bar{r} = 0$. When $\bar{r}_{\min}$ is reached, one switches from (5.253) to (5.251) with the initial conditions $E(r_{\min}) = 1$, $K(r_{\min}) = P(r_{\min})$, to avoid the singularities that P exhibits at the zeros of E. When $\bar{r} = \bar{r}_{\max}$ is reached, one can revert to (5.253) and specify the following condition at the core–cladding interface:

$$P(\bar{a}) = \mu - \frac{vK_{\mu+1}(v)}{K_\mu(v)} \equiv Q_\mu(v) \qquad (5.254)$$

where K_μ is the modified Bessel function of the second kind of order μ, and

$$v \equiv \left(k_z^2 - k_s^2\right)^{1/2} a \equiv \left(\delta^2 - B\right)^{1/2}\bar{a} \qquad (5.255)$$

k_s denotes the free wavenumber in the cladding, $\delta \equiv \delta(\bar{a})$, and $\bar{a} \equiv k_0 a$.

The first three terms of the asymptotic expansion of $Q_\mu^2(v)$

$$Q_\mu^2(v) \sim \mu^2 + v + v^2 \qquad (5.256)$$

give $Q_\mu(v)$ with sufficient accuracy, even for small values of v.

Once a modal solution $[E(\bar{r}), B]$ has been obtained, the group delay follows by application of the Hellmann–Feynman theorem (5.72). With the present notation, the relative time of flight τ, defined as the ratio of the time of flight of a pulse in the mode $E_{\mu\alpha}$ to the time of flight of a pulse of a free wave on axis, is

$$\tau = (1 - B)^{-1/2} \frac{\langle [1 - \delta^2(\bar{r})] D(\bar{r}) E^2(\bar{r}) \rangle}{D(0) \langle E^2(\bar{r}) \rangle} \qquad (5.257)$$

where the sign $\langle a \rangle$ denotes the integral of $a \, d\bar{r}^2 \equiv 2a\bar{r} \, d\bar{r}$, from $\bar{r} = 0$ to $\bar{r} = \infty$ and D is the dispersion parameter in (5.242). In the absence of dispersion, $D = 1$.

The power in the cladding, defined as the integral of $E^2 \, d\bar{r}^2$ from $\bar{r} = \bar{a}$ to $\bar{r} = \infty$, is

$$P_s = E^2(\bar{a})\bar{a}^2 \left(\frac{Q_\mu^2(v) - \mu^2 - v^2}{v^2} \right) \approx \frac{E^2(\bar{a})\bar{a}^2}{v} \qquad (5.258)$$

and the guide efficiency, defined as the ratio of the power in the core to the total power, is

$$\eta = \left(1 + \frac{P_s}{\langle E^2 \rangle} \right)^{-1} \qquad (5.259)$$

If, for example, the core is lossless and the cladding has a loss k_{si} (in nepers/unit length), the loss of the propagating mode is $(1 - \eta)k_{si}$ in nepers/unit length.

As an example of application of the previous numerical methods, let us consider a fiber whose refractive index varies by steps. The refractive-index profile approximates a square law in the limit when the number of steps tends to infinity. This profile is of practical interest because in the vapor phase deposition technique, the gases are often introduced by steps rather than in a continuous manner. Let the smooth square-law profile be

$$\delta^2(\bar{r}) = \begin{cases} 0.04\left(\dfrac{\bar{r}}{\bar{a}} \right)^2, & \bar{r} < \bar{a} \\[2mm] 0.04, & \bar{r} > \bar{a} \end{cases} \qquad (5.260)$$

$(\Delta n/n = 2\%)$, where $\bar{a} = k_0 a$, $k_0 = 2\pi \times 1.5 \ \mu\mathrm{m}^{-1}$, $a = 40 \ \mu\mathrm{m}$ (core

Table 5.8 **Relative Times of Flight $\tau - 1$ Multiplied by 10^3 as a Function of Radial Mode Number α and Number of Steps N[a]**

α	$N=1$ (step index)	$N=2$	$N=20$	$N=\infty$ (truncated square law)
0	0.019	0.036	0.14	0.00014
1	0.1	0.2	0.43	0.0012
2	0.25	0.48	0.27	0.003
3	0.47	0.89	0.42	0.006
4	0.75	1.4	0.63	0.010
5	1.1	2.1	0.1	0.015
6	1.5	2.6	0.52	0.022
7	2	3.7	0.68	0.029
8	2.55	4.8	0.22	0.038
9	3.2	5.8	0.64	0.049
10	3.8	7	0.51	0.058
11	4.6	7.8	0.1	0.070
12	5.4	−5	0.92	0.084
13	6.3	3.2	0.74	0.098
14	7.3	−4.3	0.062	0.11
15	8.3	6.2	0.78	0.12
16	9.4	0.06	0.63	0.11
17	10.5	8.2	0.051	−0.067
18	11.7	4.3	0.7	−1.5
19	13	9.4		
20	14.3	9.2		
21	15.6	8.9		
22	17			
23	17.5			

[a] To obtain times of flight in nsec/km, multiply the number shown by 5 ($\mu = 0$, $D = 1$).

radius), and the steps be uniform in $\bar{r}^2$ (each ring having the same area). That is, $\delta^2(\bar{r})$ is equal to zero from $\bar{r}^2 = 0$ to $\bar{r}^2 = \bar{a}^2/N$, where N denotes the number of steps, then takes the value in (5.260), and keeps it up to $\bar{r}^2 = 2\bar{a}^2/N$, and so on. $N = 1$ corresponds to a step-index fiber, and $N = \infty$ to the truncated square-law profile in (5.260). The relative times of flight τ for $\mu = 0$ and the radial mode numbers $\alpha = 0, 1, 2, \ldots$ are given in Table 5.8.[49] Dispersion is neglected ($D = 1$). The results for $N = 1$ (step index) and $N = \infty$ (truncated square law) agree well with expressions based on ray optics (or the WKB approximation),

$$\tau - 1 = \begin{cases} 8.7 \times 10^{-6}(2\alpha + \tfrac{3}{2})^2 & \text{(step index)} \\ 0.14 \times 10^{-6}(2\alpha + 1)^2 & \text{(square law)} \end{cases} \qquad (5.261)$$

except for the highest values of α because of the influence of the cladding.

For $N = 2$, the intermediate step is located at $a = 28$ μm. For $0 < \alpha \leqslant 11$, the time of flight is almost that of a step-index fiber having a core radius of 28 μm. For larger mode numbers $\alpha > 11$, the time of flight exhibits large and fast variations as a function of α. The ray optics technique, which ignores these fluctuations, would clearly be inapplicable. For 20 steps, large oscillations remain, which degrade considerably the impulse response of the fiber. The rms impulse width, calculated over all α, μ modes carried by the fiber, with $\Delta n / n = 0.02$, core radius $= 40$ μm, and the optimum profile, is $\sigma = 0.15$ nsec/km. For 40, 20, and 10 steps of equal area, σ increases, respectively, to 0.46, 2.4, and 8.8 nsec/km.[49] These numerical results point to the need of introducing gas flows in a continuous manner in the vapor phase deposition technique, rather than in a stepwise manner, for optimum results.

Observations of the modes of multimode fibers have been reported that agree well with theoretical predictions. The observation of the radiation pattern of fibers is usually made on a distant screen. The near field can also be observed by imaging. Single modes can be excited in multimode fibers with the help of prism couplers that select a particular value of the axial wavenumber.[50] A recent alternative technique uses stimulated four-photon mixing.[51] This technique permits the excitation of low-order modes, but it requires high-power sources.

System Considerations

In order to design a communication system using optical fibers, one must know how much loss, pulse broadening, and cross talk between adjacent fibers in a cable can be tolerated. Let us first consider the problem of loss.

Theoretical and experimental studies[52] have shown that if the source is pulse code modulated (PCM) and the detector is an avalanche photodiode (with a current gain of the order of 100), the power needed at the detector at a free-space wavelength $\lambda_0 = 1$ μm is about -62 dBm for a transmission rate of 6.3 Mbit/sec and -47 dBm for a transmission rate of 274 Mbit/sec. (Note: dBm $\equiv$ decibels above one milliwatt). The optical power delivered by light-emitting diodes (LED) and injection lasers are of the order of 1 mW.[§] Thus the tolerable loss is 62 dB for a 6.3-Mbit/sec system and 47 dB for a 274-Mbit/sec system. The optical loss consists of coupling loss, fiber loss, and splicing loss. The coupling loss from an LED to a

---

[§] To obtain a well-defined starting time, injection lasers are biased with a dc current just below threshold. The corresponding power dissipation limits the power that can be applied to the laser.

multimode fiber is sometimes as large as 15 dB. This is mainly because the number of modes (acceptance) carried by low-dispersion fibers is significantly smaller than the number of modes generated by existing LEDs.

Because the diameter of fibers is limited by requirements of mechanical flexibility to about 150 μm, and the diameter of the core of a fiber is limited to about 80 μm, the acceptance of a fiber is essentially limited by its numerical aperture (NA). For a typical LED and a step-index fiber with $NA = 0.65$, a very large numerical aperture that can be obtained only with special (e.g., lead) glasses, the coupling loss is only about 5 dB. The coupling loss would be theoretically 3 dB higher for near square-law fibers having the same numerical aperture. Lower coupling losses are possible, in principle, if the junction diameter of the LED is reduced, e.g., from 50 to 25 μm. For optimum operation the LED should in any case be designed according to the fiber considered. Low splicing losses, of the order of 0.4 dB, have been demonstrated for ribbons incorporating up to 12 multimode fibers.[53] The splicing loss is higher for single-mode fibers because the core radius is small, of the order of 12 μm. A considerable effort is currently made to find splicing techniques that can be used conveniently and reliably in the field. The cabling of single-mode fibers remains difficult. Multiple connectors have also been investigated.[54]

The normal bending loss of multimode fibers is usually negligible. However, microbending losses due to the deformation exerted by the plastic envelope on the fiber are troublesome and require special attention.[55]

The second question to be considered is how much pulse broadening can be tolerated to achieve a given bit rate. It has been established[52] that if pulse code modulation is used, the maximum bit rate is equal to $1/4\sigma$, where σ is the rms pulse width for a 1-dB signal-to-noise penalty. For a rectangular pulse of width Δt, we have $\sigma = 0.29 \, \Delta t$. Thus the maximum bit rate is just about the reciprocal of the width of the received pulse if the pulse is rectangular in shape. For LEDs and multimode near square-law fibers, the main source of pulse broadening (proportional to the fiber length) results from the nonzero spectral width of the source ($\sim 0.04 \, \mu$m) and the material dispersion $d^2n/d\lambda_0^2$. Material dispersion, as we have seen, vanishes at a wavelength of the order of 1.2 μm. This fact, combined with the fact that Rayleigh scattering varies as λ_0^{-4}, provides an incentive for developing efficient sources and detectors in the 1.2 μm region.

The third question is how much cross talk between fibers in a cable can be tolerated. For a PCM system, a -20-dB interference optical power can usually be tolerated. The coupling theory presented in Section 5.12 shows that the transfer of optical power from one fiber to another is proportional to the square of the fiber length, proportional to the fiber length, or independent of the fiber length depending on whether the fibers are identical, nominally identical but irregular, or uniform but dissimilar, respectively. In practical multimode fibers the power coupled from one fiber to the other is probably proportional to length. This point, however, would require investigations. The information transmission capacity of single-mode fibers is so large that there is little incentive to introduce more than one core within the same cladding. Thus, the problem of coupling between single-mode slabs discussed in Section 5.12 is not of major importance for the transmission of information. The problem of coupling between single-mode cores (or between cores carrying few modes) arises when one tries to increase the image transmission capacity of a fiber bundle up to the diffraction limit, each core carrying one bit of image information. As discussed in Section 5.12, cross talk (in that case image blurring) is minimized if adjacent cores are made dissimilar.

Cross talk between adjacent multimode fibers in a bundle can be reduced to tolerable values by covering each fiber with a sufficiently thick coating of lossy plastic material. For low-loss fibers, it would seem that coupling through scattering should be negligible. However this point has apparently not been investigated experimentally.

The ultimate success of communication by glass fibers rests on economic considerations. Studies have been made to see whether glass fibers may compete with pairs of wires, coaxial cables, low-loss millimeter waveguide systems, or microwave radio systems. There are, at the moment, too many unknown costs to allow accurate predictions to be made.[56] In spite of the fact that many costs are independent of the type of guide used, the future of communication by glass fibers, for both short and long distances, depends critically on the cost of the repeaters, and of the fiber. The cost of the raw materials needed to fabricate the fiber is not a negligible part of the total cost. Thus, further ingenuity on the part of optical, electrical, and mechanical engineers is very much needed in order to minimize the amount of expensive material involved in the fabrication of the fiber, the cabling and splicing costs, the repeater cost, and the number of repeaters in the link.

References

1. D. L. Mills and E. Burnstein, *Report on Progr. Phys.* **37**, 817 (1974).
2. M. Born and E. Wolf, "Principles of Optics." Pergamon, Oxford, 1965.
3. K. Artman, *Ann. Phys.* **2**, 87 (1948). J. L. Agudin, *Phys. Rev.* **171**, 1385 (1968). O. Bryngdahl, *in* "Progress in Optics" (E. Wolf, ed.), Vol. 11. North-Holland Publ., Amsterdam, 1973. H. Kogelnik and H. P. Weber, *J. Opt. Soc. Amer.* **64**, 174 (1974). T. Tamir and H. L. Bertoni, *J. Opt. Soc. Amer.* **61**, 1397 (1971). B. R. Horowitz, *Appl. Phys.* **3**, 411 (1974). J. Zagrodzinski, *Nuovo Cimento* **21B**, 129 (1974). J. Ricard, *Nouv. Rev. Opt.* **5**, 7 (1974). A. Kodne and J. Strnad, *J. Phys. A* **8**, 533 (1975).

4. O. Costa deBeauregard and C. Imbert, *Phys. Rev. D* **7**, 3555 (1973).
5. P. K. Tien, *Appl. Opt.* **10**, 2395 (1971). W. S. C. Chang, M. W. Muller, and R. J. Rosenbaum, *in* "Laser Applications" (M. Ross ed.). Academic Press, New York, 1974. L. V. Iogansen, *Sov. Phys.-Tech. Phys.* **7**, 295 (1962).
6. H. K. V. Lotsch, *Optik* **27**, 239 (1968).
7. J. A. Arnaud, *Bell Syst. Tech. J.* **53**, 1599, 1643 (1974); **54**, 1179 (1975).
8. H. Kogelnik and V. Ramaswamy, *Appl. Opt.* **13**, 1857 (1974).
9. D. P. Russo and J. H. Harris, *J. Opt. Soc. Amer.* **63**, 138 (1973).
10. J. A. Arnaud and A. A. M. Saleh, *Appl. Opt.* **13**, 2343 (1974).
11. R. Tsu, A. Koma, and L. Esaki, *J. Appl. Phys.* **46**, 842 (1975).
12. S. Kawakami and J. Nishizawa, Res. Inst. Elec. Comm. Tech. Rep., TR-25 (1967), *J. Appl. Phys.* **38**, 4807 (1967). J. Nishizawa and A. Otsuka, *App. Phys. Lett.* **21**, 48 (1972).
13. J. A. Arnaud, *Electron Lett.* **10**, 269 (1974).
14. E. A. J. Marcatili, *Bell Syst. Tech. J.* **53**, 645 (1974). V. P. Malt'sev, V. L. Mironov, and V. V. Shevchenko, *Rad. Eng. Electron Phys.* **17**, 1370 (1972). F. J. Tisher, *IEEE Trans. Microwave Theory Tech.* **11**, 291 (1963) T. Nakahara and N. Kurauchi *in* "Advances in Microwaves" (L. Young, ed.), Academic Press, New York, 1969.
15. J. A. Arnaud, *Bell Syst. Tech. J.* **53**, 1599 (1974).
16. J. A. Arnaud, *Bell Syst. Tech. J.* **53**, 675 (1974).
17. G. Biernson and D. J. Kinsley, *IEEE Trans. Microwave Theory Tech.* **MTT13**, 345, 885 (1965).
18. A. W. Snyder, *Electron Lett.* **6**, 561 (1970); *IEEE Trans. Microwave Theory Tech.* **MTT17**, 1130 (1969).
19. D. Glòge, *Appl Opt.* **10**, 2252 (1971).
20. J. B. Keller and S. J. Rubinow, *Ann. Phys.* **9**, 24 (1960).
21. J. A. Arnaud and W. Mammel, *IEEE Trans. Microwave Theory Tech.* **MTT23**, 927 (1975).
22. P. J. B. Clarricoats, *Proc. IEE* **108C**, 170 (1961).
23. E. Snitzer, *J. Opt. Soc. Amer.* **51**, 491 (1961). W. Schlosser and H. G. Unger *in* "Advances in Microwaves," Vol. 1 (L. Young, ed.), Academic Press, New York, 1966, p. 319.
24. N. S. Kapany and J. J. Burke, "Optical Waveguides." Academic Press, New York, 1972.
25. P. J. B. Clarricoats and K. B. Chan, *Proc. IEE* **120**, 1371 (1973).
26. J. A. Arnaud, *in* "Crossed Field Microwave Devices" (E. Okress ed.). Academic Press, New York, 1961. J. A. Arnaud, *Bell Syst. Tech. J.* **53**, 217 (1974). G. F. Kuester and D. C. Chang, *IEEE Trans. Microwave Theory Tech.* **MT23**, 877 (1975).
27. E. A. J. Marcatili, *Bell Syst. Tech. J.* **48**, 2071 (1969). R. Vanclooster and P. Pharisau, *Physics* **49**, 493 (1970). A. W. Snyder, *Appl Phys.* **4**, 273 (1974). R. Pregla, *Arch. Elek. Uberta.* **28**, 349 (1974). E. Voges, *ibid.* **28**, 478 (1974).
28. J. A. Arnaud, *Bell Syst. Tech. J.* **54**, 1431 (1975).
29. M. A. Miller and V. I. Talanov, *Sov. Phys., Tech. Phys.* **1**, 2665 (1956).
30. E. A. J. Marcatili, *Bell Syst. Tech. J.* **48**, 2103 (1969). L. Lewin, *IEEE Trans. Microwave Theory Tech.* **MTT22**, 718 (1974).
31. V. V. Shevchenko, *Izv. V.U.Z. Radiofiz.* **14**, 768 (1971); "Continuous Transitions in Open Waveguides." Golem Press, Boulder, Colorado, 1971. D. Marcuse, "Light Transmission Optics." Van Nostrand-Reinhold, Princeton, New Jersey, 1972.
32. J. R. Wait, *Radio Sci.* **2**, 1005 (1967). A. W. Snyder and D. J. Mitchell, *Electron Lett.* **10**, (1974); *Opt. Electron,* **6**, 287 (1974). L. B. Felsen and S. J. Maurer, *IEEE Trans. Microwave Theory Tech.* **MTT18**, 584 (1970).
33. A. W. Snyder, I. White, and D. J. Mitchell, *Electron Lett.* **11** (15), 332 (1975).

34. J. A. Arnaud, *Bell Syst. Tech. J.* **53**, 1379, 1643 (1974).
35. E. G. Neumann and H. D. Rudolph, *IEEE J. Microwave Theory Tech.* **MTT23**, 142 (1975).
36. S. Kawakami and S. Nishida, *IEEE J. Quantum Electron.* **QE10**, 879 (1974); *Electron. Lett.* **10**, 38 (1974).
37. J. A. Arnaud, OSA Topical Meeting on Int. Opt., Paper WB12, New Orleans, Louisiana, Jan. 21, 1974.
38. D. Marcuse, "Theory of Dielectric Optical Waveguides," Academic Press, New York, 1974. H. Kogelnik, *in* "Integrated Optics" (T. Tamir, ed.). Springer-Verlag, Berlin, 1975.
39. R. A. Waldron, *Radio Elec. Eng.* **43**, 751 (1973). V. G. Bezrodnyi and I. M. Fuks, *Radiophys. Quantum Electr.* **15**, 53 (1972). A. A. Zlenko, V. A. Kiselev, A. M. Prokhorov, A. A. Spithal'skii, and V. A. Sychugov, *Sov. J. Quantum Electron.* **4**, 839 (1975). A. H. Nayfeh, *J. Acoust. Soc. Amer.* **56**, 768 (1974). H. Kogelnik and C. V. Shank, *Appl Phys. Lett.* **18**, 152 (1971). J. V. Moloney, M. K. Ali, and W. J. Meath, *Phys. Lett.* **49A**, 207 (1974). V. A. Kiselev, *Sov. J. Quantum Electron.* **4**, 872 (1975). G. C. Papanicolaou, *J. Math. Phys.* **13**, 1912 (1972). H. E. Rowe and D. T. Young, *IEEE Trans. Microwave Theory Tech.* **MTT20**, 349 (1972). C. Elachi and C. Yeh, *J. Appl. Phys.* **45**, 3494 (1974). A. Yariv and A. Gover, *Appl. Phys. Lett.* **26**, 537 (1975).
40. K. C. Kao and G. A. Hockham, *Proc. IEE London* **113**, 1151 (1966).
41. S. Maslowski, *Opt. Electron.* **5**, 275 (1973). R. D. Maurer, *Proc. IEEE* **61**, 452 (1973). R. A. Andrew, A. F. Milton, and T. G. Giallorenzo, *IEEE Trans. Micr. Theory Tech.* **MTT21**, 763 (1973). M. M, Ramsay, *Opt. Electron.* **5**, 261 (1973). S. E. Miller, E. A. J. Marcatili, and T. Li, *Proc. IEEE* **61**, 1703 (1973). M. DiDomenico, Jr., *Appl. Opt.* **11**, 652 (1972). J. LeMezec, *U.R.S.I. Meeting, Lima, Peru, 8–19 August 1975,* p. 158 of digest.
42. P. Kaiser, E. A. J. Marcatili, and S. E. Miller, *Bell Syst. Tech. J.* **52**, 265 (1973). P. Kaiser and H. W. Astle, *Bell Syst. Tech. J.* **53**, 1021 (1974).
43. J. W. Fleming, *Fall Meeting Amer. Cer. Soc., Pocono Manor, Pennsylvania, Oct. 8–10, 1975.*
44. W. Eickhoff and E. Weidel, *Opt. Quantum Electron.* **7**, 109 (1975).
45. J. D. Crow, *Appl. Opt.* **13**, 467 (1974).
46. R. M. Waxler and G. W. Cleek, *J. Res. Nat. Bur. Stand., Phys. Chem.* **75A**, 279 (1971).
47. I. Camlibel, D. A. Pinnow, and F. W. Dabby, *Appl. Phys. Lett.* **26**, 185 (1975).
48. M. O. Vassell, *Opt. Electron.* **6**, 271 (1974).
49. J. A. Arnaud and W. Mammel (unpublished).
50. W. J. Stewart, *Tech. Digest Top. Meeting Opt. Fiber Transmission, Williamsburg (Virginia), Jan. 7–9, 1975,* Paper PD6.
51. R. Stolen *Appl. Opt.* **14**, 1533 (1975).
52. J. E. Goell, *Proc. IEEE* **61**, 1504 (1973). W. M. Hubbard, *Bell Syst. Tech. J.* **52**, 731 (1973). W. S. Holden, *Bell Syst. Tech. J.* **54**, 283 (1975). S. D. Personick, *Bell Syst. Tech. J.* **52**, 843, 1175 (1973).
53. P. W. Smith, D. L. Bisbee, D. Gloge, and E. L. Chinnock, *Bell Syst. Tech. J.* **54**, 971 (1975).
54. M. C. Hudson and F. L. Thiel, *Appl. Opt.* **13**, 2540 (1974).
55. W. B. Gardner, *Bell Syst. Tech. J.* **54**, 457 (1975).
56. R. J. Turner, *Post Office Elec. Eng. J.* **68**, 7 (1975).
57. M. H. Kuhn, *Arch. Elek. Uberta.* **29**, 400 (1975).
58. D. Marcuse and W. Mammel, *Bell Syst. Tech. J.* **52**, 423 (1973).
59. S. Sheem and J. R. Whinnery, *Wave Electronics* **1**, 61 (1975).

Author Index

Numbers in parentheses are reference numbers and indicate that an author's work is referred to, although his name is not cited in the text. Numbers in italic show the page on which the complete reference is listed.

A

Agudin, J. L., 334, *430*
Akimoto, T., 283(30), *324*
Ali, M. K., 414(39), *432*
Altman, C., 187(6), *218*
Anderson, I., 150(83), *161*
Andrew, R. A., 414(41), 432
Arnaud, J. A., 23(20), 30(31), 32(38), 36, 38, 41, 43(45), 44(45),45, 47, 48, 49, 50(2), 57, 72(2), 73(2), 79, 80(20), 81(2), 82(2), 86(2), 94, 97(17, 31, 33), 103, 106, 107, 108, 109(38), 111(38), 112, 113, 114, 115,119, 128(2, 39), 132, 135, 136, 140, 141, 142(16), 144(2, 36), 150(83, 84), 151, 152, 153, 157(85, 87, 88), 158, *159, 160, 161*, 184(5), 186(5), 187,190(6), 211(12), 218, 219, 239, 241, 244, 256, 260(20), 267(14), 270(20), 272(14), 274, 275, 279(20), 286(31–33), 291(34), 296(33), 297, 299, 306(20), 307, 310, 311, 312, 313, *324*, 341(7), 346, 348, 350(13), 352, 353, 354(15)355, 356, 357, 361(16), 371(21), 381(26), 384(28), 386, 390(15), 394, 397(15), 400(34), 410, 411,

415(37), 416(15), 417, 425(49), 428(49), *431, 432*
Artman, K., 332(3), 334, *430*
Ash, E., 102, *160*
Ashkin, A., 32(35), *49*
Astle, H. W., 417(42), *432*
Auston, D. N., 146, *161*
Ayers, S., 146(76), *161*

B

Berreman, D. W., 19(13), 21(13), *48*, 80(19), *159*
Bertoni, H. C., 332(3), *430*
Beyer, J. B., 147(77), *161*
Bezrodnyi, V. G., 414(39), *432*
Biernson, G., 361, 365, 377, *431*
Bisbee, D. L., 429(53), *432*
Blyler, L. L., 281(29), 282(29), 283(29), *324*
Boitsov, V. F., 146(71), *161*
Booker, H. G., 17(11), *48*, 79(18), *159*
Born, M., 2(3), *48*, 50(2), 196, *218*, 332(2), *430*
Bouillie, R., 260(16), *324*

433

Subject Index

V

Vapor phase deposition (CVD), 23, 418, 428
Variational principles, 216
Vector field, 162
Vector potential, 33
Velocities in special relativity, addition of, 138

W

Walk-off effect, 133
Wave action, 44
Wave equation, 51, 162–248
Wave function, biorthogonal, 122
Wave propagation, 44
Wave vector, 4, 221
    curve of, surface of, xx, 12, 45, 52, 175, 226
    in space–time, 236
Wavefront complex curvature, 72, 86
Wavefront radius of curvature, 55, 56, 60, 61, 73
Waveguide
    to beam transducers, 147–150
    dielectric, *see* Optical fibers
    efficiency, the dielectric rod, 362, 364, 426
    metallic, 12, 29
    system, low-loss millimeter, 430
    uniform, 42
Wavelength at resonance, 25
Wavenumber, 4, 242
    axial, 100
    effective, 14
    at resonance, 84
Weak focuser, 150–152
Weakly guiding dielectric rod, 365
Weakly guiding fiber, 366, 360
Whispering-gallery modes, 16, 17, 18, 31, 143, 369, 374, 400
    axial confinement, 398
    in dielectric rods, 367
WKB approximation, 12, 139–142, 313, 341

A 6
B 7
C 8
D 9
E 0
F 1
G 2
H 3
I 4
J 5